Dael Wolfle

# RENEWING A SCIENTIFIC SOCIETY

*The American Association for the Advancement of Science from World War II to 1970*

HIST.
Q 11
A 53
W 65
1989

Library of Congress Cataloging-in-Publication Data

Wolfle, Dael Lee, 1906—
    Renewing a scientific society: the American Association for the Advancement of Science from World War II to 1970 / Dael Wolfle.
       p.    cm.
    Includes bibliographic references.
    ISBN 0-87168-349-0 : $19.95
    1. American Association for the Advancement of Science — History.
I. American Association for the Advancement of Science.    II. Title.
Q11.A53W65    1989
506'.073—dc20                                  89-17690
                                                                CIP

Publication No. 89-14S

© 1989 by the American Association for the Advancement of Science
1333 H Street, N.W., Washington, D.C. 20005

# Contents

Foreword .................................................... v
Preface ..................................................... vii

| | | |
|---|---|---|
| One | 1945: A Setting for Growth | 1 |
| Two | Government Relations | 13 |
| Three | Annual Meeting | 41 |
| Four | *Science* and *The Scientific Monthly* | 71 |
| Five | A New Home for the Association | 101 |
| Six | The Advancement of Science | 115 |
| Seven | Changes in Governance | 127 |
| Eight | Science Education | 149 |
| Nine | Public Understanding of Science | 189 |
| Ten | International Activities | 211 |
| Eleven | Science in Society | 227 |
| Twelve | 1970: Looking Back and Looking Forward | 255 |

Appendix 1  Origins and Chronology ............................ 263
Appendix 2  Meetings and Presidents ........................... 271
Notes and References .......................................... 277
Index ......................................................... 323

# *Foreword*

*"There is no new thing under the sun."* — Ecclesiastes 1:9

Dael Wolfle has written a chronicle of the past that is also an exemplary guide for the future: a living legacy for the American Association for the Advancement of Science, which he served with such distinction and which he continues to serve through this book and in other ways. In quarrying and distilling the records and in recounting and analyzing the policies and activities of the AAAS during the 25 years beginning with the end of World War II, he has revealed mine fields and ore bodies. He tinctures facts with wisdom. Every successor executive officer, every incoming member of the AAAS board of directors, and every policy-level staff member will benefit from reading all or parts of this reliable and admirably annotated analytical history. So will many of the 135,000 members of the AAAS; and so will the officers and board members of its almost 300 constituent scientific societies.

For more than 16 years, from 1954 until 1970 (when he left, at age 64, to become professor of public affairs at the University of Washington), Dr. Wolfle was the guiding spirit, the soul, and the conscience of the American Association for the Advancement of Science. In a sense, he was the conductor of an orchestra of virtuosos. As executive officer, he was the permanent, but evolving, central nervous system of an organism whose component board of directors of about a dozen members served staggered terms of four years each. Formally, he reported to the board that hired and could have fired him. But, in fact, the board — though composed of talented, varied, and independent members — looked primarily to him for innovation and for leadership. He won and held their respect by reason of his character, judgment, wisdom, straightforwardness, progressive temperateness, imperturbability, experience, and dedication. And he was respected for those same qualities by his devoted staff, and in the scientific world, in academia generally, in the Congress, and by the press. He was — and is — a thoughtful statesman of science and an articulate spokesman for science. He helped improve the public understanding of science, both in educational innovation in the schools and, through the media, in the adult world. His book is an important contribution to the history of American science in its relationship to government, education, and public affairs.

Under his leadership, the AAAS adapted to the changed postwar world. It grew greatly in membership, in scope of activities, in financial strength, and in effectiveness to society in the promotion of human welfare. During the period of his executive officership, AAAS membership rose from 49,000 to 130,000, and the circulation of *Science* increased from 32,000 to 160,000. Early in his term of office, a new headquarters building at 1515 Massachusetts Avenue was approved, designed, and constructed. Many years later, it was outgrown and was sold at great profit in 1985.

Dael brought me into the AAAS in 1966 as a member of the Investment and Finance Committee and then, a year later, as a member of the Committee on Public Understanding of Science. It was gratifying to serve with him in these capacities, and then as treasurer beginning in 1969. As a member of the board, I helped to persuade him, uniquely qualified, to undertake the demanding task of writing this 25-year chapter of the history of the 141-year-old AAAS. And now it is my privilege to write this foreword. It is a superfluous foreword, for Dael's history speaks for itself; and his preface is its overture. Hail Dael!

Dael Wolfle's one shortcoming is also one of his virtues: modesty. He says too little in this volume about himself and his own roles and achievements. But, even if invisible, like Ariel in *The Tempest*, he is everywhere in this history; and his contributions will edify and inspire the current and future generations of AAAS boards, staff, and members. *Si monumentum requiris circumspice.*

*William T. Golden*
*March 15, 1989*

## *Preface*

Several years ago when oceanographers were planning an international congress on the history of their fields of research they invited me to prepare a paper explaining how it was that the American Association for the Advancement of Science had planned and sponsored the world's first international congress on oceanography. When it came time to give that paper the chairman of the session carefully pointed out that whereas the two preceding papers had been given by real historians, what the audience was about to hear was not by a historian but by a "participant observer" (1).

The chairman was quite right. I was not a historian, as the organizers of the program well knew. But they also knew that I had been intimately involved with that 1959 congress of oceanographers, from the first initiation of the idea to publication of the proceedings and payment of the last bills, and that I probably knew as much as anyone about how and why it was that AAAS, a national organization, had been responsible for bringing the world's oceanographers together in their first large congress.

It was for a similar reason that I was invited to write this historical account of AAAS during that expansive quarter century that started with the end of World War II and concluded around 1970. For most of that period, from the beginning of 1954 to the middle of 1970, I was the executive officer of AAAS, living and working daily at the heart of its affairs. Earlier I had served as a member of the association's council and a few of its committees. And from 1946 to 1950 I was one of the association's closest neighbors. In 1946, when AAAS purchased the property that became its new home, I was searching for quarters in which to establish the office of the American Psychological Association. The new AAAS building was larger than the association needed. Its top floor was quite adequate for the office I was establishing, and for the next four years my office was in the AAAS building, where I had a good observer's seat and worked cooperatively with some of the AAAS staff members on several matters.

Thus it was because of my first-hand experience that I was invited to write the following chapters of AAAS history. The product, however, is not a personal history or based primarily on my own reminiscences. AAAS has long maintained a set of annual "minute books" which contain the official copies of the agenda and minutes of meetings of the council, the board of directors, and the association's major committees, plus the related memoranda and correspondence. Every day in the summer of 1984 it was the task of Stacey Newton, an intern in the office of Michele Aldrich, AAAS archivist, to send a batch of that material through the copier and on to me. By the time she had finished that daily chore I had accumulated some 8,000 pages of AAAS records. Those records, together with the printed accounts of association activities published in *Science* or elsewhere, provide the documentation for the stories told in the following pages.

With all that paper as source material, there was little need to rely on personal memory. Yet not everything was recorded, and occasionally I have written "as I recall," or in some other way have indicated that on a particular point I have relied on my own memory.

With that much paper there can also be lots of reference notes, for some reader may want to see the full agenda item or minutes referred to, to know just when a decision was made, or to see more detail than is given in the text. Thus most of the notes are simply direction signs, pointing to the published record or the relevant account in one of the minutes books or elsewhere. However, some of the notes contain additional information, information that did not seem to fit into the text but that did seem to be worth recording.

The chapters are topical rather than chronological. The first chapter describes the situation of AAAS in 1945, and the second, the association's relationships with the federal government, the association's frequent supporter, sometimes ally, and always the major force determining the conditions under which AAAS planned and conducted its work. After those two chapters, the order is primarily determined by history. Before World War II the association's principal activity was conduct of its annual or semiannual meetings. Thus, the third chapter tells the story of AAAS meetings from 1945 to 1970. Chronologically, the next major activity became the publication of *Science* and *The Scientific Monthly*, so the association's publications provide the subject matter for the next chapter. From then on, the chapters are roughly in chronological order except for the one on governance of the association.

The period covered is from the end of World War II to about 1970, but some of the activities of that quarter century had started earlier, continued later, or in 1945–1970 were so different from more recent practice that it seemed desirable to say a few words about earlier history or later developments. Where that was the case I have breached the time boundaries, sometimes by a few years and sometimes by many. Yet this history is for only one 25-year segment in the life of the AAAS. For the beginning period and up to 1860 one should read Sally Gregory Kohlstedt's *The Formation of the American Scientific Community*(2). For the years between 1860 and 1945 the brief historical accounts in the annual proceedings volumes are the best source. Much of what has happened since 1970 has been reported in *Science* and other AAAS publications, but a systematic historical treatment of those years waits for some other historian — or participant observer.

For help on this volume I am indebted to several people whose assistance I gratefully recognize: first, William D. Carey, William T. Golden, and their colleagues on the board of directors who invited me to undertake what turned out to be a very enjoyable task; then Michele Aldrich and Hans Nussbaum, the association's archivist and senior staff member respectively, for always being ready to dig into the association's records for anything I wanted to see; and Stacey Newton for those daily packages from the minutes books. Several former colleagues and friends read and criticized one or more chapters in draft form: Neva Abelson, Philip H. Abelson, Water G. Berl, William D. Carey, John R. Mayor, Gerard Piel, Don K. Price, J. Thomas Ratchford, F. James Rutherford, and Edward G. Sherburne, Jr. I am grateful to all of them, and especially to Bruce V. Lewenstein and John Walsh who read the entire manuscript and made many helpful suggestions.

More generally, I want to thank the members of the AAAS staff. Some of them are named in the text, but most are not. Yet they are the people who kept the records, carried out many of the tasks that resulted from board and council decisions, did much of the work that made AAAS the effective organization that it was, and helped make the AAAS office the pleasant place in which I worked for sixteen and a half years. There were many of them, too many to list here, but let me name those who came to AAAS during the period of history covered in these chapters and who remained for at least 25 years: Philip H. Abelson, Catherine Borras, Eleanor Butz, Esther Carrico, Grayce Finger, Rose Lowery, Orin McCarley, Linda McDaniel, Ellen Murphy, Hans Nussbaum, Ann Ragland, John Ringle, Grace Smawly, James Stickley, John Walsh, and Joseph Walton. Although not directly employed by the association, five members of the advertising staff have also served AAAS for a quarter century or longer: Earl Scherago, who became advertising manager in 1955, and Herbert L. Burklund, Richard L. Charles, C. Richard Callis, and Winn Nance, all of Scherago Associates, the advertising agency for *Science*.

Let those 21 names stand for all the rest. To all of them I say "thank you" for all of your many contributions to the programs and activities described in the following chapters, and for the privilege of working with you.

*Dael Wolfle*
*Seattle, Washington*
*October 1988*

*Chapter One*

# 1945: A Setting for Growth

The most expansive quarter century in the history of American science started with World War II and ended about 1970 when federal appropriations for research and development ended their steep upward climb. That slowdown had to come, if not at the end of the 1960s then soon after, for exponential growth inevitably must slow or go through the ceiling. In retrospect, however, it is evident that science's golden quarter century was the steep part of a growth curve that started long before World War II and continued after the 1960s, but at a slower pace (1). Within that quarter century the scale of scientific and technological work grew at a rate that may never be repeated.

World War II ended with many of the nation's scientists and engineers actively involved in government programs and in the shaping of national policy for research, development, and the education and utilization of scientists and engineers. At the same time many university departments, industrial research laboratories, and other scientific institutions were conducting research for the military services or the wartime Office of Scientific Research and Development.

Recognition of the dramatic contributions those scientists and engineers had made to the winning of World War II led to widespread support for President Roosevelt's ringing affirmation of hope: "New frontiers of the mind are before us, and if they are pioneered with the same vision, boldness, and drive with which we have waged this war we can create a fuller and more fruitful employment and a fuller and more fruitful life" (2). A few weeks after the Japanese surrender in 1945 President Truman told Congress:

> Progress in scientific research and development is an indispensable condition to the future welfare and security of the nation. The events of the past few years are both proof and prophecy of what science can do. ... No nation can maintain a position of leadership in the world of today unless it develops to the full its scientific and technology resources. No government adequately meets its responsibilities unless it generously and intelligently supports and encourages the work of science in university, industry, and in its own laboratories (3).

Congress agreed and soon began providing annually increasing appropriations for an expanding program of research and development.

How the AAAS responded to those challenges was determined in part by the general temper of the times but also in part by its traditions, organizational structure, and resources and limitations. Just as the early AAAS had evolved from the Association of American Geologists and Naturalists (4), so the AAAS of 1945 and later evolved from the AAAS of 1944 and earlier. (Appendix 1 provides a brief history of the association.)

In its response and activities after 1945, as in its earlier life, the AAAS had four constituencies to serve: scientists and the members of allied professions, young people who aspired to careers in those fields, the interested public that increasingly needed to understand some of the problems and prospects of science and technology, and the political and opinion leaders who influenced decisions about the support of science and the uses and regulation of scientific and technical activities.

The postwar availability of funds from the National Science Foundation and other sources helped the association expand existing programs and develop new ones addressed to all four of its constituencies. The AAAS annual meetings placed greater emphasis on interdisciplinary topics and on matters of interest to all scientists and less emphasis on the interests of specialists. The association's journal *Science* was expanded in a similar direction. New programs on science education were undertaken. And greater attention was given to the relations of science and technology to society.

Association activities changed as conditions, opportunities, and needs changed, for the dramatic differences that distinguished pre- and postwar conditions were followed by other changes. The nation's postwar recovery of the late 1940s merged into the Cold War and economic expansion of the 1950s. In the 1960s the continued economic expansion was paralleled by the baby boom, Great Society concerns over societal problems, the trauma of the war in Vietnam, and the expanding equal rights movement. All of those forces influenced the members and officers of the AAAS and the activities that are described in later chapters. Yet it was the experiences of World War II and the new opportunities that followed, together with the historic nature of the association itself, that provided the bases for all that followed.

## *AAAS in 1945*

From its founding in 1848 the AAAS took upon itself the dual responsibility of advancing the collective interests of the whole scientific community while also serving the needs of its members in their individual specialties (4). Structurally, that duality meant that the association was both a single organization and a collection of subject-matter sections. Originally there were two sections, one for natural history and one for general physics. By 1945 there were 15 sections, one of which consisted of three subsections.

A second duality arose from the decision of the founders that membership would be open to anyone interested, scientist or not. Thus, although membership would be open to amateurs as well as professionals, many also agreed with naturalist James Dwight Dana that the association should be controlled by real scientists. As Dana wrote in 1851, "... it is of the utmost importance that those who know what true science is should strive to keep the Association in its right path"(5). To have an open society governed by a knowledgeable elite led to two classes of members and to recurring disagreements over how the association's officers should be elected. In 1945 there were still two classes of members. Most of the privileges of membership were the same for all members, but only those classed as fellows could be elected to office or serve in the association's governing council.

## Affiliated Societies

As a further complication AAAS came to be seen as two different kinds of organizations. It was founded as a single, independent membership society, but at least as early as 1891 some other societies were beginning to meet with AAAS. Some of those societies were the association's offspring, for a meeting of AAAS provided a convenient opportunity for the expanding membership of some specialized field to get together and plan a new society. By 1899 the practice of other societies meeting with AAAS was well enough established that a constitutional revision formally recognized the existence of "affiliated societies" and by 1903 the combined meetings were being called the AAAS "convocation week."

Later, a category of "associated" societies was added, and by 1945 there were nearly 200 affiliated or associated societies, including state and regional academies of science and specialist societies in many fields of science and its applications. The two categories were distinguished in that affiliates were considered to be more closely related to AAAS than were associates and — unlike associates — were represented in the AAAS governing council. (By 1965 the distinction between the two categories seemed to have lost whatever usefulness it might have had, and the two were combined into a single category of "affiliated organizations.")

## Regional Divisions

Geographic divisions became desirable as more and more scientists went to work in areas remote from the cities in which association meetings were usually held. First was the Pacific Division, which held its first meeting in 1916, and then the Southwestern Division, which first met in 1920. Daniel T. MacDougal was largely responsible for founding both. In 1913 when the association met in Cleveland, MacDougal, who was then a member of the Committee on Policies of the AAAS, pointed out that although AAAS had several hundred members on the West Coast, only two were in attendance. He knew the reason. As director of the Laboratory of Plant Physiology of the Carnegie Institution of Washington, he had his office in the District of Columbia and his Desert Laboratory in Tucson, Arizona, and it took him five days by train to get from one to the other. A West Coast scientist had to spend 15 days away from home to attend a five-day meeting on the East Coast.

AAAS was already planning to hold its first-ever meeting on the West Coast—the 1915 meeting in San Francisco that would help to honor the opening of the Panama Canal. With encouragement from the council MacDougal and other members of the local committee for the San Francisco meeting also served as the organizing committee for the association's first geographic division (6).

A few years later, as the number of scientists in the southwest continued to expand, MacDougal and Andrew T. Douglass, who was director of the Seward Observatory and had the unusual dual appointment of both professor of astronomy and professor of dendrochronology at the University of Arizona, took the lead in founding the Southwestern Division. Later, that division expanded into the Southwestern and Rocky Mountain Division (7). Appropriately, MacDougal was honored by being elected as president of both of the divisions he had helped to establish.

Much later, two other divisions were added. In 1951 the council authorized the Alaska Division, which later became the Arctic Division because it met and had members in northern Canada as well as in Alaska. The Alaska Division had an unusual beginning, for it resulted from a meeting held in Washington, D.C., rather than in Alaska and one held under the auspices of the National Academy of Science and the National Research Council rather than AAAS. It soon made up for its un-Alaska beginning, however, by meeting in the most spectacular site in which any AAAS meeting has ever been held: Mount McKinley National Park (8).

In addition to those three divisions a "south-of-the-border" division was under consideration for a long time. At the 1912 meeting the council authorized formation of a Brazilian Division, which never developed. In the 1960s there was correspondence with some of the Mexican members about a possible Mexican Division, but neither then nor as a result of the Mexico City meeting of 1973 did one materialize. However, in 1984 the council authorized formation of a Caribbean Division of AAAS and that division was established in 1985.

The Pacific and the Southwestern and Rocky Mountain divisions, and after 1952 the Alaska (or Arctic) Division, all operate as semi-independent scientific societies. Each arranges its own annual meeting, manages its own finances, and sometimes publishes symposia or other works. Yet each is an important part of AAAS, subject to oversight by the association's council and with a representative serving as a member of that council.

## Local Branches

An interest in having local branches of the association arose even earlier than did interest in regional divisions. In 1913 the AAAS council authorized establishment of local branches "in places where the members are prepared to conduct organizations that will forward the objects of the association" (9). Two decades later, in the 1930s, again when planning for the centennial meeting of 1948, and still later in 1956, the board or council sought to stimulate organization of local branches in a few metropolitan areas. A few were actually established but all but one lasted for only a few years and then disappeared for lack of leadership or because other local arrangements satisfied the members' appetite for going out to hear a lecture (10).

The branch in Lancaster, Pennsylvania, was the only really successful one. AAAS members there were enthusiastic about the opportunity, asked the council for authorization, and in 1934 quickly organized (11). At the first meeting, W. F. G. Swann of the Bartol Research Foundation in Philadelphia addressed an audience of 450 on cosmic rays and then demonstrated his virtuosity by entertaining the audience with a cello recital. In a brief business meeting Jacques Cattell was elected chairman, and the new branch was off to a fine start. With dues of a dollar a year membership grew until it reached 1,800 in 1943–1944. Membership had to be limited to 1,200, however, for an auditorium seating 1,200 was available quite inexpensively but one large enough for 1,800 was too expensive to consider (12). In a community of about 60,000 with perhaps the interest-enhancing advantage of being the city in which *Science, The Scientific Monthly,* and James McKeen Cattell's other publications were being published, the Lancaster Branch was clearly the most successful of all the local branches and is still meeting regularly, more than 50 years after its founding.

## Governance

Thus in 1945 AAAS had two classes of members, a central organization and a number of sections, two geographic divisions, one local branch, and many affiliated and associated academies of science and scientific societies.

The council was the legislative body that governed that sprawling structure, aided by the executive committee of the council (which in 1952 was renamed board of directors). For many years the executive committee was in reality the executive committee of the council, authorized to act for the council when it was not in session. It consisted of the AAAS president, the general secretary, the permanent secretary (now called executive officer), the treasurer, and eight members elected, two each year, for four-year terms. The committee annually elected one of its members to serve as chairman.

The council included all members of its executive committee, the vice president (now called chairman) and secretary of each section, one representative of each geographic division, eight members elected by the council from among the fellows of the association (those eight positions were eliminated in 1946), and one representative selected by each affiliated academy or society and a second by each affiliate that included among its members 100 or more persons who were fellows of AAAS. Because AAAS had so many affiliates their representatives far outnumbered the other members of the council.

Shortly before World War II the council and executive committee decided that the constitution adopted in 1919 needed substantial revision. A committee of three was appointed, all thoroughly experienced in AAAS affairs: Burton E. Livingston, permanent secretary from 1920 to 1930; Otis W. Caldwell, general secretary from 1933 until his death in 1947; and F. R. Moulton, who had become permanent secretary in 1937. Their proposed revisions were adopted by the council in 1946 (13).

Under that new constitution the annual election of a president who served for only one year was replaced by the annual election of a president-elect who served in that capacity for a year, advanced to the presidency, and then a year after that became retiring president and chairman of the executive committee. The general secretary, the administrative secretary (the new name for the permanent secretary), and the eight elected members of the executive committee were all retained, but the treasurer was dropped from that body, although continued as a member of the council.

Having two different secretaries has since seemed an anomalous arrangement, but for decades the association's leaders wanted two different secretaries for two different sets of duties. The permanent secretary was responsible for collecting dues, making arrangements for meetings, supervising the office staff, conducting the daily business of the association, and "such other matters as the Council may delegate (14)." The general secretary was expected to "attend to matters connected with the organization of the association, its relations to the affiliated societies, and such other matters as the Council may delegate" (15). Supporting the council, executive committee, and officers were several continuing committees and ad hoc ones as needed.

## Activities

The association's general meetings — with lectures, symposia, and other sessions arranged by AAAS, its sections, and as many affiliates as chose to meet with AAAS and arrange their own sessions—constituted the association's oldest and still major activity.

AAAS was also a publisher. A series of symposium volumes had been started in 1938. A year later AAAS began assuming responsibility for editing and publishing *The Scientific Monthly,* taking over that responsibility from James McKeen Cattell, the journal's founding editor and publisher. In 1945, under provisions of an earlier agreement with Cattell, the association also selected an editor and began preparing to assume responsibility for publishing *Science*. It was also publishing the *AAAS Bulletin*, a periodic house organ started during the war and published from 1942 to 1946 and from 1961 to 1970. At the same time it was preparing to publish the final volume of an annual series of symposium volumes that extended back to the association's beginning.

Other activities included the awarding of prizes for especially meritorious scientific work, allocation of very modest funds as grants in aid of research, and conduct of a number of week-long summer conferences on selected research topics in chemistry — the Gordon Research Conferences.

## A New Beginning

With the war over AAAS began planning for the expansion and improvement of some of those activities and for an appropriate celebration in 1948 of the association's 100th anniversary. Expanding activities could no longer be carried out in the association's small quarters in the Smithsonian Institution and a new home had to be found. The governing structure was out of date, but the committee to revise the constitution and bylaws had been interrupted by the war. Now it was clearly time for renewal, expansion, self-examination, and revitalization.

For those purposes the association had strong leadership. Among the members of the 1945 AAAS board of directors (then called the executive committee) were Roger Adams, one of the most influential and honored American chemists of his time; Anton J. Carlson, patriarch of American physiology, one of the founders of the American Association of University Professors, and president, at various times, of numerous scientific societies; James B. Conant, president of Harvard University, wartime chairman of the National Defense Research Committee, deputy to Vannevar Bush at the Office of Scientific Research and Development, and about to become High Commissioner and later ambassador to West Germany; Charles F. Kettering, for 27 years vice president of General Motors Corporation and general manager of its research laboratories, an inventor whose name still reverberates around GM as one of its founding leaders; and Burton E. Livingston, who had earlier served AAAS for a decade as permanent secretary and for three years as general secretary.

Working with those and other leaders of the association was Forest Ray Moulton, the association's first full-time executive officer. When he came to AAAS in 1937 he was 65 years old, but he did not come to that post as a retirement position; he intended to be a vigorous manager and to effect some changes in the association's activities. And well he might, for success was in his blood and in his history. He was one of five brothers, all of whom in their varied fields of endeavor had their biographies in *Who's Who in America* and one of whom, Harold Moulton, was already in Washington as the founding president of the Brookings Institution.

Moulton received his Ph.D. summa cum laude from the University of Chicago in 1899, a year after he had been appointed to the faculty. There he remained, moving up through the ranks to professor, earning election to the National Academy of Sciences and other honors, and writing many articles and half a dozen books on celestial mechanics, differential equations, and other topics in his fields of astronomy and mathematics. In 1926 he resigned from the University of Chicago faculty to enter business as financial director of Chicago's Utilities Power and Light Corporation. A few years later he became director of concessions for the Chicago World's Fair of 1933 and 1934. The theme of the fair was "A Century of Progress," and Moulton had much responsibility for developing the exhibits and buildings that centered around the scientific and technological advances of the preceding century and the hopes for the future (16).

After all that — a quarter of a century in pure science and briefer periods in commercial finance and public education — Moulton came to AAAS. Of course he was interested in science and technology and in the welfare of scientists, but he also had wider interests and plans. Soon after he came to the association he gave a talk to the Lancaster, Pennsylvania, branch of AAAS under the simple title "Science" (17) in which he emphasized the responsibilities of scientists to two other groups. One group consisted of scientists in other disciplines. AAAS could no longer do for individual fields of science what the specialized societies could do, but it could and should give more attention to the synthesis of science through excellent symposia and other interdisciplinary exchanges.

The other group he wanted the association to serve more effectively included all the interested nonscientists, a theme on which he expanded in a later talk to the Lancaster Branch. A good part of that second talk addressed the special obligations of the association to foster and encourage the large number of amateur societies in astronomy, radio, natural history, and other fields that could be found in the schools and communities of the nation. "The obligation of aiding them," he said, "falls most heavily on the AAAS, because no other society covers all the various fields in which the amateurs are interested or has members in every community in which their organizations exist" (18).

Moulton's emphasis on synthesis among scientists and aid and encouragement to amateurs illustrated perhaps the most persistent problem of the association's whole history: How should it divide its energies among several objectives? Its primary objective was the advancement of science, but "science" was conveniently left undefined, and the very first paragraph of the original "Objects and Rules of the Association" went beyond pure research: "The objects of the Association are, by periodical and migratory meetings, to promote intercourse between those who are cultivating science in different parts of the United States; to give a stronger and more general impulse, and a more systematic direction to scientific research in our country; and to procure for the labors of scientific men, increased facilities and a wide usefulness" (19).

Research was emphasized, but that final phrase went beyond the advancement of knowledge to ask for better facilities for scientists and to introduce the element of practical usefulness of their work. A still different objective, although not stated in the constitution, was in fact prominent in the organizational structure. The new AAAS was open to essentially anyone who wished to join, amateur as well as professional, those to whom science was a side interest as well as those for whom it was central. The leaders were careful to arrange matters so that they retained control of programs and activities, and there were periodic arguments about the proper place of amateurs. Yet amateurs were always accepted as members, and the fostering of their interests along with the advancement of research, the furtherance of the interests of professional scientists, and increasing public understanding of the benefits of scientific work were all among the association's goals and objectives. They were that way at the beginning and they remained that way in 1945. But always there was some uncertainty, tension, and from time to time shifts of emphasis among the three broad objectives of advancing research, public service, and education.

These three objectives coexist; they may at times coalesce; and sometimes they collide. But how the energies and activities of the association have been directed toward and divided among these three broad objectives is the continuing theme of the association's history.

## Growth, 1945 to 1970

During the quarter of a century under review the association's activities grew steadily. Table 1 shows for each year from 1945 through 1970 the membership, circulation of the association's member journals *Science* and *The Scientific Monthly* (until 1958 when it was combined with *Science*), receipts and expenditures, grant funds, and registration at annual meetings.

At the same time the nation's scientific and engineering population grew more rapidly than the population as a whole, and the scientific and technological expenditures grew more rapidly than the gross national product. AAAS membership also increased, but whether as rapidly as it "should" is hard to say. Because membership has always been open to essentially anyone who wished to join, there is no clearly defined population from which members are drawn and with which membership can be compared. However, as Table 2 shows (20), between 1945 and 1970 AAAS membership expanded in about the same ratio as did the total faculty of American colleges and universities, the number of students graduating with baccalaureate degrees in science and engineering, the total number of persons holding doctoral degrees in those fields, and the total amount of time devoted to scientific research and development.

Through the whole quarter century, association membership grew at about the same rate as did several measures of growth of the nation's scientific community. After 1958 when *Science* and *The Scientific Monthly* were combined and sent to every member of the association, the circulation of *Science* grew at a similar rate. The magazine itself also expanded during those years, publishing nine times as many pages in 1970 as in 1945 and expanding its influence by granting thousands of permissions a year to reprint material in other publications, for classroom use, or for other purposes.

Table 1. Annual records of AAAS membership, circulation of *Science* and the *Scientific Monthly*, receipts and expenditures, grant funds expended, and registration at annual meetings. (All figures in thousands except for annual meeting registration)

| Year | Members | Science | Scientific Monthly | Receipts | Expenses | *Grants | Meeting Registration |
|------|---------|---------|--------------------|----------|----------|---------|----------------------|
| 1945 | 27  | 19  | 13 | $ 374 | $ 343 |      | 2,649 |
| 1946 | 29  | 21  | 14 | 310   | 311   |      | 2,736 |
| 1947 | 33  | 26  | 16 | 386   | 310   |      | 4,940 |
| 1948 | 43  | 33  | 20 | 475   | 459   |      | 2,734 |
| 1949 | 45  | 33  | 22 | 546   | 432   |      | 7,071 |
| 1950 | 47  | 32  | 23 | 565   | 470   |      | 2,354 |
| 1951 | 49  | 33  | 25 | 585   | 456   |      | 3,702 |
| 1952 | 49  | 33  | 26 | 589   | 477   |      | 1,938 |
| 1953 | 49  | 32  | 26 | 595   | 525   |      | 3,315 |
| 1954 | 49  | 32  | 27 | 610   | 566   |      | 3,856 |
| 1955 | 50  | 33  | 28 | 603   | 593   | $ 56 | 2,636 |
| 1956 | 53  | 36  | 30 | 719   | 718   | 215  | 5,327 |
| 1957 | 56  | 38  | 30 | 834   | 825   | 230  | 3,684 |
| 1958 | 57  | 61  |    | 993   | 955   | 562  | 5,368 |
| 1959 | 59  | 62  |    | 1,184 | 1,038 | 851  | 4,636 |
| 1960 | 62  | 66  |    | 1,395 | 1,218 | 418  | 7,389 |
| 1961 | 69  | 74  |    | 1,692 | 1,482 | 523  | 4,709 |
| 1962 | 71  | 80  |    | 2,032 | 1,832 | 596  | 6,893 |
| 1963 | 79  | 94  |    | 2,390 | 2,260 | 835  | 3,660 |
| 1964 | 88  | 104 |    | 2,737 | 2,714 | 823  | 4,218 |
| 1965 | 98  | 116 |    | 3,200 | 3,120 | 940  | 7,028 |
| 1966 | 110 | 130 |    | 3,514 | 3,406 | 835  | 7,617 |
| 1967 | 117 | 138 |    | 3,754 | 3,806 | 625  | 7,279 |
| 1968 | 120 | 143 |    | 3,950 | 3,865 | 476  | 3,637 |
| 1969 | 127 | 153 |    | 4,476 | 4,282 | 316  | 7,891 |
| 1970 | 133 | 163 |    | 4,726 | 4,851 | 465  | 5,790 |

* Grants are not included in the receipts and expenses columns. In the years before 1954, funds were received for expenses of the association's Thousand Dollar Prize and the Westinghouse Science Writing Awards, and in 1951 AAAS received funds to endow the Thousand Dollar Prize, which later was given the donor's name as the Newcomb Cleveland Prize. The grants included are those from government, industry, and private foundations. Gifts and bequests from individuals are not included.

Table 2. Selected Measures of Growth, 1945 to 1970

|  | 1945 | 1970 | 1970/1945 Ratio |
|---|---|---|---|
| AAAS membership | 27,175 | 133,364 | 4.9 |
| *Total college and university faculty (full-time equivalent) | 135,000 | 574,000 | 4.3 |
| *Bachelors and first professional degrees in science and engineering | 68,576 | 334,335 | 4.9 |
| *Doctoral degrees awarded in science and engineering | 1,988 | 18,372 | 9.2 |
| Total persons in U.S. holding doctoral degrees in science and engineering | 34,549 | 182,285 | 5.3 |
| Full-time equivalent years devoted to research and development by scientists and engineers | 101,000 | 550,000 | 5.4 |
| AAAS expenditures, including grant funds | $ 374,472 | $5,316,019 | 14.2 |
| Federal expenditures for research and development (millions of current dollars) | 1,070 | 14,892 | 13.9 |
| Total national expenditures for research and development (millions of current dollars) | 1,520 | 26,134 | 17.2 |

*1945 was an abnormal year for degrees awarded, representing the low point for World War II. The figures used in the table for these measures are therefore the averages for the ten years from 1940 through 1949. The actual reported figures for 1945 were: bachelors and first professional degrees, 42,237; doctoral degrees, 1,143. No estimate of faculty size was reported for 1945; the figure given here is the average of the 1944 and 1946 estimates by the U.S. Office of Education.

*Sources:* See note 20, p. 278.

In size, therefore, AAAS expanded from 1945 to 1970 in pace with the nation's scientific and engineering community and in pace with the nation's research and development activities. The association's continuing activities and the new ones that were initiated during that period are described in the following chapters.

*Chapter Two*

# Government Relations

Efforts to influence the federal government came easily to the young AAAS. Spencer F. Baird and Joseph Henry of the Smithsonian Institution, Alexander Dallas Bache of the Coast Survey, and some other early leaders of the association were themselves government scientists, and over 40 percent of the association's first set of committees dealt with weights and measures, scientific exploration, geological surveys, wind and current charts, and other governmental responsibilities (1). Ever since, AAAS has tried to influence government policy or practice, sometimes to support and encourage, sometimes to criticize and try to change. Yet after World War II government-association relationships took on a new dimension as federal funds for research, already much larger in 1945 than they had been in prewar years, kept right on rising (after a temporary postwar dip) until the end of the 1960s. New federal agencies with new programs of interest to scientists and engineers and with fellowships for graduate students stimulated new growth in science, and after 1950 the new National Science Foundation financed a number of AAAS programs and initiated some conducted by the association.

At the same time, international tensions between Eastern and Western blocs, the Cold War, and worries about disloyalty and espionage within the country led to government actions restricting the free movement of scientists and the free flow of scientific information. Thus AAAS sometimes objected to government actions and tried to restrain them.

Overall, actions of the federal government became a primary factor in determining a number of the association's activities and how they were carried out. In later chapters of this volume, that influence will be frequently illustrated, for example, in describing the efforts of AAAS to improve education in science and mathematics (see Chapter 8). This chapter will consider more broadly the interactions and interrelationship of AAAS and the federal government, for that government was the primary influence in determining the climate in which the association lived and operated.

## Establishing the National Science Foundation

In 1945 Vannevar Bush published *Science: The Endless Frontier*, his reply to President Roosevelt's request for advice concerning the organization and support of scientific research in the years following World War II (2). A centerpiece of his recommendations was the proposal of what five years later became the National Science Foundation (NSF). Several bills calling for establishment of that new agency were quickly introduced in the Senate, with the two leading bills differing sharply on how it should be governed. One came from Senator Harley M. Kilgore of West Virginia, who had for several years been actively interested in government organization of scientific work. Dissatisfied with that bill, Bush arranged with Senator Warren G. Magnuson of Washington to introduce a rival bill that embodied the provisions he preferred.

With Senate hearings on those bills scheduled to start October 1, 1945, the AAAS board of directors appointed a committee consisting of Howard A. Meyerhoff, chairman, and Otis W. Caldwell, Kirtley F. Mather, and Walter Miles. Meyerhoff took leave of absence from Smith College and came to Washington as the association's executive secretary, a title used only for the year and a fraction he spent on that assignment. On September 12, Meyerhoff sent a questionnaire to members of the AAAS council asking their judgment, or the judgment of the officers of the affiliated societies they represented, on a number of features of the rival Kilgore and Magnuson bills (3).

All but one of some 100 witnesses at the Senate hearings supported the idea of a National Science Foundation. There was disagreement, however, on some points, especially over the form of governance of the proposed foundation (4). In an oversimplified fashion that disagreement can be illustrated by the positions taken by two AAAS presidents. Isaiah Bowman, president of AAAS in 1943, had served with Vannevar Bush and other scientific leaders of the country in the wartime Office of Scientific Research and Development. Bowman organized a committee in support of the Magnuson bill, which would have the President of the United States appoint a large board of 48 directors, primarily scientists. That board would then select a full-time director who would be responsible to the board. Supporters of this proposal considered it the safe, traditional, proven method of leaving decisions on scientific matters in the hands of scientists. Opponents considered it the conservative, elitist, "old-boy" proposal.

On the other side, Harlow Shapley (who became AAAS president in 1947), Harold Urey, and others formed a committee in support of the Kilgore bill. That bill would authorize the President of the United States to appoint the foundation director, who would, therefore, be responsible to the President. In making decisions about major matters, the foundation director would have the help and guidance of an advisory board of nine members. Supporters defended this arrangement as consistent with the U.S. form of government and as appropriate for the allocation of public funds. Opponents claimed it would politicize science; that a politician instead of a scientist might be appointed as director.

Although sides were drawn up on other issues, most of the argument was over the governing structure. Early in 1946, when it became evident that neither bill in its original form was likely to be approved, a compromise was introduced by Kilgore, Magnuson, and several other senators. It was widely supported by scientists and adopted by the Senate, but the House Committee on Interstate and Foreign Commerce decided that there was not enough time before adjournment to consider the matter adequately, so the 79th Congress adjourned without acting on the bill.

The AAAS board of directors, expecting the issue to be brought up again the next year, appointed a subcommittee consisting of Kirtley Mather (generally on the Kilgore side) and James B. Conant (generally on the Magnuson side) to plan further action when the 80th Congress convened in 1947. In addition to making a report to the AAAS council (5), Conant and Mather convened an ad hoc group that proposed that the council invite other societies to form an interdisciplinary committee of representatives of as many scientific and educational societies as cared to join. The council agreed (6) and the board of directors promptly invited affiliated societies to send representatives to an organizational meeting to be held in Washington on February 23, 1947.

Representatives of some 75 societies attended that meeting and elected Edmund E. Day as chairman. Day was an economist, president of Cornell University, and formerly a member of the faculty of Harvard University and the senior staff of the Rockefeller Foundation. Harlow Shapley, professor of astronomy at Harvard, then president of AAAS, and a supporter of the Kilgore position, was elected vice-chairman, and I was chosen as secretary-treasurer. Isaiah Bowman, supporter of the Bush Report (the Magnuson bill), Ralph W. Gerard, University of Chicago, Reuben G. Gustavson, Chancellor of the University of Nebraska, Henry Allen Moe of the Guggenheim Foundation, W. A. Noyes, University of Rochester, and Douglas M. Whitaker, Stanford University, were elected as members at large of the executive committee (7). Noyes was unable to accept; Moe did not answer the letter informing him of his election; and Bowman declined. We agreed at the first meeting that it would be desirable to have someone from the Bowman committee, and chairman Day talked with Bowman about a replacement, but nothing came of the request. The executive committee then selected C. G. Suits of the General Electric Company as an additional member (8).

Bush did not oppose organization of the Inter-Society Committee for a National Science Foundation, as the new committee was called, but he and Bowman remained aloof from it. Bush's prestige was such that he could, and did, operate individually in advising members of Congress on the pending legislation. To Bush and Bowman, the new committee sometimes seemed to represent the opposition. They did not think that the committee was truly representative of the scientific community, for it seemed to include too many social scientists. And they considered Meyerhoff of the AAAS staff to be their foe, which he sometimes was. In 1946, just before the compromise bill was passed by the Senate, Representative Wilbur Mills introduced — some people charged at Bush's urging — a bill identical to the original discarded Magnuson bill. Conflict over that action was one of the reasons that the House had adjourned the previous year without taking any action on the foundation proposal. Meyerhoff's account in *Science* started out:

> At noon, 19 July 1946, The National Science Foundation was pronounced dead by the surgical staff of the House Committee on Interstate and Foreign Commerce. The death was a homicide.
> Readers of *Science* are familiar with the promising career of the deceased, and many will mourn this untimely and unnatural passing, for the killing was done, not by politicians, but by scientists (9).

In the following year the Senate and House passed a science foundation bill of the Magnuson (Bush) style. President Truman found it unacceptable because the director would be appointed by and responsible to a group of part-time board members rather than to the President. He felt strongly enough about that feature to veto the bill and issue an explanatory statement.

For others, the decision was more difficult. From within government the President received advice both to sign and to veto the bill. William Carey, who had carried much of the Bureau of the Budget responsibility for science foundation legislation, drafted a veto message for the President and then wrote a personal memorandum "arguing vigorously against using it" (10). AAAS took no formal action, but the Inter-Society Committee officers overruled their dislike of the bill enough to urge the President to sign it. After the veto, Meyerhoff, reminding *Science* readers that the administration had earlier made it clear that the President would veto a bill such as the one actually passed by Congress, defended that position by concluding, "For the President to have no say in naming a committee chairman [chairman of the proposed Interdepartmental Committee on Science] who, in regard to scientific matters, outranks his own Cabinet officers in the executive branch of the Government is little short of preposterous" (11).

Thus Truman's veto met with praise, indignation, and disappointment, but one thing it did not do was speed up legislative action. In 1948 a bill modified sufficiently to meet the President's objections was passed by the Senate but blocked by the House Rules Committee. The next year repeated the same story: passage in the Senate and blockage by the House Rules Committee.

The Inter-Society Committee continued to work cooperatively with the House Committee on Interstate and Foreign Commerce but ineffectively with the Rules Committee — until a break was made. That significant accomplishment was achieved in a single day when the three officers of the Inter-Society Committee went to the Bureau of the Budget to review problems and possibilities with William Carey and Elmer Staats, the two staff members of the Bureau of the Budget who had primary responsibility for following science foundation legislation. Out of that discussion came agreement on a possible compromise on the issue of governing structure: to define the foundation as consisting of the National Science Board and the director and to have the President appoint the director and the other 24 members of the board. Carey and Staats thought that plan would be acceptable to the President. Chairman Day and I saw Bush that afternoon; he approved the plan and then made the prescient comment that under that arrangement someone such as Jim Conant could serve as chairman of the National Science Board and an Alan Waterman could be chosen as director; which is exactly what happened when the foundation first came into being.

With Bush's approval added to Carey and Staat's expectation of presidential approval, I took the plan to the members of the House Committee on Interstate and Foreign Commerce, who also approved it (12). With that governing structure, the bill passed both houses of Congress in 1950 and was signed by President Truman. Although the Inter-Society Committee was not a committee of AAAS, it was more closely identified with the association than with any other organization, and perhaps AAAS can claim a bit of special credit from the fact that of the five conferees at that day in the Bureau of the Budget, one (Shapley) was a former president and two (Carey and Wolfle) were future executive officers of the association.

Thus the National Science Foundation came into being but not without compromise and dilution. The five years from 1945 to 1950 saw changes in congressional priorities, and passage by the House might not have been possible without the restriction that appropriations could not exceed $15 million a year. That limit was accepted by the Senate but removed a few years later.

During those five years the Cold War had developed, fears of Communism were rampant, and McCarthyism worries about disloyalty and lapses in security were so strong that the House in 1950 adopted one amendment requiring that persons employed by the foundation or granted fellowships be certified as loyal to the United States by the FBI and that foreign nationals associated with the foundation must be cleared by the FBI. The Department of Defense and the Department of Justice both objected strongly to those unreasonable requirements, as did many scientific organizations. As a result somewhat weaker requirements were included in the final bill, but as the historian of the foundation later wrote, "Only in the irrational political climate of 1950 could anyone claim that these security requirements were in any respect loose. Indeed they revealed the effect of the Cold War anxieties on the minds of moderate men" (13). Nevertheless, the National Science Foundation was established, the National Science Board was appointed with James Conant as its first chairman, and President Truman appointed Alan Waterman as the first director. With a small initial staff recruited, the foundation was on its way toward the major scientific agency it soon became.

## Loyalty, Security, and Cold War Fears vs. Scientific Freedom and Responsibility

Moderate men and women were indeed affected by the Cold War anxieties. Communism was widely perceived as a threat to the democratic nations of the world. The Soviet Union, the temporary ally against Hitler's Germany, became the hostile enemy of the Western World. Spies and traitors were at work, but fears went beyond realities. Wild and unsubstantiated charges were made by the House of Representatives Committee on Un-American Activities. Guilt by association was a frequent charge. Applicants for NSF fellowships were required to sign a loyalty affidavit. Senator McCarthy claimed he found so much disloyalty and such threats to national security hidden in the federal government and among foreign service personnel, artists, scientists, and other intellectuals that "McCarthyism" became the name of that period.

It is difficult from this distance to appreciate the attitudes and excesses of the time, but examples were all too numerous. In congressional hearings Fritz Lanham, a former member of Congress, charged that within the proposals for a National Science Foundation there "lurked a desire of other than American origin" (14). A pamphlet from the National Patent Council described the proposed foundation as "an independent Government agency empowered to invade all research and development activities of industry and individuals, and to confiscate and pool patents for purposes of coercion and harassment in industry and perpetuation of political power ... a bill so adroitly drafted as to have fully deceived ... the naive and nonlegalistic minds of some of our greatest scientists and most patriotic citizens," and then described the hundred and more scientists and others who had testified in favor of a National Science Foundation as "members or affiliates of subversive organizations" (15).

In 1952, when the McCarran–Walter Act was passed over the President's veto, the State Department acquired authority to refuse a visa to any prospective visitor to the United States whose admission was thought to be "prejudicial to the United States." Rejected applicants could be excluded for confidential and undisclosed reasons, and that authority is still being used to exclude some writers and reporters with unpopular views as well as more dangerous applicants for visas.

Such was the national climate in 1954 when I became executive officer of the association. My initiation into dealing with McCarthyism came soon when a member of AAAS came to tell me he had been denied clearance for an industrial position because, among other reasons, he was a member of AAAS. I did not know the man and did not know whether the other charges had any more merit than that one, but at least that charge, I thought, should be removed from the record. I wrote to the industrial security clearance board to say that either they had confused AAAS with some other organization or they did not know the nature of AAAS. The letter then went on to say that the AAAS roster included Herbert Hoover, many men and women in military service, and many others with clearance to work on top secret matters, and then I offered to provide any information they needed to remove AAAS from their list of suspect and subversive organizations. I did not receive the courtesy of a reply, but neither did I see later charges that membership in AAAS was a reason for denial of clearance.

In that climate of McCarthyism, what should AAAS do as an advocate of scientific freedom and unhampered exchange of information and as an organization that believed in due process and the right of the accused to confront the charges made by unknown sources? That question was allied to a broader one: What should the federal government do about unclassified research? There was wide agreement on the need to classify some kinds of information — military codes, troop and fleet movements, characteristics of weapons not yet in extended use, research on new weapons. But at what point in the information stream, from basic research and unclassified data to the point of actual development and use, was secrecy needed? And how far back along that spectrum should workers be investigated for loyalty and security? There was no general law or act of Congress to answer that question, and the separate agencies of government were developing their own rules and requirements.

In that situation of uncertainty and nonuniformity it seemed to AAAS officers and to many other scientists that there was excessive secrecy and that persons doing unclassified research were too often being subjected to the kind of scrutiny that was appropriate for properly classified military work. For AAAS there were two possible lines of action. One was to investigate when a scientist had been discharged or refused appointment on loyalty or security grounds if that action seemed capricious or arbitrary. To do that AAAS would have to become an investigative body, operating somewhat as the American Association of University Professors had in investigating alleged cases of violation of academic freedom, with subsequent publication of the findings and, if the evidence warranted, censure of the agencies or institutions that had been found guilty (16). The board decided against that possibility: AAAS did not have and did not wish to develop the resources of an investigative agency. The second option was to issue statements, to take a stand, and to develop the principles that should be observed in balancing the needs for national security against the rights and dignity of individual citizens.

One spring morning in 1954 when I was feeling mildly ill, not incapacitated but more interested in staying in bed than in going to the office, I spent several hours drafting a statement on improving the basis for loyalty and security screening procedures while still safeguarding national security. After some polishing, I gave a draft to the board of directors. They helped revise it (17) and the statement became a collective one, with Warren Weaver playing a particularly significant role in preparing the final version that was endorsed by the AAAS council before being published in *Science*. The opening paragraph was both a summary and an indication of the tone of the document:

Those charged with safeguarding the United States have sought to minimize the danger of internal subversion through the screening of government employees and persons having access to classified information. This process is necessary, but it poses a serious dilemma: the more completely we succeed in reducing the danger that information now in our possession may leak to a potential enemy, the more risk we run of interfering with scientific progress and of reducing the technological superiority and the moral and physical strength upon which victory in the ultimate test would depend. The inherent dangers of this dilemma can be lessened and our strength enhanced by changing our basic concept of internal security from one that attempts almost exclusively to minimize our losses to one that places greatly increased emphasis on maximizing our gains (18).

That statement received substantial press coverage, but how much influence it had on government practice is quite unknown. By itself, perhaps little, but it was soon followed by other initiatives that over the next two years led to a major policy achievement.

## Eligibility for Federal Grants

In the absence of government-wide regulations, different agencies were using quite different requirements to determine eligibility to receive grants in support of unclassified basic research. In 1955 the National Science Foundation announced that decisions concerning unclassified research that did not involve considerations of security were based on the merit of the proposal and the experience, competence, and integrity of the scientist involved, as those characteristics were known to the scientist's peers. However, the statement went on,

the Foundation does not knowingly give nor continue a grant in support of research by one who is:
1. An avowed Communist or anyone established as being a Communist by a judicial proceeding, or by an unappealed determination by the Attorney General or the Subversive Activities Control Board pursuant to the Subversive Activities Control Act of 1950, or anyone who avowedly advocates change in the U.S. Government by other than constitutional means; and
2. An individual who has been convicted of sabotage, espionage, sedition, subversive activity under the Smith Act, or a similar crime involving the nation's security.
Furthermore, if substantial information indicates that a potential or actual researcher might be guilty of violating any law or regulation, the information would be forwarded to the Department of Justice for appropriate action.
The Foundation, therefore, will not knowingly support anyone who is, by admission or conviction, disloyal to this country. In the interest of science, however, it will not pass judgment on the

loyalty of an individual on the basis of unsupported charges but will rely upon the judgment of those who best know the individual and his qualifications. This position of the Foundation has been endorsed by the American Association for the Advancement of Science in a resolution passed at its annual meeting in Berkeley last winter. We believe it to be in the best interests of the Nation (19).

That policy satisfied the belief that public funds should not be granted to anyone who by admission or judicial finding had been determined to be disloyal to the country or who was a substantial security risk. It also avoided having decisions about loyalty and security made by a nonjudicial agency, NSF.

Not all agencies followed the NSF policy. A newspaper report that the U.S. Public Health Service had denied research grants to some 30 scientists was followed by a news release from Oveta Culp Hobby, Secretary of the Department of Health, Education, and Welfare, stating that the Public Health Service policy was as follows:

> We do not require security or loyalty investigations in connection with the award of research grants. When, however, information of a substantial nature reflecting on the loyalty of an individual is brought to our attention, it becomes our duty to give it most serious consideration. In those instances where it is established to the satisfaction of this Department that the individual has engaged or is engaging in subversive activities or that there is a serious question of his loyalty to the United States, it is the practice of the Department to deny support (20).

That practice put the Public Health Service into the position of being investigator and judge of the loyalty and security risk of its grantees and applicants. AAAS officers clearly favored the NSF practice, and adopted the resolution NSF cited in its statement (21). The AAAS resolution read, in part:

> The Board of Directors of AAAS and its Council urge the adoption of a uniform policy by all government agencies making grants for, or otherwise supporting, unclassified research. Specifically, they recommend adoption of the policy under which the National Science Foundation operates in making grants for unclassified research (22).

AAAS was not alone in criticizing the Public Health Service policy and at one of its meetings noted that "both individual scientists and university faculties are considering the possible desirability of refusing further grants from NIH until policies are improved" (23). The National Science Board also opposed the policy of the Public Health Service and debated what NSF could or should do to prevent that policy from being adopted widely by government agencies. Yet a government-wide policy was desirable, and in January 1955 Sherman Adams, aide to President Eisenhower, asked Detlev Bronk, president of the National Academy of Sciences, to appoint a committee to review and give advice on the matter.

Before the report of that committee appeared, Congress appointed a commission on security; Chief Justice Earl Warren joined the critics of the loyalty-security excesses; and the Civil Service Commission released figures on the extent of the problem. In 21 months, 2,778 federal employees had been dismissed for security reasons. Only one in seven had been granted hearings in which the evidence could be critically examined, and in nearly half of the cases in which hearings had been held the charges were dismissed and the employee cleared (24). Later the Association of the Bar of the City of New York came out with a recommendation for uniformity in procedures across all government departments and that clearance be required "for sensitive positions and for no others." As the report of the Bar Association was developing, the U.S. Supreme Court ruled that the intent of Congress in establishing the federal security program was to protect sensitive activities and not to include the entire Civil Service (25). AAAS was in good company in the position it had taken on the matter.

In the meantime, months went by while the National Academy of Sciences committee report was being prepared and more months before the White House responded to that report. During that period I wrote Surgeon General Scheele reminding him that a year had gone by since the issue was raised and asking what practices the Public Health Service planned to follow. In the same issue of *Science* that included the report of the National Academy of Sciences committee, an editorial on science and loyalty quoted the Surgeon General's reply that the service had adopted the practices of the National Science Foundation (26). Shortly afterward, Adams announced that government agencies would follow the practices recommended by the National Academy of Sciences committee, essentially those that had been developed by NSF. Later on it became known that the long delay was partly due to disagreements over the wording of that announcement:

> NSF wanted public recognition of the acceptance of its long-standing policy; HEW wanted an acknowledgement that it was following a similar practice but without any indication that it did so under pressure; and Bronk and Stratton [chairman of the NAS committee] wanted it clearly stated that HEW had fallen into line only after the Academy's report (27).

Quibbling over credit and blame aside, an important policy decision had been made, and government agencies avoided the fault of applying to unclassified research those review procedures that were appropriate only for dealing with classified material. AAAS probably helped in achieving that result, but most of the credit must go to the National Science Foundation, its director, Alan Waterman, and to others in government who had insisted on uniformity and a distinction between classified and unclassified research.

Not long afterward the nationally televised and widely watched Army–McCarthy hearings in the Senate discredited McCarthy and showed the nation the kind of man he was. The accusations, activities, and fears that marked the period of McCarthyism diminished somewhat and a saner and more cooperative atmosphere developed. For example, in 1956 the U.S. Commission on Government Security asked AAAS for advice on the problem of determining who needed to have access to highly classified information (28).

## Recurrence

Nevertheless, as late as 1969, the Public Health Service was informally accused of having a blacklist of scientists whose political views made them ineligible for appointment to the study sections that reviewed and rated proposals for basic research. Bryce Nelson of the *Science* staff found several scientists who had been recommended for such appointments by the National Institutes of Health (NIH) and then rejected by the Public Health Service on unspecified grounds who were willing to have their cases discussed in print (29). The review practices apparently being followed were objectionable on three grounds: (a) the reasons for rejection were veiled in secrecy; (b) those reasons sometimes seemed arbitrary and based on factors quite irrelevant to the duties to be performed; and (c) there was no provision for appeal or for confrontation of adverse charges or information. Even the NIH officers who had recommended an appointment were not told why it was rejected.

Within the Public Health Service there was a good deal of opposition to this system, including opposition from Robert Q. Marston, director of the National Institutes of Health, and other officers. Bryce Nelson quoted one Department of Health, Education, and Welfare administrator as saying, "Most of this stuff stinks, it's a lot of nonsense ... HEW is more security-minded than the Department of Defense" (30).

From outside of government the American Orthopsychiatric Association took the initiative in calling together representatives of a number of societies to consider the problem. William Kabisch, the AAAS representative to that group, kept the AAAS board of directors informed of the group's activity, and the board wrote to the HEW secretary protesting unnecessary screening procedures, pointing to the damage being done to the individuals concerned, and offering to meet with representatives of HEW to try to develop more appropriate procedures. In reply, the secretary described and defended the department's security review procedures and said that some of the statements in Nelson's *Science* article were inaccurate. The correspondence continued for a time without reaching agreement (31), but the Department of Health, Education, and Welfare modified its practices to bring them into better agreement with what AAAS and many others had been recommending (32).

Those exchanges with HEW were soon followed by urgent requests from Senators Muskie and Gravel that AAAS investigate the treatment of John Gofman and Arthur Tamplin, who believed that the Lawrence Berkeley Radiation Laboratory and the Atomic Energy Commission (AEC) had taken reprisal actions against them for their criticism of AEC radiation standards they considered inadequate (33).

The board was reluctant to have AAAS take on an investigative responsibility and agreed that it would first be desirable to decide on a general policy (34). Those discussions led to the decision to establish a Committee on Scientific Freedom and Responsibility, a committee that started out with the prestigious membership of Allen V. Astin, former director of the National Bureau of Standards; Mary Catherine Bateson, anthropologist daughter of Margaret Mead and Gregory Bateson; Walter J. Hickel, former governor of Alaska and former Secretary of the Interior; John H. Knowles, about to become head of the Rockefeller Foundation; and the Honorable Earl J. Warren, former Chief Justice of the Supreme Court (35). That committee, with gradually changing membership, has been active ever since in issuing statements of principle and in investigating and reporting selected cases of violation of the principles of scientific freedom. Since 1982 it has also made an annual award "to honor scientists and engineers whose exemplary actions, often taken at significant personal cost, have served to foster scientific freedom and responsibility" (36).

Actions of the kind taken by the association on issues and under conditions as complex as those illustrated herein must be undertaken on faith that they are right. Evidence of their effectiveness is often lacking, and experimental comparisons of effectiveness of one tactic versus another are usually impossible. The association's efforts may sometimes have had less influence than the officers and members hoped; nevertheless, AAAS could not stand by and do nothing.

## *Cooperation with Government Agencies*

When the National Science Foundation began receiving appropriations large enough to start awarding fellowships and making research grants, it used the AAAS mailing list to tell the scientific world it was ready for business (37). From then on the foundation became the primary supporter of the association's programs in science education (see Chapter 8). Some help also went in the other direction. For example, AAAS selected the science and mathematics teachers to whom NSF awarded fellowships to attend teacher-training institutes (38), and AAAS handled reimbursement and some other management details for the cooperative research agreement between the United States and Japan (39).

On one occasion AAAS was able to save the National Science Foundation from an embarrassing collapse of a popular program. The Traveling Library of Science, managed by AAAS but originally proposed by officers of NSF, had become so well regarded that the two organizations agreed on a large expansion: instead of sending sets of the books to only a couple of hundred high schools, as in 1957–1958, AAAS would send them to seven or eight times as many (see Chapter 8). That was the plan, but summer came, Congress had not yet passed an appropriation bill with funds for NSF, and prospects for early passage looked bleak. Yet if we were to expand the program that year we had to purchase some 55,000 volumes, prepare and box them for shipment, and make arrangements with over 1,300 high schools to receive and use them. Early in July we went to the bank and borrowed $300,000, purchased the necessary books and traveling-display cases, and did all the other things necessary to have the traveling libraries on the road when schools opened in September. By the end of September NSF had its appropriation and AAAS had its expected grant and had paid off the loan (40). We were confident that the grant would come, for the whole program was a mutually planned one and the foundation's educational program had strong congressional support. When NSF could not provide the up-front money, we could, and the board agreed that if a similar situation arose in the future we should again come to the rescue.

## The National Science Board

The act that established the National Science Foundation instructed the President to consult with certain named organizations before selecting members of the National Science Board. AAAS was not one of those named organizations, but each biennium, as members were to be reappointed or new ones named, AAAS was invited to submit nominations. That invitation was always accepted with interest, and AAAS had a rather good record of recommending men and women who were selected. Sometimes as many as half of the names submitted for appointment or reappointment were in fact appointed or reappointed and as each new term began the board was pleased to find some of its nominees included (41).

## Military Service and the Scientific Manpower Commission

The military buildup that followed the outbreak of the Korean War in 1950 depended partly upon selective service and partly upon recall of reservists from World War II and the following years. As usual, difficulties sometimes arose in deciding whether a particular reservist or draftee would be of greater usefulness in military service or in a civilian position. Engineers had the help of the Engineering Manpower Commission in dealing with such cases, but scientists had no similar organization. The American Chemical Society took the initiative in inviting other societies to join them in forming a committee to cooperate with government agencies in establishing a program for the proper training and use of scientific personnel in time of national emergency.

The AAAS board of directors was hesitant to join in that arrangement but did appoint a small committee to keep watch over the situation and to advise the board (42). Individually, Howard Meyerhoff, the association's executive officer, thought more aggressive and positive action was needed, and he quickly became effective in working with military and selective service authorities. Together with

Alden Emery and Walter Murphy of the American Chemical Society, he called together representatives of other societies for a meeting that eventually led to formation of the Scientific Manpower Commission (SMC).

That commission might have come into existence sooner had the AAAS board been more enthusiastic. They still hesitated, however, and decided not to send a representative to the proposed organizational meeting. They did tell Meyerhoff that he might attend as an observer for AAAS (43).

By the following spring the board had reversed itself and agreed to become one of the organizations exercising policy control over the planned Scientific Manpower Commission. Meyerhoff and Detlev W. Bronk became the association's representatives. Organizing the SMC took several months. By March of 1953 the SMC was incorporated and the board was willing to support it. Although they agreed to contribute $5,000 to help the SMC get started, board members warned that they were not setting a precedent and added, "It would therefore not be proper for the Scientific Manpower Commission to assume that the AAAS will necessarily continue its support" (44). Meyerhoff then left AAAS and accepted an appointment as administrative officer of the SMC (45), a position he held from 1953 until 1962 when he left to become chairman of the Department of Geology at the University of Pennsylvania.

In addition to its original function of dealing with military and selective service officials on individual cases, SMC broadened its program and became a widely useful collector and disseminator of information about scientific and engineering personnel trends in the United States through its periodical *Manpower Comments* and by other means.

In its early years the SMC was given free office space by the Carnegie Institution of Washington, but when Carnegie needed that space for its own use the commission had to rent commercial office space. That strain on its limited budget was relieved in 1962 when M. H. Trytten, then one of the two AAAS representatives on the SMC board of directors, offered space adjacent to his own office, the Office of Scientific Personnel of the National Research Council. Shortly after that move Betty Vetter succeeded Howard Meyerhoff as executive director of the commission, and 10 years later she and AAAS reached agreement that SMC would become a participating organization of AAAS. Under that arrangement SMC was given a new title, Commission on Professionals in Science and Technology, moved into AAAS office space, and began to serve as the association's office of scientific and engineering personnel, although retaining its separately incorporated status (46).

## Testimony and Advice

From time to time officers of the association appeared before congressional committees with testimony presented in the name of the association. The council frequently passed resolutions concerning matters of national policy and sometimes asked to have those resolutions sent to the appropriate government agencies. Occasionally a phone call from a committee staff member asked advice about a pending bill, requested the names of people who might be invited to present testimony at a committee hearing, or sought information about an individual. As one example, when the National Advisory Committee on Aeronautics was being transformed into the National Aeronautics and Space Administration, one of Senator Lyndon B. Johnson's aides phoned to ask whether scientists would approve appointment of T. Keith Glennan as NASA's first administrator, a position in which he then served from 1958 to 1961.

Some government officials, members of congressional staffs, and occasionally members of Congress attended AAAS annual meetings. The seminars for members of Congress or their staff aides (see Chapter 9) never dealt with matters of pending legislation, but the friendly relations established through those evenings sometimes resulted in informal requests for information and advice. The association's tax-exempt status meant that never more than a minor amount of time or effort went into attempts to affect legislation. Yet courtesy and the self-interest of the scientific community called for responding to requests for names or advice, and in that informal way as well as in formal statements or testimony the views of the scientific community, or at least of its AAAS representatives, were given to members of Congress.

Sometimes AAAS initiated the advice. In a board of directors discussion of the association's political responsibilities, the members agreed that education concerning issues important to science was clearly a responsibility of the association. As one means to that end, Paul Scherer suggested that when such an issue arose the association might invite a small group of persons to come together to discuss the matter and that we then publish in *Science* whatever statement the group decided to make, with the statement clearly identified as that of the participants in the discussion and not as a statement by AAAS or its officers (47).

One example was an informal seminar held on January 8 and 9, 1960, to consider the proposition then being advanced by Senators Hubert Humphrey, Estes Kefauver, and other members of Congress that the United States should bring a number of its scientific agencies together in a Department of Science. Lloyd V. Berkner, then president of Associated Universities, A. Hunter Dupree, University of California, James McCormack, vice president of MIT, James A. Mitchell, director of the conference program of the Brookings Institution, Emanuel R. Piori, director of research at IBM Corporation, Don K. Price, dean of the Graduate School of Public Administration (later called the John F. Kennedy School of Government) at Harvard, and I spent those two days holed up in a Washington, D.C., hotel debating the advantages and disadvantages of such a department. Rather than try to arrive at a "yes" or "no" decision, we agreed to consider the several kinds of science department that had been proposed and the criteria by which their merits might be judged. The published report of our discussions (48) did not settle the basic question of centralization versus decentralization of responsibility for scientific matters. That question had arisen periodically from the time of the Allison Commission of 1884–1886 (49) to 1985 when President Reagan's Commission on Industrial Competitiveness recommended bringing civilian research and development agencies together into a Department of Science and Technology (50).

## Political Meetings

In addition to its scientific meetings and research conferences, AAAS held two meetings whose major purpose was to influence national policy: the Parliament of Science in 1958 and the Symposium on Basic Research in 1959.

### The Parliament of Science

Clarence E. Davies, secretary of the American Society of Mechanical Engineers and a constructive and devoted member of the AAAS council and several AAAS committees, several times proposed that in addition to its annual scientific meetings the association should also hold an annual parliament of science, a meeting to consider current policy issues concerning the status and problems of science and technology and their contributions to the national well being. Only once, however, was a meeting with that title actually held.

The shock that followed launching of the first Soviet satellites led to extensive political debate and much media attention to questions of what the United States should do to "catch up" or "stay ahead." It was in that atmosphere in 1958, while Congress was planning the National Defense Education Act of 1958, that AAAS held its Parliament of Science. A "parliament" should of course have a representative membership, and to that end all of the association's sections and affiliates were invited to nominate participants. From those nominees a small committee under the chairmanship of former AAAS president Warren Weaver selected somewhat over 100 working scientists as parliament members and invited an additional 60 observers from industrial, governmental, academic, and other organizations.

The report of the three days of lively discussion (51) started off with a statement of basic principles, such as the tentative nature of scientific knowledge; the fact that science is an activity of the human race and not of political subdivisions; that science is only one expression of human curiosity and one part of human knowledge; and that the nation therefore needed an educational system strong at all levels and in all fields of knowledge.

After those statements of basic principles came recommendations in five areas. Those areas and their discussion leaders and recorders were:

- the support of science—William V. Houston, president of Rice Institute, and Mina Rees, dean of Hunter College;
- organization and administration of science in government—W. Albert Noyes, dean of the Graduate School, University of Rochester, and Emanuel R. Piori, director of research, IBM Corporation;
- communication among scientists and communication of scientific ideas—J. Murray Luck, professor of biochemistry, Stanford University, and Mary I. Bunting, dean of Douglass College, Rutgers University;
- the selection, guidance, and assistance of students—Henry Eyring, dean of the Graduate School, University of Utah, and Harry C. Kelly, assistant director, National Science Foundation; and
- the improvement of teaching and education—George W. Beadle, chairman of the Division of Biology, California Institute of Technology, and James F. Crow, professor of zoology and genetics, University of Wisconsin.

After an opening plenary session each section met in at least two extended sessions to develop recommendations in its area. Section reports were then brought to two sessions of the parliament as a whole, where most of the sectional recommendations were approved by the parliament. Some of those recommendations endorsed existing government practice. Some opposed a current proposal, for example, a federal Department of Science. Others recommended new or expanded programs. Some were addressed directly to government agencies and others, although addressed to scientists or educators, had implications for possible government action.

There is no need to repeat here the recommendations that were made decades ago in a different political and international climate, and no similar meeting has been held since. The 1958 Parliament of Science did demonstrate, however, that under conditions of national stress the association could bring together a reasonably representative group of scientists from industry, government, and academia to debate and to agree on the procedures and priorities they considered most useful for sustaining the nation's scientific enterprise and for making that enterprise most useful to the nation as a whole.

## Symposium on Basic Research

The 1959 Symposium on Basic Research was a special, one-time event proposed by Alfred P. Sloan, who had retired from the General Motors Corporation and was devoting most of his time to the private foundation that bore his name. Five years earlier General Lucius Clay, a trustee of the Sloan Foundation, had proposed a program of research support to selected young scientists to help them move forward on whatever research problems they found most interesting and challenging. Experience with that program had been so favorable that Sloan wanted to stimulate more widespread societal support for basic research.

Accordingly, he and his associates at the Sloan Foundation invited AAAS and the National Academy of Sciences to work with the foundation in organizing and conducting a symposium on basic research. An arrangement committee with a representative of each of the three organizations was chaired by Warren Weaver, who was then about to retire as vice-president for the natural and medical sciences of the Rockefeller Foundation to become vice-president of the Alfred P. Sloan Foundation.

The symposium was held in New York, with all sessions but one in the Caspary Auditorium of the Rockefeller Institute, which was soon to become the Rockefeller University. Attendance was by invitation, with nearly 250 present for the first two days of formal papers and panel discussions and with 100 invited and present for a final day of discussion. Participants heard addresses by prominent scientists and by leaders of government and industry, including William O. Baker, Robert Oppenheimer, James R. Killian, Crawford Greenwalt (president of E. I. du Pont de Nemours Company), and Robert L. Wilson (former chairman of the board of Standard Oil Company). All of the addresses and a summary of the concluding discussions were published in the proceedings volume, *Symposium on Basic Research* (52).

The one exception to the Rockefeller Institute site was a banquet at the Waldorf-Astoria Hotel attended by an additional 200 guests — mostly major officers of industrial firms, publishing houses, or banks — selected by Sloan because he thought they should be influenced to have a more supportive attitude toward basic research. The principal speaker was President Eisenhower. A few months earlier he had declined the association's invitation to address the 1958 annual meeting in Washington, D.C., but an invitation to address the Sloan–AAAS–NAS symposium was more persuasive. Under the title "Science: Handmaiden of Freedom," the President described government responsibility for the support of basic research, called for increased support, and made the first announcement that he was recommending to Congress that the federal government finance construction of the Stanford Linear Accelerator. In conclusion the President stressed a point that was made repeatedly during the three days: "We cannot improve science and engineering education without strengthening education of all kinds. America must educate all the varied talents of our citizens to the limit of their abilities."

In order to reach a larger audience than the 250 invited participants, a public relations firm saw to it that the sessions had good coverage by the print and electronic media. A clipping service recorded over 800 resulting editorials and news stories from all over the country.

There is no way of knowing how much was accomplished by the symposium. Press coverage was extensive; the symposium volume sold well; some very influential leaders of business, education, and government heard strong pleas for more favorable conditions for the pursuit of basic research. However, scientists among the participants probably heard little new. When the AAAS board of directors held a post mortem discussion of the symposium, they "considered ways in which similar meetings might be made more effective. It was generally agreed that, to as great an extent as possible, discussions should avoid mere repetition of the obvious and the well agreed upon and should focus upon problems on which there is not agreement. ... Interest-arousing titles such as 'Is the Day of the Small College Over?' and 'A Trap for Talent' were suggested" (53).

The criticism implied in that quotation seems to miss the point, however. The symposium was intended to help scientists interested in basic research, but they were not the target audience. The symposium was an "event" directed toward societal leaders. The summary volume included this justification:

> When the authors of *The Pursuit of Excellence*, the Rockefeller Brothers Fund report on American education, looked back over the span of history, they reached the provocative generalization that a society gets the kind of excellence it understands and appreciates. The raw, rebellious, revolutionary American Colonies produced great and enduring political theory and political institutions, but not the great art being produced in Europe at the same period (54).

The objective of the symposium was to help a number of leaders of the nation gain a fuller understanding and appreciation of the role of basic research in acquiring and retaining the world leadership the nation then enjoyed. If it helped in that respect, the symposium was a success.

## Financial Relationships

The National Science Foundation first-year appropriation was only $25,000, just enough for hiring a staff and to begin planning future activities. As those plans developed, the budget request for fiscal year 1952 was for $14 million, nearly all for research grants and graduate fellowships. In August of 1951, the House of Representatives approved an appropriation of $300,000, little more than had been appropriated the year before. The Senate had yet to act, and Howard Meyerhoff wrote to all members of the AAAS council urging them to write to their senators and urge appropriation of the full requested amount (55). Although NSF did not get the desired $14 million, the House and Senate did agree upon $3.5 million.

Meyerhoff's action was representative of the mutually cooperative and helpful relationship that developed between NSF and AAAS. The foundation became the major supporter of the association's programs in science education, and the association performed a number of services for the foundation. Whether a particular program was initiated by one organization or the other, there was usually a good bit of give and take along the way as continuing plans were developed and modified. Initially a relaxed and friendly relationship extended even to the matter of indirect costs on grants and contracts.

## Overhead Rates

During World War II the Office of Scientific Research and Development wrote contracts with industry and universities designed to pay all of the expenses, direct and indirect, but no more. Those "no loss–no gain" contracts were intended to mean that contractors neither lost nor gained on work for the government. That policy was one of the parents of the postwar grant programs supporting research paid for by government funds but conducted by nongovernmental organizations. The other parent was the policies of the Carnegie, Rockefeller, and other private foundations that supported academic research. Their grants were normally of a stated amount, based upon a budget submitted by the grantee, often not differentiating between direct and indirect costs, and sometimes not even recognizing the indirect costs involved.

As the new National Science Foundation staff debated the merits of not paying indirect costs (more money for research) versus paying those costs (less drain on the institution's other funds), the staff decided that the foundation should pay at least part of the overhead costs. The decision still left "the practical question of whether to pay the full amount, which would require detailed inquiries into the varying practices of institutions, or to pay at some uniform rate which might or might not approximate the full overhead costs. For 'simplicity and ease of administration' the staff chose a uniform rate, later fixed at 15 percent of the total direct costs" (56). That was the original policy, but as grantees later found, NSF began to require cost sharing and federal auditors lost all reluctance to make "detailed inquiries into the varying practices of institutions."

AAAS also had a policy decision to make. Led by its treasurer Paul Scherer, the board decided that AAAS should not request support for any activity for which it was not willing to devote some of its own money. The easiest way to do that was to request modest or no reimbursement for the overhead costs on grants. Officers of NSF and other supporters knew that was AAAS policy, and with that understanding, financial negotiations could concentrate on the substance and scale of the work to be done, and overhead costs were not a source of friction.

In 1958, during a general review of the association's policies and practices, the board again considered the matter of overhead charges and confirmed the earlier decision. On grants or contracts of substantial size to support new activities—for example, the Traveling Science Libraries—we had requested an overhead rate of about 10 percent of direct costs, an amount somewhat less than true costs. For some smaller grants no indirect costs had been collected. For example, the Parliament of Science, held earlier that year, was the association's idea and would have been held with or without supporting grants. An overhead charge of 15 percent might have been taken from the Sloan and Rockefeller Foundation grants that supported the meeting, but the board decided to seek reimbursement only for the direct costs (57).

In the 10 years from 1955 through 1964 the association received a total of $4,649,143 for direct costs under grants and contracts, of which $3,639,521 came from government funds. In addition the association received $435,943 for indirect costs, of which $348,727 was from government funds. That meant that overall the association had been paid an indirect rate of 9.4 percent on all its grants and contracts, and 9.6 percent on those from the federal government. If one assumes that the 15 percent that NSF had originally decided to provide for overhead represented the true cost, the difference between 15 and 9.4 percent represented a AAAS contribution of about $260,000 to its grant-supported activities. If the "true" cost to AAAS was 27.74 percent of direct costs — the average rate determined by government auditors for the years 1961 through 1964 — the association's contribution to those activities was $853,653 (58).

From the standpoint of the board that was quite satisfactory, but the relaxed treatment of overhead could not continue. As federal grant programs continued to increase, government agencies had to pay closer attention to indirect cost rates. And as the amounts grew, financial aspects came more and more under the control of auditors rather than program managers. So it was that in 1965 auditors for NSF reviewed in complete detail all contracts and grants from NSF that had been in effect in 1964, and with less completeness, all during 1961, 1962, or 1963. They disallowed one item of $2,000 and another of $240; asked for more complete information on charges totalling about $5,000; and recommended that all indirect charges for off-site expenditures under the grant for the Commission on Science Education be disallowed. During summer writing sessions members of the education staff and a number of scientists and teachers worked together at a university that could provide housing and work space for a large team. The auditors did not think those sessions involved any indirect expenses for the association!

At the same time the auditors calculated that the true indirect costs to AAAS had ranged from 26.46 to 29.06 percent of direct costs for the four years under review and recommended that NSF pay additional amounts for part of the unreimbursed indirect costs. As the board of directors reviewed the auditors' findings and recommendations, they decided to abandon the practice of asking for a lower overhead rate than needed to pay all indirect costs but to continue the policy of cost sharing. On future grants, they therefore decided, AAAS would ask for full allowable overhead reimbursement at the rates determined by government auditors, with the budget for direct expenses showing the portion being borne by the association.

From then on, the audited rates for indirect expenses seemed reasonable. The auditors came to recognize that it did indeed cost the association something to establish, support, and monitor a group of people working somewhere outside of the association's headquarters building. AAAS continued to bear part of the costs of activities initiated by the association, and the federal agencies paid indirect costs as determined by government auditors (59). Yet the later practices took more time by both granting agency and AAAS and at least in that respect were more costly than the earlier system that relied more on mutual trust than on detailed auditing. However, occasional irritation and disagreements over whether certain costs should be allowed or disallowed were far outweighed by the benefits of the programs being supported.

## Tax Reform

In November of 1956 *Science* carried two articles by Paul Klopsteg on the coming financial difficulties of higher education. The first article developed the following argument. Most of the research support coming from federal agencies was for applied research rather than for the basic research that was traditionally the hallmark of university work. So far there was little federal money for the general support of higher education, but the impending influx of the baby boom generation would require much expansion of campus facilities. Much of the necessary new money might have to come from the federal government. There had not yet been much federal control of higher education, but increasing dependence on federal support would bring increasing danger of that control and the centralized bureaucracy needed to administer it. To avoid that danger Klopsteg proposed that scholarly research in universities and colleges be supported "to the greatest possible extent by gifts and grants ... derived from many private and some public sources" (60).

A week later the second article explained how that might be done: Change the income tax laws in such a fashion as to make gifts to colleges and universities (and other selected and appropriate institutions) as attractive and inexpensive to citizens of modest means as they were to the very wealthy (61).

That was not the first time *Science* had published such a recommendation, nor was *Science* the only source of that advice. Ten years earlier, when the postwar upsurge of federal support for academic research was just beginning, Robert W. King had proposed that gifts to the prospective National Science Foundation and other approved scientific agencies be made costless to the donor by having the donor's income tax reduced by an amount equivalent to the gift (62). A few months after Klopsteg's article appeared, the National Science Foundation published *Basic Research: A National Resource*. That report, which had the same objective as the association's later Symposium on Basic Research, argued that universities would be better able to perform their traditional functions of education and research if methods were adopted to achieve substantial increases in philanthropic giving (63).

Klopsteg's proposal received immediate attention from AAAS officers and others, and later from Congress. Specifically what Klopsteg proposed was to equalize, more or less, the net cost of a gift to the giver regardless of the giver's level of income. That equalization could be accomplished in various ways; the one he used to illustrate the principle was this:

(a) First, compute the income tax in the usual way, without deducting gifts to approved institutions from gross income;

(b) subtract the amount of qualifying gifts from the computed tax as a tax credit; and

(c) add a small "gift surtax" ranging from perhaps two percent for those with low incomes up to nine percent for those in the 82-to-91-percent bracket (64). (In those years individual income tax rates did go as high as 91 percent for single taxpayers with taxable incomes above $200,000 or for married taxpayers filing joint returns above $400,000.)

Some such arrangement did seem "fairer" than the then-current practice under which a dollar given by one with a low income cost the giver 80¢, or even the whole amount, while a dollar given by a donor with a large enough income cost the giver only 9¢.

In the weeks immediately following publication of those articles there were so many requests for reprints that AAAS relieved Klopsteg of the burden of replying (65). Three months later reports that the Bureau of the Budget was conducting a study of the implication of the proposal for federal tax revenue prompted the board of directors to decide that a private study of the proposal would also be desirable (66).

To conduct that study the association engaged the services of Stuart Rice and Associates, a consulting firm in Washington, D.C. Rice had been professor of sociology and statistics at the University of Pennsylvania before coming to Washington where he had served for a number of years as assistant director of the Bureau of the Budget monitoring statistical standards for the federal agencies (67). Later in the year the Ford Foundation made a grant of $20,000 that covered most of the costs of the study. At the foundation's request the study was expanded to consider not only the particular changes proposed by Klopsteg but also several alternative changes of tax regulations that would accomplish approximately the same end (68).

Libert Ehrman of the Stuart Rice and Associates staff analyzed the probable consequences of several methods of equalizing the cost to the donor of gifts to higher education or other qualifying recipients. His report included historical data on the amounts of philanthropic contributions by taxpayers of different income levels, changes in philanthropic giving following several earlier changes in income tax rates and regulations, and comparative estimates of the probable consequences of several alternative methods of achieving the objectives that prompted Klopsteg's original proposal (69).

With the full report in hand the board authorized me to have prepared a congressional bill "embodying what appear to be the most desirable adjustments in federal income tax laws to stimulate philanthropic gifts for educational and other purposes, and to seek means of having the bill introduced into Congress" (70). The proposal soon gained support from representatives of the National Industrial Conference Board (71), and over the next year several implementing bills were introduced into the House and Senate. They varied somewhat in their provisions, with H.R. 2440 introduced by Representative Frank Thompson of New Jersey being the one that gained the most attention. Thompson's bill limited the proposed equal-cost provisions to gifts to educational institutions and did not include the churches, hospitals, and research institutions that Klopsteg had originally suggested should have the same treatment. Wilbur Mills, chairman of the House Ways and Means Committee, held hearings over several weeks covering a variety of matters concerned with income tax legislation, asked Klopsteg to testify (72), and said that after Congress reconvened in 1960 he planned to hold further hearings specifically on Thompson's bill and the others stimulated by the Klopsteg proposal.

That time at the end of 1959 marked the peak of interest in the proposal. As the House hearings were being held AAAS received a number of replies to a letter Klopsteg had sent to college and university presidents telling them of the bill, the hearings, and the opportunities for increased philanthropic support. Most replies were supportive, and many included copies of letters to senators or representatives urging adoption of the proposed legislation.

But as heightened attention aroused support, it also aroused opposition. The American Council on Education (ACE) announced that its Committee on Taxation and Fiscal Reporting to the Federal Government had decided that its support of the Thompson and related bills would be inadvisable. The reason, as I learned in discussions with members of the ACE staff, was that although they liked the idea, it was opposed by the American Association of Land Grant Colleges and Universities and the National Association of State Universities (which later combined as the National Association of State Universities and Land-Grant Colleges). Those colleges and universities had had extensive experience with direct federal support under the Morrill Land Grant Act and some other programs. They liked that direct support better than they expected to like the more diverse and uncertain support that might come through many private gifts, and they were not afraid of federal control coming as a consequence of greater federal support (73).

In response to that negative position the AAAS board urged the American Council on Education to get together with other interested groups to consider how higher education would be financed through the coming decade of rapid expansion and recommended that the question of support be examined in terms of total needs rather than from the limited perspective of supporting or opposing a particular proposal. ACE did hold such a meeting, but the results were mostly negative.

In a parallel effort to revive or maintain interest Paul Klopsteg and I met with representatives of the Bureau of the Budget, the Department of the Treasury, and the Joint Economic Committee of Congress in a meeting arranged by Stuart Rice.

That session ended in a puzzling disagreement: Estimates of how much money the federal government needed to collect in order to provide one dollar in grants or appropriations to higher education ranged from $1.07 to nearly $3.00 — that is, that it cost as little as seven cents to as much as two dollars to collect as taxes, process, appropriate, and deliver one dollar in a grant or appropriation to higher education (74). Someone's figures seemed to be sorely amiss.

The focus of the Klopsteg proposal was institutional independence: How can we increase financial support for colleges and universities without also increasing government control of those institutions? The extent to which the practical interests of federal government agencies were shaping university research programs made that question a timely and important one to many scientists. But the proposal raised other issues and the Bureau of the Budget reinterpreted the proposal to ask: What reduction would there be in tax revenues? The land-grant colleges and state universities shifted the question to: Would we rather receive one large grant from the Department of Education or a lot of small grants from alumni and other friends? They opted for the large grant, but Klopsteg and other supporters of his proposal chose the many smaller gifts and the reduced threat of external control.

Some real differences in values were involved. In 1959 Don Price, a new member of the AAAS board, broadened the questions to consider some of the political aspects of the proposal. In a letter to Klopsteg he raised two issues concerning the more widespread effects that might follow adoption of H.R. 2440 or a similar bill.

The first issue concerned the size of the deduction for charitable gifts. Klopsteg had proposed equalizing the deduction upward by giving to all taxpayers essentially the same percentage savings in income tax payments as were then being enjoyed by taxpayers with the largest incomes. But, Price pointed out, the very existence of the right to deduct for charitable gifts was being questioned. If the Treasury and the Ways and Means Committee gave detailed consideration to the Klopsteg proposal one of the first questions they would consider would be the amount by which tax rates would have to be raised to produce the same revenue as before. And an outcome of that debate might well be to flatten the cost of charitable gifts at 20 or 30 percent instead of at the low figure Klopsteg had proposed. That change would have little effect on small gifts from taxpayers of modest means, but it would surely discourage the large gifts to research universities that only the wealthy could afford. More fundamentally, Price speculated:

> Your plan proposes to transfer the arena in which the distribution of tax money will be fought over from inside the administration (which you want to avoid in order to prevent political control over education) to the arena of legislative debate and political pressure, where I think the political danger is even greater.

Once the principle of the plan is admitted a great many institutions will want to get in on the privilege. Is higher education distinct enough from primary education as a national goal to be able to deny this privilege to private schools? What share of the money would then go to parochial schools? What share to private schools set up to avoid racial integration? What share to hospitals and community chests and the Red Cross? As I think about the political melee that would follow the adoption of this principle I believe I would almost prefer a system of grants in aid from the Office of Education than the plan which I might call (if I were not afraid of competing with Margaret Mead as a phrase-maker) taxation without administration (75).

Klopsteg replied that the principle of allowing individual taxpayers to determine where some of their money would go as gifts instead of as tax payments had already been adopted, and that his proposal was intended not to make a basic change in the principle but to increase the amount of money flowing to appropriate institutions through that choice. As for possible extension of the principle to many kinds of recipients, he was not worried about what might happen to institutions that already were receiving support of an appropriate amount, but he was concerned about colleges and universities that were entering into a period when greatly increased size and responsibility would require greatly increased support (76).

The exchange of letters was a friendly one. Each recognized the problems raised by the other, but they differed in how they weighted those problems. I sent copies of both letters to all board members to help them think about the issues involved. But in the pejorative sense in which the term is frequently used, the exchange of letters was merely an "academic" argument, for the whole proposal was moving off stage. Discussions with the American Council on Education had not resulted in support from that central organization of American colleges and universities. The congressional hearings enshrined Klopsteg's proposal in the *Congressional Record* (77), but neither H.R. 2440 or any similar bill got out of committee. In 1960 the idea dropped out of board discussions and off its agenda. It did not disappear, however, and in 1988 Burton Weisbrod advanced a similar proposal as a means of helping to meet the mounting needs of universities at a time when large deficits in the national budget and international trade would restrict federal funds for research and educational programs (78).

As AAAS was attempting to change income tax regulations it was also deciding not to request changes in its own tax liability. As a tax-exempt organization under Section 501(c)(3) of the federal tax code, AAAS was not required to pay federal income taxes, and under the laws of the District of Columbia it was not required to pay sales taxes on its purchases. It was, however, required to pay property tax on its land and buildings and District of Columbia personal property tax on its other assets. To obtain exemption from the property tax would require an act of Congress adding AAAS to the list of institutions to which Congress had already granted that exemption. To secure exemption from the personal property tax would require reincorporation in the District of Columbia.

As those possibilities were discussed by the board of directors after the association had purchased the property at 1515 Massachusetts Avenue, two questions had to be decided: Should the association attempt to secure exemption from either or both of those taxes and, if so, before or after the proposed new building had been erected? Warren Magee, the association's legal counsel, advised postponing a request for exemption from the property tax if the association was likely to erect that new building in the near future. The board accepted that advice, but disagreed with Magee's recommendation that AAAS not reincorporate but rather continue to pay the relatively small personal property tax (79).

On reconsideration, however, the board came to agree with Magee and with Howard Meyerhoff who recommended retaining the original 1874 incorporation in Massachusetts. The incorporators included Joseph Henry, James D. Dana, F. A. P. Barnard, Asa Gray, and other distinguished pioneers of American science. The new incorporators in the District of Columbia would not attain comparable seniority and distinction for a century or so, and the modest annual savings to be gained from not paying the District's personal property tax was a small price to pay for the distinction of the old charter granted to such illustrious forbearers.

Nevertheless, the original charter did have to be amended, for its limitation on the amount of property the association could hold had become completely out of date. To take care of that matter the Massachusetts Legislature on May 16, 1952, adopted a special act raising to $5.5 million dollars the value of real estate the association might hold and granting authority to hold other assets of unlimited amount (80).

After the new building was erected in 1955 the assessed value naturally increased and the association was charged with a property tax of about $12,000 a year. Arthur Hanson, the association's new legal counsel, and his father had earlier secured congressional legislation adding several Washington-based organizations to the tax-exempt list and in 1956 thought there would be a better-than-even chance of persuading Congress to add AAAS to that list. At that time several other organizations also wanted relief from the property tax. Thus the board approved an agreement with Hanson that he would seek to have an appropriate bill introduced into Congress and that the costs would be divided among the several organizations included in that bill. Nothing came of that decision; congressional opposition to extending the tax-exempt list proved to be too strong to allow realistic expectation of adoption of the proposed bill (81). The more favorable opportunity of earlier years had been lost.

Tax rates gradually increased so that in 1970 the association paid a tax of $20,303 on the headquarters building and land and a personal property tax of

$3,153. By that time, however, some nonprofit organizations that were not required to pay local property taxes were making voluntary contributions to compensate for fire and police protection and other services they received from city governments. Perhaps the decision not to request exemption from the property tax was evidence of good corporate citizenship.

*Chapter Three*

# Annual Meeting

Ever since 1848 AAAS meetings have brought American scientists together to discuss research findings, interdisciplinary problems, relations with government agencies and with society at large, and other matters of concern to scientists. And ever since its founding the association has had two audiences to satisfy: men and women directly engaged in scientific and technical work and amateurs, students, and members of the public who were interested in science but not — or not yet — engaged in its advancement, application, or teaching. This duality of audiences plus the diversity of interests of AAAS members resulted in a basic conflict in planning meetings: How much attention should be given to reports and discussion of new research findings and how much to matters of wider interest such as cross-disciplinary discussions, consideration of educational, methodological, or policy issues that affect all or most of science, or to public information and education about scientific affairs? Some sessions on those wider topics were surely appropriate, for AAAS was comprehensive in its coverage. Yet while science-wide topics had merit, many of the association's members were specialists and also wanted programs in their own scientific specialties.

## Affiliated Societies

As the number of specialist societies increased, as many of those societies became affiliates of AAAS, and as affiliated societies chose to meet with AAAS, the program of the annual meeting came to be a product of more and more different planning bodies. As a further consequence of the growing number of affiliates, the planning of program sessions became one of the central issues in relations between the association and its affiliates. Early in the 1900s, the AAAS council decided that whenever an affiliated society met with AAAS the corresponding association section could dispense with the usual sectional program and give program responsibility entirely to the affiliated society. In fact some of the sections thought that arrangement should be mandatory (1).

The issue was complicated by the divergent nature of the affiliates in different fields. Most of the affiliated societies that chose to meet with the association were in the biological sciences, for biologists organized themselves rather differently than did other scientists or engineers. Chemists, physicists, social scientists, and engineers usually founded only a few national organizations. But not biologists — there were too many different kinds of biologists. In fact, biology as an integrated science resulted from the coming together of many different groups interested in different life forms or processes, rather than being the parent discipline from which the specialties arose. Thus instead of only a few large organizations of biological scientists, many were organized to serve the interests of scientists specializing in agronomy, bryology, comparative anatomy ... and on to zoology. Most of these specialized and often relatively small professional societies became affiliates of AAAS (2), and many of the affiliates found it advantageous to meet at the time and place of the AAAS meeting and to have AAAS handle some of the details of arranging meeting facilities and publishing programs. Thus AAAS meetings came to have a fairly heavy biological emphasis and included numerous sessions for reports of original research.

However, there was also a counter trend. The growing number of meetings of specialized societies provided many opportunities for reports and discussions of new research findings, so it became less necessary for AAAS to meet that need. Thus, in 1946 a small conference of Albert S. Blakeslee, Detlev Bronk, Sewall Wright, Robert M. Yerkes, Paul Weiss, and a few others called upon AAAS to devote more of its energies to symposia, conferences, and publications dealing with the objectives, methods, and broader issues of scientific inquiry (3).

This thinking became sharply crystallized in 1947 when Harlow Shapley, the association's new president, drawing partly on discussions he had with Moulton concerning the roles and functions of AAAS and of the specialized societies, proposed some drastic changes in AAAS activities and structure (4). He recommended abolishing some or all of the sections, changing the basic nature of the annual meetings, and giving substantially greater emphasis to international relations of science.

The sectional organization, he pointed out, had been adopted when there were few specialized scientific societies in the United States and when sectional meetings of AAAS provided prime opportunities for reporting new findings. That situation had changed so much that he proposed eliminating some of the sections. Using his own field as an example, he wrote that the section on astronomy "has been ... of slight importance for the past thirty years; in fact it has to many been more of a nuisance than an asset." In contrast, he added, "symposia involving the joint efforts of two or more sections have been pretty uniformly successful at past meetings of the A.A.A.S."

This observation led directly to the next recommendation: "I venture the suggestion that the whole purpose of the annual meetings of the A.A.A.S. be radically redirected in the following way: (a) that large professional societies be encouraged *not* to meet at the same time and place as the A.A.A.S.; (b) that the week-long series of programs at annual meetings ... be devoted wholly to interfield conferences, symposia and addresses; (c) that the open sessions, operated by a widely-drawn committee on program structure and management, be built almost exclusively on invited contributions for the various interfield symposia; [and] (d) that the aim, spirit, and slogan of the meetings of the A.A.A.S. be integration."

Concentrating exclusively on interfield symposia, conferences, and addresses would, he thought, probably reduce attendance substantially. However, he added, "after one century of existence the A.A.A.S. is presumably modest enough and dignified enough to prefer quality to quantity. There is no great advantage to the society in rivaling the attendance records of the American Legion, or even of the American Chemical Society. It might be to our credit to be able to report that the attendance is only sixty percent as great as ten years ago and that in consequence much good work has been accomplished."

As a start in these directions, Shapely proposed that in 1948 the centennial meeting serve as a transition from the old type of meeting to the type he was recommending and suggested that all of its sessions be planned around the theme "One World of Science."

The board rejected Shapley's proposal to abolish some of the AAAS sections, but the members liked the idea of a special kind of meeting for the centennial year. The resulting Centennial Policy Committee (5) soon began to plan a three-pronged program: (a) the centennial meeting would be planned along the lines proposed by Shapley; (b) a special membership drive would seek to gain 15,000 new members during the centennial year; and (c) efforts would be pushed forward to arrange for construction of a permanent new home for the association.

As things turned out, progress on building plans was slow, and construction of the association's new building did not actually begin until 1955 (see Chapter 5). The membership drive was more successful, gaining 12,000 instead of the hoped-for 15,000 new members. And arrangements for the 1948 meeting had some unintended consequences.

On uncomfortably short notice — only a few months — section officers and officers of affiliated societies were told that at the 1948 meeting, no provision would be made for any of their meetings or for any sessions for the presentation of individual research reports. Instead, the entire program would consist of symposia and addresses arranged by a central planning committee. The abrupt departure from the association's long-established meeting pattern led to a good bit of grumbling in some of these sections and affiliated societies and gave a boost to what soon became a new annual meeting for biologists. The American Institute of Biological Sciences (AIBS) was then being formed (6), and under that new banner, the American Society of Zoologists, the Botanical Society of America, and several smaller societies quickly arranged to meet together in Washington, D.C., in the days just preceding the centennial meeting of AAAS.

## The Centennial Meeting, September 1948

To arouse interest in the centennial meeting *Science* published a number of articles describing scientific institutions in or near Washington that would be open for visits during the meeting. Ten days before the meeting, a special issue of *Science* included brief histories of AAAS, its Pacific and Southwestern Divisions, Silliman's *American Journal of Science*, the National Academy of Sciences, and the National Research Council. Appropriately, the cover bore pictures of two patriarchs of the association: 90-year-old Liberty Hyde Bailey, who had served as president in 1926, and 91-year-old Leland O. Howard, whose honoring was particularly timely since he had served as permanent secretary (executive officer) for 22 years,—longer than anyone else before or since—and in 1920 had become president (7).

The centennial meeting opened on the evening of September 13 with a plenary session in Constitution Hall. A brief address by President Harry S Truman was followed by the retiring presidential address of Harlow Shapley. With the convention theme being "One World of Science," Shapley entitled his address "The One World of Stars," a talk full of examples and advantages of international cooperation in astronomical studies (8).

That evening, however, most of the attention was focused on President Truman. It would have been difficult for the President to give an address more to the liking of the audience. He called for greater financial support of research, repeated his endorsement of the proposed National Science Foundation, and urged the Congress to pass the enabling legislation. He described the conditions of freedom and open communication needed for progress in science and belabored those in government who thought the nation's security could be better guarded by excessive secrecy concerning what was already known than by creating conditions leading to new advances in knowledge (9). Two days later, the AAAS board of directors adopted a resolution thanking the President and endorsing his statement. Well they might; it would not have been more favorable to scientific progress if the directors themselves had written the speech.

As a matter of fact, the first draft was written by a future president of AAAS. Truman had not accepted the invitation to speak until just before the meeting. On Friday night, three days before the meeting was to open, he was sailing down the Potomac River on the presidential yacht with Clark Clifford, his special counsel, and John Steelman, who headed the President's science policy staff. They discussed the AAAS invitation and agreed that the President should accept it. Saturday morning Clifford telephoned George M. Elsey, one of the President's assistants, and asked him to prepare a suitable address of about 10 minutes length. That request was an unexpected interruption, for Elsey was busy preparing outlines for the many short talks President Truman was planning to give on his "whistle-stop" campaign trip to the West Coast, a trip that was to start only a few days later. Putting that task aside, Elsey set to work preparing the President's address, starting with a draft that had earlier been submitted by Edward U. Condon, director of the

Bureau of Standards and soon-to-be president of AAAS. Between Saturday morning and Monday evening the address went through five drafts and was reviewed or edited by David Bell, Clark Clifford, Charles Ross, and a few others, including Truman himself, for the fifth draft shows changes in his handwriting, as does the final reading copy (10).

Condon must have been pleased as he heard the address proceed and recognized the ideas being expressed (11). The audience, too, was pleased, not only by the content but also by the President's manner. Early in the address, as one attentive listener recalls, Truman misread a sentence, giving it exactly the opposite meaning from what was obviously intended. "He stopped abruptly, in a sense grinning at himself, and laughed, as the audience did ... [then] said 'That reminds me of a story!'" (12). He had the audience with him.

Newspaper reports the next day emphasized the President's attacks on some of the witch hunts then being conducted by the House of Representatives Committee on Un-American Activities (the Thomas Committee). The *Washington Post* headlined a first page account "Truman Hits Smears on Scientists" and quoted the President's statement that the indispensable work of American scientists "may be made impossible by the creation of an atmosphere in which no man feels safe against the public airing of unfounded rumors, gossip, and vilification." That attitude toward the House of Representatives Committee on Un-American Activities had been more succinctly expressed two days earlier in the Sunday *Post* by the cartoonist Herblock. His cartoon showed Chairman Thomas asking the other members of the committee "How did atomic energy information leak out to the damned scientists in the first place?"

The *New York Times* headlined its account "Truman Charges Smears and Gossip Hinder Scientists: Hits Politicians: President Says 'Red Herrings' Curb, Not Help Security" (13). Neither Edward Condon nor any other targets of the smear campaigns were mentioned by name in the President's address, but many members of the audience were familiar with the House Committee's — and especially Congressman Richard Vail's — characterization of Condon as "the weakest link in our atomic security chain" (14). William Laurence, science editor of the *Times*, concluded the story by writing that as Truman finished speaking he went down the front row of officers and members of the association who were seated on the platform and "smiled broadly as he shook Condon's hand." He probably did not know that Condon had written the first draft of the address he had just given (15), but he surely knew that Condon was one of the smeared victims he had been talking about.

Although no one knew it at the time, that address ended a long tradition. At every previous Washington meeting of AAAS — in 1854, 1891, 1902, 1911, and 1924 — the President either spoke or recognized the association in some other way, such as a reception at the White House. In 1958 President Eisenhower was reminded of that long tradition and invited to address the association, but the invitation was not persuasive; golf in Georgia was more attractive than staying in Washington for Christmas week. President Eisenhower did, however, come to New York the following September to address the banquet session of the association's Symposium on Basic Research (see Chapter 2).

For the four days following the gala opening evening of the centennial meeting, mornings were devoted to several concurrent symposia. Typically, each included papers by three speakers from different fields of science or different types of institutions. Representative topics were genes and cytoplasm, human individuality, interactions of matter and radiation, sources of energy, the upper atmosphere, and waves and rhythms. *Science* published abstracts of all of the talks and the full texts of some (16), and all were included in the *Centennial Volume* that appeared a year later (17).

Afternoons were free of sessions. Many scientific societies with headquarters in or near Washington held open house during those afternoons, and tours were available to a number of research laboratories. Some 300 participants toured the U.S. Department of Agriculture's Agricultural Research Center at Beltsville, Maryland; over 500 visited the National Institutes of Health in Bethesda; 156 went to see the U.S. Navy's David Taylor Model Basin; and others took tours to other research establishments (18).

Evenings were for addresses open to the public. Except on the first evening when President Truman and Harlow Shapley spoke, there were about four speakers each night. Those addresses were also abstracted in the proceedings issue of *Science* and published in the centennial volume.

AAAS meetings have regularly received substantial attention from the media, and the centennial meeting was no exception. Some 180 reporters wrote accounts for their newspapers or magazines. All three radio networks and the new field of television reported on the meeting. CBS gave it the largest amount of time, reporting several daily features and ending each day with a late evening commentary devoted exclusively to the meeting. On one evening CBS staged its popular *Town Meeting of the Air* in Constitution Hall, broadcasting a discussion on "What Hope for Man" by Brock Chisholm, director general of the World Health Organization, Fairfield Osborn, president of the New York Zoological Society, Harlow Shapley, chairman of the AAAS board of directors and professor of astronomy at Harvard, and Edmund W. Sinnot, AAAS president and director of the Sheffield Scientific School at Yale. Irving Gitlin of CBS concluded that the hours that CBS had devoted to reporting the meeting "represent coverage of an organizational meeting second only to that devoted to political conventions" (19).

## What Kind of Future Meetings?

The AAAS board of directors had already decided that in 1949 the association would go back to the kind of meeting held in 1947 and earlier years. For advice on the longer future, however, council members were polled to determine their preference. Twenty-six percent voted for the traditional type of meeting, 21 percent preferred the centennial-year type, and 54 percent compromised by voting to alternate the two types (20). But before this advice could be followed, a new factor was introduced by the change of administrative secretaries.

Forest R. Moulton retired at the beginning of 1949 and was succeeded by Howard A. Meyerhoff, a geologist on the staff of Smith College. Meyerhoff had earlier had substantial experience with AAAS as chairman of the section on geology and in 1945 and 1946 as the association's staff member who followed legislative consideration of the proposed National Science Foundation and wrote frequent reports on that topic for *Science*. In 1946 he chose to return to geology, initially as a member of a party of geologists surveying mineral resources in the Argentinean Andes. At the beginning of 1949 he came back to AAAS as the board of directors' first choice as Moulton's replacement. He considered the vote of the council on the preferred type of future meetings to be inconclusive and argued that, in any event, the affiliated societies rather than the council should have been polled. Pending later decisions about future years, he proposed that the 1950 meeting be a combination of the two types.

When Moulton retired, John Hutzel, the assistant administrative secretary, resigned to assume a new position in the federal government. His place was taken by Raymond L. Taylor, who had been serving as chairman of the biology department of Sampson College in New York. His primary responsibilities were membership recruitment and the planning and management of the association's meetings. A botanist, he hoped to persuade some of the new AIBS societies to resume meeting with AAAS.

As longer-range plans were being debated, the 1949 meeting in New York was a great rebound from the 1948 meeting in Washington. In contrast with a registration of 2,734 at the centennial meeting, there were 7,014 registrants in New York, far surpassing the previous peaks of 4,940 in Chicago in 1947 and 4,206 in Washington in 1924. Instead of the centrally planned symposia and addresses of the centennial meeting, sessions in New York were arranged by the AAAS, by 17 of its sections, and by 62 affiliated societies, of which many were biological (21). Then as earlier, one of the features of the meeting was the annual Biologists' Smoker, an evening gathering with no programs, just an opportunity to meet and chat with old friends and to make new ones. Raymond Taylor happily reported that those in attendance at the Biologists' Smoker of 1949 had consumed 7,200 bottles of Coca Cola, 1,200 cans of Pabst beer, and uncounted Nabisco pretzels and Cheese Ritz, all donated for the occasion.

The 1949 meeting also had a new feature. Then, and for several following years, a committee consisting of scientists and others from in or near the meeting city selected one or more topics for association-wide symposia. These symposia on generally important topics drew substantial audiences, but they did not end the arguments over the amount of attention to be given to individual research reports. At first, AAAS tried to forbid any competing programs at the time of the association symposia, but objections from some of the affiliated societies soon led to withdrawal of that ban.

It was 11 years before registration again reached the level of the 1949 meeting, for from 1950 on, AIBS met regularly in the late summer on a university campus. Automobile travel was easier then than at the time of AAAS meetings, and campus dormitories were less expensive than city hotels. The AIBS meetings quickly became popular, and AIBS and AAAS agreed not to meet in close proximity in any year. In an attempt to repair the slightly strained relations, the AAAS board of directors adopted a resolution affirming that the association welcomed participation of the biological societies in the AAAS annual meetings and that any expression of opinion to the contrary did not represent the thinking of the board, either then or at any earlier time (22). Raymond Taylor continued to try to woo back some of the biological societies, and in some years some did meet with AAAS. With AIBS meeting annually, however, AAAS meetings were no longer so heavily biological in content as they had been.

## The Arden House Statement

In June 1951 Howard Meyerhoff received a letter signed by A. J. Carlson, A. C. Ivy, and Ralph Rohweder that began with this bald assertion: "The AAAS has to find another function or die." Going on, the letter contrasted past and current conditions for science and the number of scientific societies, mentioned the increasingly troublesome restrictions on science that were developing out of fears of the Cold War, and concluded that the changed status of scientists in modern society required thorough reconsideration of the whole purpose and function of AAAS (23):

> Once the scientist had the freedom of the ivory tower, but when the products of the ivory tower had changed the world, the world of nonscientific men took over the tower. There is no secure retreat for the scientist today. He is inevitably involved in problems of public understanding because the public has involved itself in his work.
> Obviously this is no new thesis. The Committee of Atomic Scientists and the National Society for Medical Research are concrete embodiments of the rather widespread realization among scientists that they must accept responsibility for conveying to the public information which is critical to the making of public policy.
> The Committee of Atomic Scientists and the National Society for Medical Research cannot touch the central problem, however. The concepts of free enquiry and of nonauthoritarianism, of integrity and pragmatism — to mention a few principles of scientific activity — must be conveyed to the public or there can be neither substantial public understanding of science nor wider adoption of its

effective approaches to man's problems. An effort broad and strong enough to make progress toward so great a goal would have to be undertaken by an organization comparatively great, such as the American Association for the Advancement of Science.

The letter concluded:

We would like to have you consider this not merely as a suggestion for a new undertaking for the AAAS but as a plan for a total analysis of the work of the AAAS and what must be done to advance science from the point at which it stands in June 1951. Such an analysis may produce conclusions and program decisions in no way similar to any mentioned in this letter. It is not our purpose here to suggest answers. We hope simply that we have performed a more important function of scientists, the pointed formulation of a question. Is the AAAS living up to its great name in 1951?

Of the three authors of that letter, Carlson was a recent president of AAAS and the American Association of University Professors and the long-time chairman of the Department of Physiology at the University of Chicago. Ivy, a physician and physiologist, was head of the department of clinical sciences and vice president of the University of Illinois and the recipient of many national and international honors. Rohweder was executive secretary of the National Society for Medical Research, an organization defending and supporting the use of animals in research, an organization in which Carlson and Ivy were active.

As was customary for serious proposals that came to the association, that letter was submitted to the board of directors at their next meeting. The authorship and content of the letter would have assured careful attention, but in addition it met with an especially interested reception from Warren Weaver, a relatively new member of the Board of Directors who was director of the Division of Natural Sciences of the Rockefeller Foundation and a vigorous contributor to efforts to improve public understanding of science. Discussion of the letter led the board to invite several consultants to join them for a couple of days to debate the implications of the letter and the ways in which AAAS should respond to its challenge.

The outcome of their discussions was the "Arden House Statement," so named because the meeting was held at Arden House, Columbia University's conference center on the Harriman estate above the Hudson River. The Arden House statement became a permanent addition to the association's policy documents reprinted every year in the *AAAS Handbook*. As it was intended to do, that statement set off a serious and sustained search for more effective ways to meet some of the association's responsibilities, but it also led to uncertainties and conflict over the nature of the annual meeting and to such severe disagreement between Howard Meyerhoff and some of the elected officers that his services as administrative secretary were terminated.

The Arden House statement, of which Warren Weaver was the principal author, was unanimously approved by the participants in the Arden House meeting, and after the guests had departed was formally adopted by the board of directors. The statement did not recommend specific actions; rather, it called for a thorough reassessment of the association's objectives and activities. However, it was unambiguously critical of the annual meetings, calling them "outmoded," and asserting that they "must, first of all, serve scientists and science in such a way as to command the confidence and backing of scientists," the statement went on to call for a general reexamination of programs and activities. Then it became specific about meeting content:

> We have reached the stage where one over-all organization cannot effectively deal with the intensive and specialized interests of individual branches of science. The technical papers that present detailed results in chemistry, in physics, in mathematics, in zoology, etc., can more properly be presented before meetings sponsored and arranged by the appropriate professional groups.
> 
> It is thus clear that the AAAS should not attempt to hold to a pattern of annual meetings that was natural and effective many years ago, but which is now outmoded.
> 
> This is, in fact, only one aspect of an important general principle. In view of the present size and complexity of science, in view of the seriousness and importance of the relations of science to society, and in view of the unique inclusiveness of the AAAS, it seems clear that this organization should devote less of its energies to the more detailed and isolated technical aspects of science, and devote more of its energies to broad problems that involve the whole of science, the relations of science to government, and indeed the relations of science to our society as a whole.
> 
> This increased emphasis on broad problems should lead to new activities in wider fields, but it also requires a modification of what the AAAS tries to do within and for science. Thus it seems clear that a major present opportunity for the AAAS within science is to act, in all ways that promise useful results, as a synthesizing and unifying influence. As an obvious example, this indicates meetings at which one branch of science is interpreted to the other branches of science, meetings at which are stressed the interrelationships between the branches of sciences, meetings which cultivate borderline fields, and meetings at which the unifying theme would be central problems whose treatment requires the attack of several disciplines (24).

The statement then recommended that the association give greater attention to one of its constitutional objectives: "to increase public understanding and appreciation of the importance and promise of the methods of science in human progress."

Copies of the statement were sent to members of the council with an invitation for comment. Most of the 85 replies were supportive, although there were diverse and sometimes contradictory suggestions for change. At its 1951 meeting the council approved the statement in principle and asked the chairman of the board "to appoint an overall committee and subcommittees to study and analyze the activities of the association and make specific recommendations ... with reference to the future programs of the AAAS" (25). Kirtley Mather, then chairman of the board of directors, promptly prepared a long memorandum proposing that the board itself be the overall committee authorized by the council and that several subcommittees be organized to deal with association meetings, association publications, the association and the general public, services of the association to American scientists, the association and the United States government, cooperation with other scientific organizations in the United States, and international affairs (26).

That was too elaborate a structure for the board, which decided to appoint only two subcommittees, one on internal AAAS functions and the other on the relations between science and society. Each was asked to analyze the implications of the Arden House statement in its area and to define the functions of other committees to be appointed later. The board also authorized employment of a staff assistant to serve full-time on Arden House plans. Douglas Cornell, a physicist, was their first choice for that post. He was ready to move at that time, but chose the more attractive invitation to become executive officer of the National Academy of Sciences.

The two committees never really became active, and further committees were not appointed. Nor was a special secretary for Arden House analysis and planning ever engaged. Instead, dissension began to appear among some of the AAAS officers responsible for the association's activities and planning.

To some of the program arrangers, the Arden House characterization of AAAS meetings as "outmoded" and the suggestion that sessions for reporting individual research results be abolished were red flags. In planning for the 1952 meeting, to be held in St. Louis, Raymond Taylor warned the directors that attendance would be low because many of the biological science societies were meeting with AIBS. He then went on to explain that "an additional ... negative factor is the impression in many quarters that the association already has decided to deemphasize, if not to discontinue, its annual meetings" (27).

Taylor's warning of low attendance was disappointingly accurate; there were only 1,938 registrations. At Philadelphia in 1951 registration had been more satisfactory, but the 1950 meeting in Cleveland had drawn only a few more than had St. Louis in 1952. The annual meeting was clearly in trouble, and in that troubled time Edward U. Condon, who was about to succeed to the presidency, and Warren Weaver, who had just been chosen as president-elect, had what turned out to be an inflammatory joint interview with Earl Ubell, science editor of the

*New York Herald Tribune.* With quotation marks around some of the derogatory terms, Ubell reported that the association was intellectually bankrupt, that its annual meetings were "outmoded," that its programs had grown "thinner," that Weaver headed a group to "revitalize the association," and that the Arden House revival had really just gotten started with the election of Weaver as president-elect (28).

Meyerhoff was furious and responded with an editorial in *Science* that started out by denying the alleged rumor that the AAAS staff took a holiday after the annual meeting. Then he got to the real point of the editorial, his rebuttal of the Condon–Weaver editorial Ubell had published: "If there is no substance to the rumor of post-convention relaxation for the staff, there is even less to the unfounded impression that the Association meetings are 'outmoded,' that its 'programs have grown thinner.' Neither facts nor figures bear out these defeatist statements" (29).

Detlev Bronk, chairman of the board of directors, Condon, and Weaver replied with another editorial, published in *Science* and *The Scientific Monthly*, in which they recognized, they said, that some members feared that change would be too abrupt and revolutionary, while others were disappointed that there seemed to be no progress. They assured readers that they they had no intention of introducing disruptive changes; that although changes might not yet be apparent, they were truly under way; that implementation of the Arden House statement should be slow and gradual; and that for the time being it was desirable to include in the annual meeting programs "a certain body of short reports of current research in specialized fields" (30).

That peace-making effort came too late. *Science* had already announced that "Gladys M. Keener, executive editor of *Science* and *The Scientific Monthly*, and Howard A. Meyerhoff, administrative secretary of the American Association for the Advancement of Science and chairman of the Editorial Board, will voluntarily discontinue their duties with the Association on March 31, and will subsequently submit their resignations" (31). That announcement explained their departure as due to serious disagreements with the president and president-elect over management of the association's meetings and publications.

Those reasons were given more fully in the last agenda Meyerhoff prepared for the board of directors, a meeting he did not attend. He reviewed some reasons for the low attendance at the St. Louis meeting of 1952 and then added:

> Other facts are less tangible but no less significant; most conspicuous is the hesitation felt by several societies that have met with the AAAS in the past, but which now think it inadvisable to make long-range plans to meet with the association because of the uncertainties resulting from the Arden House statement of policy, which forecasts drastic changes in the type of meeting, and which thus discourages participation on the part of those societies that wish to have technical sessions. Correspondence and conferences with

society officers have demonstrated that this is an increasingly serious factor in curbing attendance at the annual meetings of the association.

After a brief summary of Earl Ubell's report of his interview with Condon and Weaver, Meyerhoff's own "Boston 1953" editorial, and the Bronk–Condon–Weaver reply, Meyerhoff concluded: "So far as the administrative secretary is aware, the president-elect has not indicated at any time what kind of meeting would meet with his approval; and until some objectives other than those currently being followed have been defined, the association cannot operate in a vacuum" (32).

Meyerhoff's disagreement with Weaver was disruptive enough, but he was also engaged in a rancorous quarrel with Condon over management and editing of *Science* and *The Scientific Monthly* (see Chapter 4). Perhaps one fight could have been resolved, but two were too many.

The departure of Gladys Keener and Howard Meyerhoff was the sad outcome of those disagreements, for the association lost the services of two capable and devoted staff members, although Meyerhoff later served as a constructive and useful member of the council and of several AAAS committees. The departure of Keener and Meyerhoff helped push Arden House considerations off center stage, for their departure gave the board more urgent business to consider.

The board had to start searching for a replacement for Meyerhoff, in fact, for two replacements. When Meyerhoff succeeded Moulton he asked for, and was given, the additional duty of editing *Science* and *The Scientific Monthly*. As both administrative and editorial responsibilities increased, that arrangement seemed undesirable to the directors, and they decided to return to the pre-Meyerhoff arrangement by separating the editorial responsibilities from those of the administrative office. The board also had another urgent issue to consider. The plan to hold the 1955 meeting in Atlanta was beginning to raise arguments as to whether the association should meet in any city that practiced such race discrimination.

## *The Atlanta Meeting: Problems of Discrimination*

Prior to World War II, AAAS had occasionally met in the Old South — New Orleans in 1905 and 1931, Atlanta in 1913, Nashville in 1927, and Richmond in 1938. But not since 1938 had the association returned to that section of the country. Members of the board had two good reasons for deciding to meet in Atlanta in 1955: a meeting there would recognize the growing scientific activities of the southern states and would make attendance easier for scientists, including black scientists of those states. Because of Atlanta's segregation laws, however, the board insisted that all sessions be completely open regardless of color and that all feasible measures be taken to overcome the city's discriminatory practices in hotels, restaurants, and taxi systems.

Early in 1954 the board received a thoughtful memorandum from the Council for the Advancement of Negroes in Science. After briefly mentioning the distinguished contributions of a few Negro scientists and discussing the segregated school systems, the inadequate facilities of Negro colleges, and the discriminatory practices that made it difficult for Negroes to earn advanced degrees or find employment in science, the memorandum concluded with four recommendations. Three urged AAAS and its affiliates societies to work toward equality of opportunity for Negroes in higher education and scientific employment. The fourth recommendation was that AAAS and its affiliates "hold their meetings and conventions only under those circumstances where nonsegregated facilities are or can be made available to all attending" (33). The recommendation certainly did not endorse an Atlanta meeting, but neither did it specifically ask for cancellation of that plan. In fact, it pointed out that more Negro scientists could attend a meeting in the South than one held elsewhere and that despite the great complexity of local conditions, it was possible to arrange for unsegregated meetings and some social events.

At the same meeting the board also received a report from Raymond Taylor, who had recently visited Atlanta to survey possible arrangements and accommodations. He reported that although hotels and restaurants were strictly segregated, there would be no such problems about scientific sessions, evening lectures, or other scheduled events of the meeting, whether held in hotels, the convention center, or state institutions, and that the president's reception, the Biologists' Smoker, and a luncheon bar at the civic auditorium would be unsegregated. He also reported that the National Association for the Advancement of Colored People had recently met in Atlanta, had held some mixed meetings, and had adopted a resolution of appreciation to the city (34).

On the other side of the argument were supporters of the position that it was wrong for AAAS to meet in Atlanta or any other city in which segregation rules obtained. Yet organized protest did not develop until April 1955 when one member sent a letter to all members of the council protesting the Atlanta site. In response, AAAS received eight letters or telegrams supporting that protest (35). W. Montague Cobb, chairman of the Section on Anthropology, opposed an Atlanta meeting, as did Gabriel Lasker, secretary of that section. Lasker had earlier asked the advice of Margaret Mead, a new member of the AAAS board of directors and a former chairman of the Section on Anthropology. She in turn wrote to several black leaders, describing the situation and asking their advice. After receiving replies, Mead wrote to Lasker that "the advice I have obtained — exclusive of my own opinion — stacks up four to one in favor of going through with the meeting, collecting material on the spot, using the local opportunity to protest, and mobilizing these materials for future protests against meeting in any city where all members of the AAAS are not treated with equal dignity" (36).

The Atlanta meeting turned out to have a small and atypical attendance. As expected, there were more blacks than would have attended a meeting in a northern city. In total, there were fewer registrants from the region of the meeting than normal (37).

Atlanta's segregation laws were not changed, but there was no interference with the unsegregated scientific sessions or social events of the meeting. Warren Weaver's address as retiring president was delivered — as he had requested — at Atlanta University. The dilemma of whether it was "right" or "wrong" for the association to meet in a city that practiced segregation was still undecided, but not undebated.

Four days before the Atlanta meeting opened, Montague Cobb, chairman of the Section on Anthropology, sent a letter to all members of the board and council directly challenging one of the reasons for the Atlanta meeting: "It is my hope," he said, "that the Council will issue some declaration which will affirm for the world that American science does not place a regional educational mission above fundamental respect for the person and personality of human beings" (38). Looking to the future, the Resolutions Committee recommended to the council that it endorse a resolution stating that AAAS membership was open regardless of race or creed and that, because the objectives of the meeting "cannot be fulfilled if free association of the members is hindered by unnatural barriers," meetings should "be held under conditions which make possible the satisfaction of these ideals and requirements."

Some members of the council objected that this resolution did not give as clear and unambiguous guidance for the future as they wanted. Some thought it inappropriate to adopt any such resolution. Some pointed out that the members of the council actually present were not representative of the total membership of that body. During a recess in the council meeting the board of directors, led on this matter by Margaret Mead, decided to recommend that the resolution be submitted to the entire council for a mail ballot (39).

When the resolution was later mailed to the entire council, 220 members voted to adopt it, 28 voted "no" and only three failed to vote. The resolution then adopted remains in effect, and despite the changes that have occurred since 1955, the association did not schedule another meeting in the Old South until the 1990 meeting in New Orleans (40).

## Change of Management

By early autumn of 1953 the board's search for one of the successors to Howard Meyerhoff had been successfully completed with the appointment of Duane Roller as editor of the two journals. A few weeks later I accepted appointment as administrative secretary (41).

With the two searches over, the board returned to Arden House matters by appropriating $10,000 for the expenses of a committee to undertake some of the studies recommended by the Arden House statement. At the same time, Warren Weaver requested that his special responsibility for Arden House matters be transferred to the new administrative secretary. In accordance with those decisions, and at my first meeting as one of its members, the board of directors unanimously agreed to

> request the Administrative Secretary, with such staff and other assistance as he considers appropriate, to study the assignment placed upon the Board of Directors by the findings of the Arden House Conference, and present to the Board suggestions for necessary or desirable changes in a) meetings; b) publications; c) facilities and methods for services of AAAS to members and constituent societies, including the integrating influence of the Association among the sciences; d) facilities and methods for the AAAS for improving the general public understanding of science.
>
> Moved, further, that the Administrative Secretary in the light of suggestions on the four points listed above determine whether other steps are desirable, such as appointment of special committees or the undertaking of other activities by the Board of Directors, in order that the greatest possible benefit may be derived by the Association from the Arden House study (42).

That was my initiation into Arden House responsibilities.

At the next meeting, working from a long memorandum that outlined options for the four issues involved (43), the board adopted two basic recommendations concerning Arden House generally and four concerning specific operations. As general principles they agreed that the association should move forward on specific tasks — such as meeting arrangements — without waiting for some global "Arden House" policy and that activities in the several areas involved would be responsibilities of the association's normal governing structure, staff, and committees, rather than of a special new structure as had earlier been contemplated (44). With agreement reached on those two general principles, the board then approved the following four recommendations concerning annual meetings and *Science*:

- There would be no active discouragement of sessions devoted to reporting research results. Such sessions might diminish in number as other sessions were more attractive and as other meetings satisfied the need for reporting new research findings, but they should not be banned from AAAS meetings.
- Special efforts would be undertaken to improve and strengthen symposia and other programs of wide scientific interest.
- Lectures and sessions addressed to the general public, including high school students, would be given greater emphasis. The National

Geographic Society's annual lecture and the Junior Scientists' Assembly were well-established features of AAAS meetings, but the association should develop more programs that would be attractive to high school students and the interested public.

- Although *Science* should continue to depend primarily upon voluntarily submitted papers and other communications, the editorial staff should take a more active role in writing and in inviting editorials, news material, and feature articles on such topics as science policy and education. These instructions approved the first steps toward what later became the "News and Comment" section of *Science*.

With those decisions made, the Arden House statement continued to guide association objectives and policy, but "Arden House" activities were no longer treated as something in addition to or parallel with the rest of the association's responsibilities, structure, and activities. Instead, they became integral parts of AAAS planning and activity.

## Improving the Annual Meeting

Early in 1953 the Section on Engineering sent Meyerhoff a memorandum on the association's meetings. It arrived just as the Meyerhoff–Condon and Meyerhoff–Weaver disagreements were resulting in Meyerhoff's departure, and it was given little attention at the time. We resurrected it the next year and one of its recommendations was readily adopted: "We suggest that the Board of Directors establish a Program Planning Board (coordinate with the Board of Editors) charged with the responsibility to supervise the program of the Annual Meeting. This responsibility would include the statement of policies and session requirements and the establishment of a planning schedule"(45).

In accordance with that recommendation the board established a standing Committee on AAAS Meetings, authorized the committee to plan such sessions as it considered desirable, and asked it to select and schedule the sessions it approved from among those proposed by affiliated societies, AAAS sections, and other AAAS committees (46).

As the new committee on meetings started to organize, it found plenty of advice already on hand. In addition to the Arden House statement and the memorandum from the Section on Engineering there was a paper prepared by John Behnke, assistant administrative secretary, who had come to AAAS in 1952 after 23 years in book publishing, most recently with W. H. Freeman Co. in San Francisco. Behnke had assumed responsibility for analysis and planning of AAAS activities before and during the Condon–Meyerhoff–Weaver upheaval. He recommended giving greater attention each year to a few top-notch symposia on major topics with nationally recognized leaders as participants (47).

Further advice came from Jacob Bronowski. Early in 1954, on a trip to Europe, Warren Weaver had discussed AAAS affairs with Bronowski and had asked his advice. Bronowski replied, in part:

> What has slowly taken shape as I have been thinking is, I now see, a simple conception: that the key to all these problems must lie in the word "science" in your title. What holds the plurality of sciences together in your Association is the conviction that there exists, embracing them all, the idea of science in the singular. Without this belief, the specialist sciences would (and indeed do) go their own ways apart, hold their own conventions, and become business clubs, at which people give good advice on new techniques. If the AAAS, or the British Association, or for that matter, the Royal Society, are to have a meaning, it must be because they represent not sciences but science (48).

He went on to suggest that sessions of wide interest to scientists from many fields could be planned around such topics as common elements that unite all of science, what the different branches could learn from each other, or the question of whether there is an ethic common to all science.

This emphasis on broad issues of wide scientific concern was, of course, supportive of the Arden House recommendations, and of Harlow Shapley's earlier call to change the basic nature of the annual meetings along the lines that were tried out in the centennial meeting. But the pressures were not all in that direction: Meyerhoff and Taylor were not the only ones wanting to continue to hold sessions for the presentation of new findings and the discussion of technical details.

In 1955 an Ad Hoc Committee on Membership Development concluded its work with a memorandum covering several aspects of AAAS activities. With respect to the annual meetings it made two strong points. As for the on-going argument over the inclusion of sessions for reports of current research, the committee advised caution. It recognized the desirability of programs on general and interdisciplinary topics, but also argued that when scientists decided which sessions to attend, they chose strongly in favor of "sessions heavily loaded with technical science." That caution was followed by the suggestion that the association try to obtain good information from scientists as to what they wanted in AAAS meetings and the suggestion that if the technical sessions should gravitate to other meetings such as those of AIBS, the association could "move smoothly into more complete emphasis on the all-embracing aspects of science and the public" (49).

The committee's other recommendation was that AAAS should follow the example of the British Association in making each annual meeting a great event for the host city. There should therefore be sessions of interest to "students, teachers, industrial leaders, government officials, and the interested public should be participants in the program and should become a great audience for presentations of public interest by scientists with something to say" (50).

The committee's statement that attendance figures showed a strong preference for sessions "heavily loaded with technical science" may have been a bit of an exaggeration, for detailed attendance records were seldom kept. However, at the 1954 meeting the median attendance at symposia was 95, while the median attendance at sessions of contributed papers was 60, and there were a few more sessions for contributed papers than symposia (51).

The larger number of sessions of contributed papers was partly due to the fact that many sessions were still being planned by affiliated societies. From 1950 through 1955, 85 different affiliated societies participated in some way in the annual meetings, ranging from holding multiple sessions wholly planned by the affiliate to cosponsoring one or more sessions arranged by some other organization. As AIBS meetings became popular, attendance began to overcrowd the facilities of the campuses on which they met, and AIBS asked member societies to consider meeting with AAAS at intervals in order to avoid the undesirable crowding of AIBS meetings (52).

## *Meeting Policies*

As the new Committee on AAAS Meetings settled down to work, the members agreed that before considering individual programs they should decide how they wanted to exercise the considerable amount of responsibility and authority they had been given. Several basic decisions were made (53):

- That the major function of the Association — and therefore of its meetings — was to facilitate communication among scientists, particularly among representatives of different disciplines.
- That the basic pattern of traditional AAAS meetings consisting of different types of sessions arranged by a variety of program planning units should be preserved. Improvement should be made within that pattern rather than by switching to a quite different form of meeting, such as the one held at the association's centennial.
- That, beginning with the 1956 meeting, it would be desirable to set aside a block of time, free from competing programs, for association-wide addresses or symposia to be presented under the continuing title "Moving Frontiers of Science." For those sessions the committee planned to select speakers who "would give addresses explaining to a diverse scientific audience recent developments and current problems in a selected scientific discipline or problem area. ... These addresses should be both reviews of recent and current work and explanations of how that work could affect developments in other fields of science."
- That the association's second major objective was the presentation of scientific information to the public, and that in carrying out this objective there should be some public sessions as regular parts of the annual

meeting and that the staff should also seek to secure widespread television coverage of appropriate parts of the meeting.
- That the committee and the AAAS staff should take a substantially more active hand in deciding which proposed programs would actually be included in the program and when each would be scheduled so as to minimize conflict or overlap and so that the total would constitute as coherent and effective a program as they could organize.

Later on, the committee withdrew somewhat from this last decision, agreeing that they did not want to veto plans made by sections or affiliates. Nevertheless, it was clear that the committee intended to be aggressive and forceful in attempting to improve the quality of the meetings. When the board of directors met a few days later they endorsed the committee's proposed methods of operation, encouraged it to go ahead as planned, and said they did not want to restrict the committee and that it should come back again whenever it wanted advice or further recommendations concerning its own functioning (54).

In selecting members for the committee, the board chose active scientists, science administrators, and members who were well acquainted with the problems and characteristics of association meetings (55). Some of those who served on the committee were Leonard Carmichael, Harry C. Kelly, Preston Cloud, William Steere, Francis O. Schmitt, Gerard Piel, Fred Singer, Walter Sullivan, Philip Handler, Robert Jastrow, Norman Hackerman, and Derek J. deSolla Price. When he became editor of *Science* in 1962, Philip Abelson resigned from the committee and its chairmanship on the grounds that it was not appropriate for a staff member of the association to chair such a committee.

One of the innovations and principal responsibilities of the committee was to plan the annual "Moving Frontiers of Science" sessions. In some years all of the papers under that title were organized about a single theme. For example, in 1956 the central theme was the fundamental concepts and units of science, with addresses by Jerrold Zacharias for the physical sciences, Ralph Gerard for the biological sciences, Robert Macleod for the social sciences, and a concluding address by Michael Polanyi of Victoria College, Manchester, England, that tried to synthesize the concepts and units of the three broad areas. A discussion panel of Paul Lazarsfeld, Paul Weiss, and Jerome Wiesner completed the program. As another example of a single theme, in 1958 the focus was on the organization of scientific activities in several countries. E. S. Hiscocks, director of the United Kingdom Scientific Mission in Washington, spoke on "The Organization of Scientific Activities in the United Kingdom." Robert Major, director of the Royal Norwegian Council for Scientific and Industrial Research and one of the most respected "wise men" of European science policy councils, described the organization of scientific activity in his country. B. G. Ballard, vice president of the National Research Council of Canada, gave a similar account for that country. And Don K. Price, dean of the Graduate School of Public Administration at Harvard, compared the organization of scientific activities in the United States with the organizations found in other countries.

In other years the "Moving Frontiers of Science" sessions included addresses on four different topics. For example, in 1962 Homer B. Newell, from NASA, spoke on "Space Science," Sterling B. Hendricks of the United States Department of Agriculture on "Biological Timing," William O. Baker of the Bell Telephone Laboratories on "Coupling of Industrial Research and Modern Science," and Sydney Brenner of the Medical Unit at Cambridge University on "Perspectives in Molecular Biology."

"Moving Frontiers of Science" sessions, presidential addresses, and other events selected by the Committee on AAAS Meetings drew large audiences, partly because of their intrinsic interest and partly because many of them were scheduled at times when the committee permitted no competing programs. In general, those times, after some negotiation with section secretaries and program representatives of affiliated societies, were the evenings throughout the whole meeting period and all day on December 28 (56).

Evening addresses, including the address of the retiring president, were always a popular feature and were arranged under a variety of auspices. The annual Sigma Xi lecture, presented first in 1922, was sometimes delivered by the recipient of the year's Proctor Prize for Scientific Achievement. An annual address arranged by Phi Beta Kappa, started in 1935, usually featured a scientist or historian of science. In 1944 that lecture started Harlow Shapley on a two-year cycle; he delivered the Phi Beta Kappa address in 1944, the Sigma Xi lecture in 1946, and his presidential address in 1948. The annual lectures sponsored by the National Geographic Society regularly drew large audiences of AAAS members attending the annual meeting and of local members of the society. In 1959, the year of the Darwin Centennial, the Committee on AAAS Meetings also scheduled a "distinguished lecture" intended for a general audience. That address, given by George Gaylord Simpson under the title "The World into Which Darwin Led Us," was so successful that the committee decided to devote one evening at each meeting to an address by a distinguished speaker for a general audience. In 1960 that address was given by Sir Charles Snow on "The Moral Un-Neutrality of Science," an address followed by commentaries from Father Theodore Hesburgh and William O. Baker.

In addition to the "Moving Frontiers of Science" session and the public addresses there were more numerous sessions of more specialized interest. Some of these — usually selected from among those arranged by related groups of sections or affiliates — were considered by the committee to be of sufficiently general interest to be scheduled for presentation on December 28, the one day of the Christmas week meeting wholly controlled by the committee. Others had to compete with larger numbers of alternatives.

Under the general auspices of the Committee on AAAS Meetings several kinds of special programs were arranged. Some were annual events. For example, the Conference on Scientific Manpower provided an annual forum for discussion of issues concerned with the supply of scientists and engineers and the demand for their services. Financed and later published each year by the National Science Foundation, those conferences were planned by a committee representing AAAS, NSF, and other interested organizations. Another example started in 1960 when the George Sarton Memorial Foundation provided $200 to finance the first annual George Sarton Memorial Lecture on the history of science. Some other special programs were one-time affairs, such as the 1964 International Symposium on Communication and Social Interaction in Primates. Six sessions with speakers from four continents provided a special incentive for primatologists to attend the 1964 meeting in Montreal. Under such auspices or as other one-time events that took advantage of special opportunities, AAAS audiences heard addresses by Jacob Bronowski, Jane Goodall, astronaut Edward White, and other speakers of unusual interest.

From the 1940s through the 1960s AAAS and the British Association (BA) for the Advancement of Science arranged for a number of exchange lectures (see Chapter 10). Thus American audiences had the opportunity to hear A. V. Hill, Lord Brain, Dame Kathleen Lonsdale, and other presidents of the British Association (BA) or other speakers selected by that organization. In other years those in attendance at BA meetings heard addresses by American scientists. Financing was always a problem, however, especially for the BA, and the exchanges never became solidly established and disappeared after 1972.

Another type of special program was a symposium on science in some other country. A 1951 symposium on Soviet science became a popular volume in the association's series of published symposia (57). It was followed in 1960 by the symposium on science in Communist China (we had not yet learned to call it the People's Republic of China). Primary stimulation for that one came from the Office of Science Information of the National Science Foundation. Scott Adams of that office called together representatives of AAAS and a number of other scientific societies and 10 Chinese-American scholars to consider the possibility (58). Adams' timing was excellent. There was still fairly free exchange between the two countries; for example, the Library of Congress was receiving 150 scientific and technical journals from the People's Republic of China. Moreover, 10-year summaries of progress in major fields of science had been prepared under government auspices. All of this information was available, and in preparation for the symposium, a committee of representatives of several scientific societies selected topics and authors, and 25 specialists went through some 200,000 pages of Chinese scientific literature. Then, before the symposium was presented, the People's Republic closed access and clamped down on the export of literature. Library of Congress receipts fell from 150 journals in 1959 to a handful in 1960 and none at all in the early part of 1961 (59). Interest was therefore high and the papers presented at the symposium became a successful addition to the association's series of symposium volumes (60).

The symposia on the sciences in the Soviet Union and the People's Republic of China were supported by grants from the National Science Foundation, and their success led some of the NSF staff members as well as some members of the AAAS to want to extend the series to other countries. It was even suggested that we might do a different nation each year. Japan was not closed to visitors and Japanese scientists published a good deal of their work in English. Yet the Japanese language was a barrier and, moreover, AAAS was providing administrative support to the effort of the Department of State and the National Science Foundation to encourage exchanges and cooperation with Japan. It therefore seemed appropriate to select Japan as the third country for a special symposium even though that country did not have the special aura of secrecy that brought substantial audiences to the Soviet and Chinese symposia. The 18 Japanese scientists who came to present their well-prepared papers must have been greatly disappointed at the small audience that came to hear them at the Cleveland meeting in 1963. Fortunately the published papers (61) reached a much larger audience.

## When to Meet

When to meet was a frequent question and the answer was always a source of some complaint. For its first half century the association usually met in August. Early in the 20th century the week after Christmas became the standard time, with a second and usually smaller meeting in the summer. Meetings were suspended during most of World War II, and after their resumption the association met only once a year. Because the first AAAS meeting in 1848 was held in September, it seemed appropriate to celebrate the centennial in the same month.

But there were always objections to meeting the week after Christmas. Winter weather sometimes made travel difficult. The holidays were for family gatherings, not a time for the scientist member of the family to rush off to a convention. So there were periodic searches for alternatives. In 1958, and again in 1964, Raymond Taylor searched the calendar; every week of the year was examined for likely travel difficulties, availability of hotel space and ability to secure favorable hotel rates, conflicts with other major scientific meetings, and conflicts with such events as the Fourth of July or the start or finish of the academic year. The most attractive alternatives seemed to be the days following Thanksgiving and the last week in January, but the board decided that neither was sufficiently better than Christmas week to justify a change (62).

A study committee of the council then took up the question. Polls of a small sample of AAAS members, of registrants at the 1965 meeting in Berkeley, and a straw vote of council members resulted in no majority for a particular time of year. However, late June drew a few more votes than Christmas week and the committee proposed that for several years the association meet at both of those times. Then, depending upon attendance records, a choice could be made of June or December or both. When the poll results and the recommendation were reported to the board, they decided that the greater availability of hotel space at lower rates during Christmas week outweighed the slight preference for a June date and that the association would therefore continue its traditional meeting time (63).

Two years later a changed meeting time was adopted. Following discussions with section secretaries, chairmen of several AAAS committees and representatives of some affiliated societies, Walter Berl, the meeting editor, recommended that the Christmas week period be continued through 1972, that 1973 be kept open for a special meeting tentatively being planned for Mexico City, and that beginning in 1974 the association try for several years a meeting starting on the last Monday of February. The board agreed and that new schedule was adopted (64).

The meeting time was changed but the problem was not solved. One meeting was actually held at the new February time, then one in January, several in February, several in January, and then several in May, before settling on the January or February period.

## Social Events

When the association held its first meeting in Washington, D.C., in the spring of 1854, "the members were elegantly entertained, on different evenings, by Franklin Pierce, President of the United States; Jefferson Davis, Secretary of War; James Guthrie, Secretary of the Treasury; and William W. Corcoran, Esq. [founder of the Corcoran Art Gallery in Washington]" (65). AAAS, in turn, invited the President of the United States, members of his Cabinet, and members of the Senate and House of Representatives to be present at the sessions of the association. Good feelings at that meeting resulted in adoption of resolutions of thanks to the hosts and other Washingtonians, and to the 14 railroads that had offered the facilities of free return tickets to the members in attendance at the meeting (66).

Seldom have the hosts been as distinguished as at that 1854 meeting, but social events have always been a part of annual meetings. During the period under review the two major ones were the President's Reception and the Biologists' Smoker. The former followed the address of the retiring president and everyone who attended that address was invited. The Biologists' Smoker was intended especially for biologists, the largest group of members, but was really open to anyone who cared to spend part of an evening in informal conversation with old friends or new acquaintances. The President's Reception is still an annual event, but the Smoker is no longer held.

Other social events came and went as different committees or groups decided. One that became permanent was the annual dinner for past presidents and former members of the board of directors. That dinner was the invention of Chauncey Leake and Margaret Mead. In 1961, the year of his chairmanship of the board,

Leake proposed that we invite all former presidents to join the board for dinner one evening during the annual meeting. Margaret Mead added that we should also invite former board members who had not been elected to the presidency. I agreed and the now traditional Past President's Dinner was started at the 1961 meeting in Denver. Understandably, former presidents and board members were never all present, but those dinners quickly became a welcomed opportunity for current officers to honor a few special guests and to meet or renew acquaintance with James B. Conant, Harlow Shapley, Roger Adams, or other earlier officers of AAAS (67).

## Extending the Audience

From 1949 until 1967 primary responsibility for coordinating planning for AAAS meetings was carried by Raymond L. Taylor. In 1967 when he retired and we had to find a successor we followed the precedent of the 1962 search for a new editor of *Science*. In that case Philip Abelson had been appointed with the agreement that he would serve as editor while also continuing his own research activities as head of the Geophysical Institute of the Carnegie Institution of Washington. That agreement had worked very satisfactorily and, in a similar fashion, Walter Berl, a member of the research staff of the Applied Physics Laboratory of Johns Hopkins University, became meeting editor — a title that was deliberately chosen to indicate the nature of his responsibilities. AAAS and the Applied Physics Laboratory agreed that he would divide his time between planning and handling meetings for AAAS and continuing his own research at the laboratory. To assist him with many of the responsibilities of arranging and conducting the meetings, Daniel Thornhill, a retired Navy commander then serving on the staff of one of the National Academy of Sciences committees, was appointed as meeting manager.

Working with the Committee on AAAS Meetings, the section officers, and officers of affiliated societies, Berl and Thornhill continued the same basic meeting patterns that had developed under the auspices of the Committee on AAAS Meetings but Berl soon began to put his own impress on those meetings. He had three objectives to emphasize: (1) to present an "Annual Report" on the state of science, its successes and concerns, through in-depth discussions of a wide variety of interdisciplinary topics, including topics that extend beyond the conventional boundaries of science; (2) to involve the citizens and institutions of the cities in which the meetings are held; and (3) to report the content and conclusions of the discussions to as large a public as possible by all suitable communication devices (68).

To accomplish those objectives he introduced several changes without altering the general pattern of the meetings. For one, the number of illustrated lectures or other addresses intended for large scientific or more general audiences was increased, to between 9 and 15 at each of the annual meetings for which he was primarily responsible. The list of speakers included Kathleen Lonsdale from

London, Vadim S. Semenov from Moscow, A. Doxiadis from Athens, Louis S. B. Leakey from Kenya, Norman Borlaug from Mexico, and Sol Linowitz, B. F. Skinner, Loren Eisley, Robert Merton, Kingman Brewster, Margaret Mead, Edwin Aldrin, Philip Handler, and others from the United States.

As interesting reminders of where science had been, he sought out or endorsed programs commemorating the work of important pioneers or using anniversaries of their birth or work as occasions for examining progress in their fields. Scientists so remembered included Michael Faraday, George Ellery Hale, John Wesley Powell, and Dmitri Mendeleev. Events whose anniversaries were celebrated with special programs or exhibits included the 200th anniversary of the *Encyclopedia Britannica*, the centennial of the founding of Yellowstone National Park, and 75 years of scientific films.

Going back to an earlier type of program that had been abandoned when the annual symposium committee was replaced by the Committee on AAAS Meetings, he introduced the "General Chairmen Symposia," each planned under the sponsorship of the meeting's general chairman. Their contents were intended to be of special interest to the city in which the meeting was being held.

In some respects the annual meeting became more of an intellectual smorgasbord than it had been earlier, offering public lectures on several topics, symposia on many topics, special conferences, anniversary celebrations, the annual exhibit of books, instruments, and services, a substantially enlarged showing of films, tours to scientific institutions or other places of interest, special exhibits, and an occasional concert on, for example, electronic music or early Scott Joplin jazz. As with any festival, there was always much more to see and hear than anyone could take in.

The number of attendees grew. The worrisomly low attendance figures associated with the early post-war years and the formation of the American Institute of Biological Sciences were no more. Attendance always depended somewhat on the location of the meeting; high along the Washington-Boston corridor and lower in most other cities. Yet as Figure 1 shows, the general trend was upward, and later meetings in any city were almost always larger than earlier ones in the same city. In the six years from 1965 through 1970, four of the meetings had registrations above 7,000 and the only small registration for that period was in Dallas.

Registration figures never gave a full count of attendance, but over a span of years trends in registration offered the best measure the association had for judging how well the meetings were satisfying the expectations and needs of the members. The upward trend was satisfying.

Whatever the registration count, the people actually in attendance constituted only part of the audience Berl was seeking to reach. Earlier meetings had extensive press coverage and some had also had some coverage by radio and television. In the 1960s, however, Berl and his staff made special efforts to assure coverage by television. For the 1967 meeting New York's Channel 13 received a

Extending the Audience

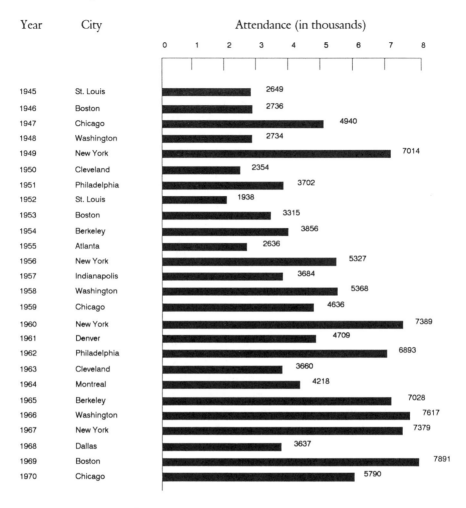

Figure 1. Registration at AAAS Annual Meetings, 1945–1970

grant from the Ford Foundation that enabled the station to broadcast live interviews and selected symposia and panel discussions for three and a half hours on December 26 and for eight hours on each of the following four days. Channel 13's coverage concluded on December 31 with 11 hours of material that had been taped on earlier days of the meeting. Almost all of the 35 and a half hours of live coverage were also broadcast in Washington, Boston, Philadelphia, and nine other cities of the Eastern Educational Television Network. After it was all over, the station reported that although it had not received a staggering number of letters of comment it had received a respectable response. The reactions to the coverage by those who did comment was consistently articulate, sincere, and appreciative.

One New York viewer was even prompted to contribute $1,000 to Channel 13 as a direct result of the broadcast (69).

In 1968 television coverage at Dallas was similar to that in New York the year before, but with less feeding to other stations. In 1969 in Boston, 9:00 am to 4:00 pm daily coverage of selected sessions and interviews was broadcast by Station WGBH and several other stations of the Eastern Educational Television Network. In addition, one-hour "Convention Magazine" programs provided evening prime-time audiences with summaries and discussions of some of the major events of the day's sessions. Those programs were broadcast by 17 of the Eastern Educational Network stations and were taped and broadcast early in 1970 by several others. Thanks to the coast-to-coast lines of the Corporation for Public Broadcasting they were also broadcast in 40 cities outside the area of the Eastern Educational Network (70).

A novel feature of the broadcasting from Boston was a satellite program in New York. Through the cooperation of Channel 13 in New York the concluding session of the symposium, "Is There an Optimum Level of Population?" was projected in color on a 16-by-22-foot screen in the Caspary Auditorium of Rockefeller University for viewers gathered there. After the broadcast ended those viewers held their own discussion of the symposium topic.

In the days following the 1969 meeting the individual television stations, the National Educational Television office, and AAAS all received many appreciative letters commenting on individual broadcasts, expressing the hope that similar broadcasts would be available next year, asking for transcripts, or otherwise making the staff members who worked on the taping and broadcasting happy. A few letters brought subscriptions to *Science* and a letter from Vice President Hubert Humphrey asked for a transcript of a session he had found particularly interesting.

The cost came to about $100,000 per year. For 1967 most of that cost was borne by a grant from the Ford Foundation. The same source paid half the cost for 1968. In 1969 the Ford Foundation provided only $10,000; it was a scramble to get about twice that much from a variety of other supporters and AAAS had to carry most of the cost. Foundations do not like to give continuing support for operating expenses and the cost seemed too great for AAAS to bear. Under a variety of arrangements some programs were broadcast in later years, but 1967 to 1969 was the short period of extensive coverage for the whole period of the annual meeting.

A different kind of extension of the audience for AAAS meetings was secured in 1968 and later years by audiotaping selected sessions. An order desk at the meeting and published announcements later made copies available for purchase by anyone who wanted to use them for classroom instruction, to initiate discussion, or for other uses. Unlike the extensive television coverage the audiotapes have survived and new ones continue to become available each year. Some of those symposia are published and for that purpose papers can be revised, the discussion summarized, or more papers added. The resulting volumes provide fuller and more polished reports, a year or more later. In contrast, the essentially unedited audiotapes are available immediately after the actual event.

Early in his tenure as meeting editor, in an editorial in *Science*, Berl described his plans for the annual meeting with a quotation from Caryl Haskins, who became a member of the board of directors a few years later. After acknowledging that the primary task of scientists working at the frontiers of knowledge was to enlarge and extend those frontiers, Haskins — and Berl — went on to emphasize that scientists also had two other obligations: to explain their work to scientists in other fields and — a more difficult obligation — to explain to the interested public "the nature, the purpose, the rationale, and the intense social relevance of the scientific way" (71).

To explain developments in one science to scientists in other fields. To explain science to nonscientists. Those were the two major objectives of the meetings. Neither was new. Both were expressed in the Arden House statement of 1951, earlier in Harlow Shapley's 1947 recommendations for changing the whole character of AAAS meetings, and in the thinking and writing of other AAAS leaders. Both were primary, and both need frequent reaffirmation. Symposia, public addresses, films, audiotapes, newspaper and magazine reports, television, and radio have been the media for those efforts. At the end of the 1960s they were effectively reaching large audiences. Yet there is no permanently satisfactory formula for the "ideal" AAAS meeting. Changing times, conditions, problems, and resources have always provided motivation and opportunity to try to make the next meeting better than the last one.

*Chapter Four*

# Science and The Scientific Monthly

Successful publication of a magazine requires an editor who is knowledgeable in and has good judgment about the quality of written material in a particular field; who has enough money — from subscriptions, the sale of advertising space, a subsidy, or some combination of those sources — to pay the bills; and — most basic of all — who has a clear idea of why the magazine is being published and to whom it is addressed. When AAAS started publishing *The Scientific Monthly* and soon afterwards *Science,* it was suddenly faced with all three of those requirements. Learning how to meet them did not come easily; in fact, there was a good bit of stumbling before a steady course was reached, before the association learned how to become an effective publisher of periodicals.

## The Cattell Years

*Science* came to AAAS in 1945 and *The Scientific Monthly*, its sister publication, in 1939 (see Appendix 1, p. 267). Both came from James McKeen Cattell, who had started the *Monthly* in 1915. *Science* had a longer history. It was started in 1880 by Thomas A. Edison and John Michels (1). It soon foundered, was resurrected by Alexander Graham Bell and his father-in-law Gardiner Greene Hubbard, was partially supported by AAAS when it ran into financial problems, and in 1894 was sold to Cattell when Bell and Hubbard wanted to found *National Geographic*.

Both *Science* and *The Scientific Monthly* had been AAAS magazines before the association owned either. In 1900 Cattell and L. O. Howard, then permanent secretary of the association, made a risky deal. AAAS agreed to give Cattell two of the three dollars of annual dues paid by each AAAS member and in return Cattell agreed to send copies of *Science* to every member of the association. For Cattell that meant that instead of collecting the nominal subscription price of five dollars a year he was getting only two, and for AAAS it meant giving away two-thirds of its annual dues income. Both were gambling and for both the gamble paid off. AAAS membership, which for 20 years had been coasting along at the 2,000 level, took off and within the next 15 years reached 10,000. With its greatly increased circulation *Science* became a more attractive advertising medium and so increased that source of revenue for Cattell. After he established *The Scientific Monthly*, a similar arrangement was made with the association for those members who preferred to receive that magazine rather than *Science*.

Thus from the beginning of the 20th century *Science* was the official magazine of AAAS even though the association did not own or edit it, and later the *Monthly* gained a similar status. The magazines were "official," for under the arrangement Cattell published the official papers of AAAS and frequent news items of its plans and activities.

In 1925 Cattell, aged 65 and the retiring president of AAAS, made another agreement with the association: At his death AAAS would become the owner of *Science* and in compensation would pay an annual royalty to his widow, Josephine Owen Cattell, for the rest of her life. In 1936 a similar agreement was made concerning *The Scientific Monthly*, and in 1938 the two agreements were combined so that the payment for each magazine would be a sum equal to one-half of the excess of income over expenses for the five years preceding the date of transfer. That sum was to be paid in 10 annual installments, and the canny Cattell had written into the contract a clause stipulating that the annual payment would be increased by an amount sufficient to compensate for any decrease in the purchasing power of the dollar (2).

Under an optional clause of that contract, *The Scientific Monthly* came to AAAS in 1939, and in 1945, after Cattell's death, *Science* followed. Despite the association's close relationship with both magazines and the long-anticipated transfer of ownership,— and perhaps because of the forceful Cattell's dual role of publisher of the magazine and influential member of the association's governing hierarchy, — AAAS was poorly prepared to assume its new publishing responsibilities.

Before AAAS learned how to manage its new assets or formulated what later became its basic editorial and publishing policies, it wavered for awhile on some policies and options, made some poor editorial appointments, and was the victim of ill fortune. In 1939 when AAAS assumed responsibility for the *Monthly*, Ware Cattell, one of Cattell's sons, was appointed as editor. That arrangement did not work out satisfactorily, and in 1943 F. R. Moulton, permanent secretary of the association, fired Cattell for not doing his job. In retaliation Cattell sued the association for $17,500 and eventually collected $7,500 (3). Frank Campbell, an entomologist on the faculty of Ohio State University, was then appointed as editor and continued in that capacity until 1949 when Howard Meyerhoff succeeded Moulton and assumed editorial as well as administrative responsibilities.

When James McKeen Cattell died on January 20, 1944 (4), AAAS was caught without an editor or editorial staff for *Science* and without sufficient space in its crowded offices to make room for such a staff. Arrangements were therefore quickly made to have *Science* edited by Mrs. Cattell, with help from her sons Ware and Jacques. She was thoroughly experienced, for although her name had never appeared as an assistant or associate editor, she had helped edit the magazine for nearly 50 years. During 1944 the board of directors debated several alternatives and then asked Mrs. Cattell and Jacques to continue as editors through 1945.

That arrangement was forced partially by collapse of another possibility. Late in 1944 retired admiral Charles S. Stephenson was appointed, — or at least he and some of the members of the board thought he had been appointed, — to the editorship of *Science*. However, his plans for editing the magazine were not acceptable to the board of directors, who unanimously voted to thank "Admiral Stephenson for his analysis and recommendations in connection with the editing and publishing of *Science*. The Committee does not find itself in a position to adopt the plan suggested now, and therefore, regretfully feels obliged to accept Admiral Stephenson's resignation as tendered orally to the Committee at its session on October 22, 1944."

Admiral Stephenson did not agree that he had resigned. He claimed he had been fired and brought suit against the association for $35,000. Warren Magee, the association's legal counsel, served AAAS well in that case; over the next couple of years he persuaded Stephenson to reduce his demands to $10,000, then to $7,500, $5,000, $2,500, and $2,000. At that point the directors decided it was worth $2,000 to end the nuisance with an out-of-court settlement. Magee charged $500 for his services, and the matter was ended (5).

With Stephenson no longer under consideration and Mrs. Cattell editing *Science* there was time for a proper search for a new editor, one that ended with the appointment of Willard Valentine, professor of psychology at Northwestern University and treasurer and publications manager of the American Psychological Association (6). His appointment started four months before January 1, 1946, the time when he would succeed Mrs. Cattell, in order to give him an opportunity to recruit staff, find space, and make necessary arrangements and plans.

## Editorial Turnover

A new staff was needed and, as Moulton had told the board of directors, the minimum necessary staff would consist of "an editor, one assistant editor, one highly competent stenographer-typist and one typist clerk" (7). By November Valentine had hired Mildred Atwood as assistant editor and two stenographic and typing assistants. There was no room at the Smithsonian Institution to house the editorial staff, but that problem was nicely solved by the generosity of Paul Douglas, president of American University, who offered free space for the editorial and advertising staffs on the main campus of the university. As a return courtesy Valentine agreed to teach a course in psychology for the university. No AAAS editor since Valentine has personally had to pay the rent on the magazine's office space (8).

Valentine revised the format, discontinued advertisements on the front cover, started using cover pictures, rearranged sections, and intended "to make the journal maximally useful to American scientists and the general public interested in scientific news and advance." Sadly, his editorship lasted only 15 months. On the Saturday afternoon of April 5, 1947, he suffered a heart attack while working on his home property and died within an hour (9). Mildred Atwood, his assistant, took over as acting editor and in 1948 was given the new title of executive editor when George Baitsell became chairman of the editorial board and in that capacity was de facto editor. A year later when Howard Meyerhoff became administrative secretary he also succeeded Baitsell as chairman of the editorial board. With the assistance of Gladys Keener as executive editor he was responsible for both *Science* and *The Scientific Monthly* until early in 1953 when he and Mrs. Keener suddenly left AAAS.

Bentley Glass was then pressed into service as chairman of the editorial board and part-time editor of *Science* and the *Monthly*. Because he had prior commitments to be in Europe for part of that summer he persuaded his Johns Hopkins University and editorial board colleague William Straus to take his place for several weeks. The directors hoped Glass would continue indefinitely, but he did not want to leave Johns Hopkins. The wider search that thus became necessary resulted in appointment of Duane Roller, then assistant director of research at Hughes Research and Development Laboratories in California. Roller was the founding editor of the *American Journal of Physics* and in 1945 had been high on the list of possible editors when Willard Valentine had been chosen.

Roller stayed for only a year and departed on such short notice that there was not time to make a proper search for a replacement. Consequently I served as acting editor for the two journals through 1955, and during that year we selected Graham DuShane, professor of biology at Stanford University, as editor (10). His six and a half years in that position was the longest continuing editorship since Cattell's death in 1944. In the intervening years there had been Josephine Owen Cattell and Jacques Cattell, Willard Valentine, Mildred Atwood, George Baitsell, Howard Meyerhoff and Gladys Keener, Bentley Glass and William Straus, and Dael Wolfle — eight editors or acting editors in 12 years! No wonder that John Walsh later described that period in the life of *Science* as less than satisfactory (11). In contrast, the next two editors, DuShane and Abelson, served for 28 years.

## The Two Magazines: Differing Philosophies and Audiences

Although both magazines were intended primarily for scientists, *Science* was clearly both the more technical and the more timely. It published reports of recent research and other scientific developments, accounts of events of interest to scientists, personal news about scientists, and major addresses and other official papers of the association.

*The Scientific Monthly* was more leisurely, more philosophical, and intended to interest a less technically specialized audience. As Frank Campbell, the second editor under AAAS ownership, explained, "I have privately regarded *SM* as the *Atlantic Monthly* of science" (12). It published more than *Science* did on education, social science, the social aspects of science, and the history and philosophy of science. It included some poetry and occasionally published a quite offbeat piece, of which the one that elicited both the most applause and the most criticism was a spoof entitled "The Schuss Yucca," a description of a rare variety of yucca found only in a small area near Pasadena, an area "noted for its queer flora and fauna." In common with other yuccas the schuss yucca flowered only at long intervals, but when it did, watch out, for the large stem and blossom shot up to a height of 10 to 20 feet in a few seconds instead of the few weeks of other varieties of yucca. The earlier report that a cattle rustler and his horse had been impaled in midair while jumping over one of the plants was surely an exaggeration, but as evidence of the speed of growth the article included five successive pictures of a young woman looking down, directly at, and then up to the flower-headed shoot (never mind that the length of shadows belied the author's statement that the photographs had been taken at one-second intervals).

The schuss yucca spoof was published in 1952 (13). When I came to AAAS in 1954 there were still occasional letters complaining that publishing such trash was beneath the dignity of a scholarly association of serious scientists; a few angrier members even used the article as a reason for resigning. But the mails also brought letters to the effect that "I'll keep on paying my dues as long as you continue to publish an occasional article that is as much fun as the yucca one."

But then there came a letter from an Austrian botanist who took the article seriously. Unable to locate some of the references to earlier German articles about the variety, he asked if the author had made some errors in his citations. I took the letter to Hans Nussbaum, our best German scribe, and asked him to reply. Nussbaum's kindly explanation brought another letter shamefacedly admitting that the botanist had been taken in by the hoax but admonishing us for not having published it in an April 1 edition of the magazine.

*Science* could not have accepted such an article but *The Scientific Monthly* could (14). Because the two magazines differed, so did their competition. *Science* could be compared only with the British journal *Nature* (15). *The Scientific Monthly* was compared with *American Scientist* and, after 1948, with the revived and greatly improved *Scientific American*. The *Monthly* usually came out second best in those comparisons, and as early as 1947 the board of directors asked the Publications Committee to consider the advisability of continuing or suspending it "in view of the present status of the *American Scientist*" (16).

## Merger with *Scientific American*?

The *Monthly* was continued, but soon the association had an opportunity to merge it with *Scientific American*. In 1948 that venerable title began appearing on the cover of an excellent new magazine being published and edited by Gerard Piel and Dennis Flanagan. It was well edited, had a format under which it quickly became widely known, and was reaching a larger audience than the *Monthly*. Circulation and advertising trends were promising, but the magazine was exhausting its initial capitalization before income was large enough to meet expenses.

In that situation Piel came to AAAS with an offer from the owners to give all of the stock of the company to AAAS if the association would then merge *Scientific American* with *The Scientific Monthly* and use the staff of *Scientific American* to produce the combined magazine. Many details were left to be worked out later, but the proposal was generously favorable to AAAS. By proposing that the combined magazine be published by a new corporation wholly owned by AAAS, the association's financial interests would be protected in the unlikely event that the combined magazine did not become financially successful. And by providing that the majority of the members of the new corporation's board of directors be appointed by AAAS the association's interests would be further protected (17). The owners of *Scientific American* expected that as an official publication of AAAS the magazine would gain enhanced prestige and would quickly reach the break-even point and become financially self sufficient. And although it was not yet a money maker, the association was being offered a very promising magazine and the services of an able and enthusiastic editorial staff.

By the time the Publications Committee met to consider that offer Meyerhoff had examined seemingly favorable trends in *Scientific American* circulation and income and knew of the imminent demise of *Science Illustrated*, an unsuccessful would-be competitor. Moreover, he had discussed the matter with Warren Weaver at the Rockefeller Foundation and had Weaver's encouragement to expect that the Rockefeller and Ford foundations would provide initial support to AAAS as owner, support they could not give to a commercial publisher (18).

Correspondence among members of the Publications Committee and board of directors, internal memoranda, and correspondence with Piel all seemed to indicate that the offer would be accepted. But when it came time to vote, George Baitsell, chairman of the editorial board and in that capacity editor of *Science* and *The Scientific Monthly,* moved that the proposal not be accepted. That was the vote by the Publications Committee (19).

When the board of directors met on the following day they had that negative recommendation standing against Meyerhoff's favorable one. The board, which included all members of the Publications Committee, voted "that the acquisition of the *Scientific American* not be further considered at this time, and that Piel be informed that the Association declines with sincere regret to undertake publication of this periodical." Meyerhoff's letter to Piel explained that "The Committee feels that it has not yet mastered all of the problems connected with its publications, and is unwilling to take over the major problem which acquisition and management of the *Scientific American* would present." And so the opportunity was lost (20).

Fortunately for *Scientific American*'s now greatly enlarged list of readers, the owners provided enough additional capital to get the magazine over its temporary financial problems. AAAS had not accepted their offer, but AAAS interest may have helped give the owners the courage to keep their magazine afloat a little longer. Within a few months the magazine had achieved a circulation of 100,000, had increased its advertising revenue, and had sufficient capital to assure continued publication (21).

The record does not explain why the AAAS officers were so timid or negative. Perhaps the financial risk seemed too great. Meyerhoff had just that year been the board's first choice to succeed F. R. Moulton, but he was still new to the position. And perhaps they were influenced by a surfeit of publishing opportunities. They did agree to join with the National Institutes of Health in publishing a special series of symposium volumes (22), but as they rejected Piel's offer they also turned down an invitation to publish an American edition of the British magazine *Discovery* (23) and shortly declined two other — and very different — offers: one from Jacques Cattell, who hoped AAAS would purchase and publish *American Men of Science* (24), and the other from the International Academy of Proctology, which wanted AAAS to publish *Proctological Journal* (25).

In October of 1954 Duane Roller resigned to accept a new appointment with Ramo-Wooldridge Company and suggested that his resignation become effective upon completion of arrangements for his successor (26). Early in December, when we had not yet selected a new editor, he wrote that a matter he was not free to explain "makes it almost imperative that I leave at the end of the present month" (27). The board accepted that request and asked me to serve as acting editor, a title we selected to make it evident that the positions of administrative officer and editor were not being recombined (28).

In part my assumption of editorial responsibilities was possible because of the excellent staff Roller was leaving. Valentine's three assistants in 1946 had increased to 10 in 1953, but the magazines were still understaffed and early in 1954 the board had authorized three additions. For those positions Roller selected Charlotte Meeting, who served as managing editor until her resignation in 1957; Ellen Murphy, who in 1988 was still chief production editor; and Robert V. Ormes, successively assistant editor, managing editor, and assistant publisher until his untimely death in 1984. They and the generous members of the editorial board gave me a cram course in how to handle my new responsibilities as acting editor of the two magazines.

At that time one very live possibility was a possible joint venture with *Scientific American*. In earlier conversations with Gerard Piel I had learned that he and his colleagues were considering bringing out a weekly magazine of science as their next publishing venture. When Piel learned of Roller's resignation he offered to cooperate with AAAS in publishing *Science* (29) and proposed a contract along the following lines. *Science* would continue to be owned by AAAS and would be published under that title as an official journal of AAAS. Editing, advertising, and promotion would become responsibilities of the *Scientific American* staff. AAAS would contribute all of its advertising contracts, inventories, and prepaid and future subscriptions (and a portion of each member's dues) and would provide a loan to serve as working capital. His company would contribute the time and services of *Scientific American* staff members and sufficient risk capital to expand operations of the magazine. Although responsibility for management would rest entirely with *Scientific American*, operation would be overseen by an advisory committee appointed by AAAS. Excess of income over expenses would be divided equally between the two parties (30).

That proposal looked attractive. We needed a new editor and we also needed a new advertising representative to replace the current one who was unsatisfactory. So the board took Piel's proposal seriously, and with that encouragement he submitted the proposal to the other owners of *Scientific American*. By that time they were in a quite different position than they had been five years earlier when they had appealed to AAAS for rescue. Some of them insisted on altering Piel's proposals in a fashion so unfavorable to AAAS that the board's subcommittee (Thomas Park, Paul Scherer, Paul Sears, Warren Weaver, and Dael Wolfle) quickly agreed that the conditions were completely unacceptable. Indeed, they were so unfavorable that my first account of them was in a call from Piel to say that the deal was off: we could not possibly accept what his co-owners were offering (31).

In 1949 it had seemed to me, as an uninvolved outsider, that AAAS made a mistake in not accepting ownership of *Scientific American*. In 1955 it seemed to me, as a much-involved insider, that the owners of that magazine missed a fine opportunity in not making an acceptable offer for joint responsibility for *Science*. In the long run, however, both decisions worked out well. *Scientific American* soon reached the financial break-even point and was on the way to success, and there is no reason to think that it would have become better under AAAS ownership. As for *Science*, the refusal of *Scientific American* to offer acceptable terms forced AAAS officers and staff to try to overcome the troubles of *Science* and to put it on course to becoming a much-improved magazine.

## Improving Advertising and Typography

When negotiations with *Scientific American* terminated, the board agreed that we should start immediately to try to make *Science* into a better magazine. To that end, the board authorized adding staff and incurring a deficit if need be. My recommendation that we terminate the contract with the unsatisfactory advertising agent and seek a new one was approved, as was the recommendation that we increase the page size of *Science* from its atypical 7½ by 10¼ inches to the more standard size of 8¼ by 11¼, which was used by *American Scientist, Scientific American, Chemical and Engineering News*, and the news magazines. The larger page size permitted advertising plates made for other magazines to be used in *Science*, and *Science* thus became a more attractive advertising medium than it was when special plates and press work were necessary to fit its unusual page size (32).

In making these changes the cordial relationships that had been developed with members of the *Scientific American* staff were frequently helpful. Piel and Martin Davidson, advertising manager of *Scientific American*, offered to screen candidates for appointment as the new advertising representative. We had two good ones from which to choose and selected Earl J. Scherago, who still heads Scherago Associates, the firm that sells advertising space in *Science*. After earning a degree in chemistry from Cornell he started graduate work with special interests in electron microscopy, but then decided he did not want to follow in the academic steps of his father, who was chairman of the bacteriology department at the University of Kentucky. Instead he became an instrument salesman and then a salesman of advertising space in the American Chemical Society's *Analytical Chemistry*. At the time we interviewed him he was being offered promotion to the position of sales manager for that journal, but he chose *Science*, to the benefit of AAAS. As he established his office in New York, began recruiting a staff, and planned his sales strategy, our friendly relations with *Scientific American* continued to be helpful, for Davidson gave Scherago useful advice and help in his early days with AAAS.

Scherago's selection proved to be an excellent one for AAAS. The original contract stipulated that he would work only for AAAS (that restriction was later removed); that he would work on a commission basis; and that at least a stated portion of his commission had to be used to cover the costs of selling and the costs of promotion, the remainder of the commission being his personal compensation. With an eye on the future Scherago continued to take less than he could have as personal compensation and used more than the required amount to promote the journal as an effective medium for advertising scientific instruments and services. That strategy paid off; advertising revenue — which had been running at less than $150,000 a year — soon started moving sharply upward. For the next 15 years, advertising income increased at an average rate of 20 percent per year, a compound rate of increase that brought the annual total to $2.2 million in 1970.

As we decided to increase page size and try to attract more advertising, Charlotte Meeting, the managing editor, and I decided to improve the format and clean up the typographical hodge-podge that had gradually accumulated as different type styles and sizes had come to be used for different purposes. Again *Scientific American* was helpful as its art director, James Grunbaum, became our generous advisor. The result was that *Science* acquired a cleaner and more readable page. We made one mistake, however, as readers quickly told us, and we continued another.

Our mistake was to go too far in simplifying typography. We stopped using boldface type for the names of individuals in personal new items or lists of deceased members. Letters from readers soon reminded us that boldface type made it easy to scan for familiar names and that mistake was soon rectified.

The other mistake was on the cover. When Willard Valentine had become editor he stopped selling the front cover as advertising space and started using pictures of people, scenes, new building, or events of scientific interest. In 1949 Howard Meyerhoff dropped the use of cover pictures and used that space to list the issue's contents. In the redesign of 1955 we continued that practice. Four years later, however, with Graham DuShane as editor, *Science* returned to the use of pictures of scientific interest on the front cover (33). Since then under cover editor Grayce Finger's careful selection there has been a new picture every week. Some became popular posters or were used for the association's annual calendars that were printed during the 1970s.

## *Merging* Science *and* The Monthly

With more money from advertising, with *Science* redesigned and being published on a faster schedule, and with Graham DuShane in the editor's office, the time seemed right to try to decide a nagging issue: Should we continue to publish two magazines or should the two be combined? Should the *Monthly* be revised to become more attractive to a different readership or was some other option more desirable?

There was no uncertainty about *Science*. Whatever its shortcomings it was the American weekly newsmagazine of science and clearly should be continued and improved.

The *Monthly*, however, had two strong competitors: *Scientific American* with a circulation of 130,000-plus and *American Scientist* with 50,000. *The Scientific Monthly* had only about 27,000 subscribers (34), most of whom seemed reasonably satisfied despite occasional derision of its poetry (or verse), some complaints over publication of the schuss yucca spoof, and a couple of angry articles about schools of education. Nevertheless, there was a problem of quality; on average, papers submitted to the *Monthly* were of lower quality than those submitted to *Science* and new editors sometimes reported that previously accepted articles were giving them trouble (35).

The question of whether the *Monthly* should be continued had not been resolved when the board raised that issue in 1947. The question kept on being raised (36) until, at the end of 1956, Graham DuShane and I recommended combining the two magazines under the title *Science*. However, we added, one issue a month should be planned as "the 'Scientific Monthly' edition of *Science*" and consist mostly of material of the *Monthly* character. The board approved and presented the plan to the council which also approved (37).

Financial consequences of the merger were expected to be unfavorable in the short run. Combining the two magazines would lose subscription income from those libraries and individuals that subscribed to both. There would probably be a small initial loss in advertising revenue. And printing costs would increase, for it would cost more to print some 22,000 additional copies of 52 issues a year of the combined magazine than to print the same number of copies of 12 issues of the *Monthly*. In the long run, however, we thought both subscription and advertising income would increase if — and this was the justification for the merger — the combined magazine could be improved enough to earn the larger circulation that we hoped to achieve. The board accepted that reasoning, was willing to risk a temporary deficit if necessary, and decided to make the merger effective on January 1, 1958 (38).

Circulation of *Science,* already recovered from the doldrums of the early 1950s, jumped from 38,000 to 61,000 in the year of the merger, rose to 66,000 by 1960, and then soared to 163,000 in the next 10 years, an average increase of nearly 10,000 a year during the 1960s.

*Science* flourished, but the promised "Scientific Monthly" issues of *Science* never appeared. Some members disliked the loss of their favorite magazine and said so in the most convincing fashion — by resigning. The editorial staff worked hard to increase size, improve and broaden coverage of its articles and reports, and add new features such as the News and Comment section, and *Science* did include some articles of the type that would formerly have appeared in *The Scientific Monthly*.

Some members who had not liked the merger began to write saying they were pleased with the changes in *Science* and that increasingly they found it filling the place of the former *Monthly* (39). Nevertheless, as a magazine, the *Monthly* died and its death left unsatisfied those members and officers who thought AAAS should be publishing a second magazine. The nature of that magazine was not clear, however. Some thought it should be for students, some for the general public. Some wanted a magazine reviewing the history and philosophy of science. Ware Cattell tried to interest AAAS in reviving the *Monthly* under its old title and with him as editor. It was easy to say no to that proposal (40), but more difficult and only after extended discussion to decline to approve Daniel Greenberg's carefully prepared recommendation of a new magazine on science policy (41). To some minor extent the gap created by the demise of the *Monthly* was filled by arranging with other publishers to offer AAAS members reduced subscription rates for *Daedalus* (published by the American Academy of Arts and Sciences), *The Advancement of Science* (published by the British Association for the Advancement of Science), and a few other magazines. But it was not until appearance of the new and highly popular magazine with the annually changing title, *Science 80, Science 81* ... that AAAS met what a number of its officers had long considered an association responsibility — an association magazine of high quality intended for an interested general audience (42).

For *Science,* however, the merger was a complete success. Larger circulation brought greater advertising revenue; more income financed improvements in the magazine; those improvements attracted more member-subscribers. Within slightly over a year the typical issue increased from 48 to 64 pages and circulation grew by more than 50 percent. In 1960 *Science* reached the publisher's landmark of having enough advertising income to pay for all of the printing and paper costs. A larger part of dues and subscription income could thus be used for editorial improvements.

## News and Comment

*Science* has always been a news magazine and for a time paid Science Service to provide news items. When Bentley Glass was editor he employed Bethsabe Pederson to expand and systematize the news section under such headings as News Briefs (some of which were signed by members of the editorial board or other scientist-contributors); Scientists in the News; Necrology; Grants, Fellowships, and Awards; Education; In the Laboratories; and sometimes a further category — Miscellaneous. But further improvement seemed desirable and achievable, in both speed of printing and depth of analysis.

In 1954 the normal time between preparing the news columns and their appearance was about two weeks. Early in 1955 Charlotte Meeting and I decided to show the staff that *Science* could handle a newsworthy event much more rapidly. The occasion was an invitation to a General Electric Company news conference to announce an "important" but undescribed achievement. From knowledge of some of the work going on at GE, Meeting guessed that the successful making of artificial diamonds would be announced. The timing of the press conference fitted so neatly into the printing schedule for *Science* that we agreed to save an appropriate amount of space in the news section. She would then attend the press conference, write an account to fit the saved space, phone that account back to Washington so it could be inserted in the corrected page proofs and be sent immediately to the printer. Her guess was correct and with newsmagazine speed *Science* reported how H. Tracy Hall and his colleagues at GE had made synthetic diamonds by applying high pressure and temperature to graphite (43). The editorial staff stood a little taller that week.

Few items required that speed, but as the news columns moved increasingly from personal to topical news it became more desirable to shorten the time between writing and publication. A faster printing schedule and the increased press run that followed the merger of *Science* and *The Scientific Monthly* took the magazine beyond the capacity of the Science Press Printing Company in Lancaster, Pennsylvania. With regret at having to leave the firm that James McKeen Cattell had founded to print his journals, in 1959 *Science* printing was moved to the National Publishing Company in Washington, D.C.

With faster presses closer at hand, news items could be published on a much shorter schedule. In 1963 when President Kennedy was assassinated on a Friday, Jerome Wiesner, Kennedy's Special Assistant for Science and Technology, used the weekend to write a commemorative and appreciative account of the slain

President's relations to science; Daniel Greenberg wrote a companion article; and selected quotations from the address that President Kennedy had given a month earlier at the centennial of the National Academy of Sciences were substituted for the previously planned editorial. This sadly necessary new material was set in type on Monday, corrected, and in the mail for distribution on Tuesday night and Wednesday morning (44). The United States Information Agency sent a thousand copies to individual scientists in the Soviet Union.

Doing the same thing more rapidly made the magazine more timely, but we also wanted to publish a different kind of science news. In a kind of reverse way, the model of what we wanted was the newspaper science writer who reports scientific developments in terms that interest and inform nonscientists. We wanted analyses of political, economic, societal, and governmental actions and their implications for the work of scientists. No other journal was publishing such material in a regular fashion so there was no specific model to use as an example, and we tried two or three writers before finding one who got the idea. The first who did was Howard Margolis, a Harvard graduate with a major in government and a minor in physics who had been writing for the Food and Drug Administration and the Federal Trade Commission. He came to AAAS in 1960 and in 1961 was followed by Daniel Greenberg from the *Washington Post*. In 1962 John Walsh left a congressional staff position to come to *Science*. Those first three were the pioneers of News and Comment.

Reporters for the staff were not scientists and were chosen with other qualifications as criteria. In the first decade or so of the News and Comment section every reporter had graduated from college with a major in history, economics, government, or a similar field. Each had written for a daily newspaper and had worked on Capitol Hill, either as a staff member or as a reporter. Several had been Rhodes Scholars. They were an able group and they wrote the kind of interpretive articles Graham DuShane and I originally had in mind.

But they also wrote some articles of a kind that we had not anticipated. As they worked they were necessarily looking at the scientific community to which their articles were addressed and they began to write articles about that community.

Some of those articles — and also some editorials — were critical of government actions or policies, and in a few cases News and Comment pieces included uncomplimentary personal statements about a member of Congress or of the Executive Branch of government. All of that made some members of the board and some readers quite uneasy. We were told that we were damaging the whole scientific enterprise when we criticized government officials and were accused of "washing dirty linen in public." Some critics opposed the practice of open debate and of publishing individual or divergent views. Frederic Seitz, president of the National Academy of Sciences, told the editor of *Science* that it would be better practice for AAAS to have a small group of scientists discuss a policy issue and then "use *Science* to summarize the results of these deliberations after they have been digested." (45).

Complicating the whole matter was a continuing problem for the board: When and under what circumstances was it appropriate to release a statement for or against a practice or policy issue? AAAS members' attitudes and views differed widely and never did they vote to determine majority views on anything. Thus statements made by the board or council were carefully labeled to make their source clear and evident.

As for News and Comment, some clashes were probably the inevitable result of having a group of able and energetic journalists write for a magazine owned by and published for a scientific community whose members varied widely in their views, opinions, and political and societal attitudes. Open and personal controversy is generally more attractive to journalists than to scientists; what made a good story from the writer's point of view was sometimes objectionable to a reader with quite different views. In the board's view, a disturbing aspect of the resulting arguments was the danger, and sometimes the fact, that statements published in News and Comment or in editorials were interpreted as representing official positions of AAAS or its officers. Of course the board wanted members of the staff to use good judgement and good taste in everything they wrote, but they did not want to tell the editor what should or should not be said or what topics were or were not appropriate for discussion in *Science*. However, the board did want it to be clearly recognized that individual authors were not speaking for AAAS. To give greater emphasis to that point, they decided to position immediately adjacent to the weekly editorial the statement that all articles of whatever kind published in *Science* "including editorials, news and comment, and book reviews are signed and reflect the individual views of the authors and not official points of view adopted by the AAAS."

That statement helped on one aspect of the problem, but in 1970 internal conflict over what could be published cost the association one of its ablest, most admired, and sometimes most criticized reporters.

In 1970 Glenn Seaborg had agreed to run for president-elect of AAAS. In 1963 and 1968 he had declined similar invitations because he felt that his duties at the Atomic Energy Commission (AEC), would not allow him adequate time for the AAAS position. In 1970, however, he decided that if he were elected he could adjust his AEC duties to do justice to the AAAS responsibilities. All of that was strictly in accordance with due process of the Committee on Nominations and Elections, but after ballots had been mailed controversy arose. Some members of the board thought it would be inappropriate to have the chairman of the Atomic Energy Commission also serve as president and later chairman of the AAAS board. The question was a personal one; nobody doubted Seaborg's ability or scientific standing, but some did worry about conflicts of interest. No such fears had been expressed a decade earlier when Paul Klopsteg, associate director of the National

Science Foundation, was nominated, or later when Wallace Brode, science advisor to the Secretary of State or Alan Waterman, director of the National Science Foundation, were elected to the same office. But 1970 was a different year and attitudes had changed. Bentley Glass, then chairman of the board, accepted the unpleasant task of informing Seaborg of the concerns of some members of the board — not of all, for the Seaborg candidacy was supported as well as questioned within the board. Seaborg considered the issues, consulted with some friends, and decided to leave his name on the ballot. The council elected him and as far as I know there were no later charges that conflict of interest had affected his service to AAAS.

The whole matter led to an exposé article in the *Washington Post* while the balloting was going on. Daniel Greenberg, news editor of *Science*, wrote a piece that Philip Abelson, the editor, refused to print. Greenberg took the position that the controversy about Seaborg's nomination was a legitimate news story and that "secrecy is anathema to the well-being of the scientific community." Abelson took the position that Seaborg's chairmanship of the Atomic Energy Commission was public knowledge; that the electorate had received biographies of the two nominees; that *Science* had already published articles about some of Seaborg's other interests; and that for *Science* to publish an account of adverse attitudes toward one of the two nominees would constitute meddling in the election by the association's own journal.

Abelson prevailed, Greenberg resigned in protest, and AAAS lost the services of an excellent reporter who had helped pioneer the News and Comment section of *Science*. There was a further result: In no way could AAAS avoid unfavorable criticism over the whole matter and the press seemed to delight in publicizing it. The account in *Newsweek* was headlined "The Squabbling Scientists" and the *Washington Evening Star* headlined its account "Science Association's Internal Rot" (46).

Despite some troubles, *Science* helped create a new kind of journalist and writers for the magazine began to get other and sometimes more attractive offers. Howard Margolis went to Defense Secretary Robert McNamara's office as speech writer and then to the political science department of the Massachusetts Institute of Technology. On a fellowship at Johns Hopkins University Daniel Greenberg wrote *The Politics of Pure Science*. He was invited to write special articles for many other periodicals and when he left *Science* in 1970 he began publishing *Science and Government Report*, his own newsletter reporting Washington, D.C., developments of interest to scientists and technologists. Victor McElheney, our first staff member stationed in Europe, became science editor of the *Boston Globe* and then director of the Vannevar Bush fellowship program at MIT (Now renamed the Knight Fellowships). Philip Boffey went to the *New York Times*. Bryce Nelson left for the Los Angeles *Times* and then became director of the journalism school at the University of Southern California.

The quality of their work was also recognized by honorary awards. In 1970 President William J. McGill of Columbia University conferred on Daniel Greenberg the University Medal for Excellence; the citation described him as

> the biographer, conscience, and political mentor of the scientific community. He combines intelligence, imagination and wit with the best tools of the journalistic craftsman in fashioning an account that sparkles with integrity and credibility. There are few scientists in the nation whose self-understanding has not been immeasurably enhanced by Greenberg's reporting of the best and worst of our folkways and frustrations ... [He does] for the nation what we have otherwise failed to do for ourselves. He has sought by induction to discover a national scientific policy from the innumerable determinations and doings of presidents, congressmen, foundations, universities, corporations, and the activities of the mandarins of America's research establishment. (47)

Another award that gave us particular pleasure came from the American Medical Writers' Association, which in 1967 conferred on *Science* its Honor Award in Medical Communication and explained that the award was for "excellent general medical science reporting, especially by Elinor Langer" (48). Elinor Langer had been employed not as a reporter but as an assistant to search out information and to aid the staff writers. But she quickly demonstrated that she could write well and with some initial coaching from the more experienced members of the staff she became a regular contributor to the News and Comment section.

The best award, however, was the readers' judgment. Responses to periodic reader surveys regularly showed News and Comment to be one of the favorite sections of the magazine and frequently the section read first.

## Policy, Management, and Authority

As AAAS gained experience in publishing *The Scientific Monthly* and, especially, *Science*, policy had to be developed on five major issues of responsibility and authority: 1. For whom is the magazine published? 2. Who should be responsible for its contents? 3. To whom is the editor responsible? 4. To whom is the executive officer responsible? 5. What are the appropriate relations between the executive officer and the editor?

Probably the most fundamental of these is the question: For whom is the magazine published? "The reader" is the obvious answer, but not always the accurate one. Some magazines are published for the advertisers. Sent without charge to a selected list of recipients, they provide sellers of the advertised products or services a direct line to a targeted group of prospective purchasers. Other magazines are published primarily for the benefit of the authors. Periodicals with this purpose often have names such as "Annals of the ..." or "Bulletin of the ...". Still other magazines are published for the benefit of the organizations that own them. For example, a company publication informs stockholders, staff members, and interested others about new products or other developments.

But a magazine published by a voluntary membership organization as diverse as the AAAS must be for the readers, not the advertisers or the authors or the owners. *Science*, even before it was owned by AAAS, had a minor house organ function. It published the presidential addresses, annual reports, and other official papers of the association, and it continued to do so in years when the *AAAS Bulletin* was not published. Yet benefit to the association was not the primary reason for publishing. The primary objective was to benefit the readers.

That emphasis was early recognized and early stated. In 1945, the first year of AAAS ownership, some changes in advertising content were announced, but readers were assured that those changes would have no impact on editorial content (49). And even advertising was judged by that principle. The general standard for acceptability has long been that products or services advertised in *Science* should be of interest to readers as scientists. Other kinds of advertising are inappropriate.

The principle of reader control became established not as the result of a specific policy document or in the resolution of a major conflict, as did some of the principles discussed below, but more gradually as specific cases were considered. Should all vice-presidential addresses be published, as one member of the board of directors insisted? Should reports of prominent AAAS committees be published, as some committee chairmen claimed? Not unless the editor decided that publication would be of interest and value to a sufficient number of readers.

Thus readers' judgment was an essential guide for the editorial staff, and readers' judgment was frequently and sometimes vigorously made available to the staff. One channel was through periodic questionnaires sent to random samples of individual members or subscribers. Responses told us which sections they read first or which they would like to see expanded, contracted, or left about as they were (50). This information was helpful in allocating space as the magazine continued to expand.

Guidance also came from unsolicited letters. Some dealt with an individual article or report and many of those were published in the letters column of the magazine. Others concerned the News and Comment section, usually the most controversial section of the magazine. Some were complimentary, and those I wanted to make certain were seen by the appropriate staff members. Others were critical, and those I wanted to see for myself. I have no illusions that I saw all of the complaints, but "the complaint" was a valuable feedback, whatever the topic, and I wanted to know what the problems were.

My all-time favorite complaint concerned a matter of editorial style. Duane Roller was interested in rules for forming and using symbols. He argued that units of time should be treated as were amp, ohm, curie, or other physical units, and he insisted that they be represented by their proper symbols such as min for minute, or wk for week. One reader who objected was George W. Corner. After complimenting Roller for recent improvements, he added that his expression of general satisfaction with *Science* entitled him to protest one specific policy:

> When I have seen the abbreviation yr for *year, years* in the text of other people's articles in *Science*, I have merely shuddered; but now to see it in my own galley proofs — in this 45th year of my attempt to write clear, dignified, pleasant-looking English on scientific subjects — and to realize that an article which will be seen by cultivated men in many countries is going to represent me as using this barbarism, has really upset me.

Then with an obvious debt to "Always," Irving Berlin's love song to his wife, Corner concluded (51):

> This protest, Sir, is registered
> Not for just a da
> Not for just a yr
> But alw.

That protest did not persuade Roller, who won the skirmish. The last things he wrote during his year of editorship were an editorial and an article on symbols of units of measurement, including measurement of time (52). But Corner won the war. After Roller left, the editorial staff quietly went back to the use of ordinary English in writing about days, weeks, or years.

Corner's letter was my favorite complaint. My favorite response to a complaint was, egotistically, one I wrote myself. A letter addressed to me personally started with the blast: "You undoubtedly run the most inefficient office in Washington, D.C." Investigation showed that we had indeed messed things up pretty badly. I wrote a letter of apology, explaining what had gone wrong and what we had done to rectify matters. Then on an impish impulse I added a short final paragraph: "For my consolation I hope you will tell me who runs the second most inefficient office in Washington, D.C." By return mail there was a contrite letter of thanks and from a mutual acquaintance I later learned I had acquired a strong supporter.

## Responsibility for the Magazine's Content

The principle that benefit to the reader should be the guiding policy in editing *Science* was generally accepted without argument. Adoption of some of the other principles was traumatic. Willard Valentine, the first AAAS editor of *Science*, was given a substantial amount of authority to plan and manage the magazine, but editorial latitude rapidly diminished after his death and during the temporary arrangements with Mildred Atwood and the part-time editorships of George Baitsell and Howard Meyerhoff. Because the original AAAS system of sending manuscripts to outside referees was thought not to be satisfactory, the association created an editorial board whose members were paid $1,000 a year to review manuscripts in their fields and to recruit other reviewers when necessary. The members soon went beyond that function, however, and tried to manage *Science* and *The Scientific Monthly* by committee, ruling, for example, that any one member could approve a technical report, any two a lead article, and any three an editorial (53).

The editorial board apparently liked committee management, for the members "instructed" Howard Meyerhoff, their new chairman, "to inform the Publications Committee of the Board's conviction that the present method of directing the publications through the medium of an Editorial Board is proving sufficiently satisfactory and promising to warrant continuation for another year" (54).

The board of directors accepted that recommendation (55), but within a year Meyerhoff had had quite enough. He told the board of directors that it was time for a clear statement of how authority was assigned among the Publications Committee, the editorial board, and the editor. He then went on to urge that as the actual editor he should have full authority and responsibility for deciding what was to be published in the two magazines, with the understanding that he would accept the judgment of other members of the editorial board on papers in fields other than his own and that he would consult them on other relevant matters. Moreover, he added, in publishing a weekly, decisions often had to be made too quickly to wait for the next board meeting or replies from a quorum of the members. Then he came to the clincher: Concentration of authority as he requested would give the Publications Committee and the board of directors "complete control of the journal. If they don't like the chairman's performance, they can remove him" (56).

The board of directors agreed and ended the festering question of allocation of authority by ruling that (a) all matters of policy concerning the association's journals resided in the Publications Committee (which was made up largely or wholly of members of the board of directors); (b) the responsibility for executing those policies rested with the editor, then the chairman of the editorial board; and (c) the responsibilities of the editorial board were to assist the chairman in carrying out established policies, to review papers in their respective fields, to solicit manuscripts, and to give advice to the chairman (57).

The editorial board was not altogether happy with those decisions and the editor's authority was challenged by at least one member, but as Meyerhoff later explained he refused to let the matter come up for further discussion because if he were to be able to achieve the improvements he wanted to make he needed the authority the board of directors had given him (58).

Fortunately for his successors in the editor's chair he prevailed. The issue of relations among editor, editorial board, and the Publications Committee and board of directors was settled and has remained so. However, it took three more years and a major upheaval to clarify relations between the editor and the association's elected officers and directors.

## The Meyerhoff-Condon Correspondence

On New Year's Day of 1953 Edward U. Condon, who was about to finish his year as president-elect and become president, wrote Meyerhoff a long, 18-point letter. On several points he wanted information about what was expected of him as president of AAAS or how the association handled certain matters. On several points he expressed dissatisfaction with *Science* and *The Scientific Monthly*, named several articles he wanted Meyerhoff to publish quickly, or requested changes in the normal publishing schedule of the magazines. Two of the 18 points dealt with the handling of letters or other communications either addressed to him personally or concerning him. He had for several years been one of the choice targets of the House of Representatives Committee on Un-American Activities (see Chapter 3, especially note 14) and a few days earlier at the annual meeting in St. Louis had been angered by a newspaper account of what Meyerhoff had said to a reporter concerning Condon's support and opposition within AAAS (59).

Much of Condon's letter could be described as the queries of a new officer not well acquainted with how matters were handled and wanting to be informed, but some queries seemed strange coming from one who had been a paid member of the editorial board for several years, and some could be described as interfering with the responsibilities and authority of the association's editor and chief executive officer. In fact, the letter was later described in all of those ways (60).

Meyerhoff replied on January 6 by supplying the requested information on one of Condon's 18 points and saying he would reply to the other points in succeeding days as time permitted. The next day he sent Condon four letters. Two were factual and informational; one explained why some of Condon's request for changes in *Science* were impractical or departed from established custom; and the fourth was a sharp rejection of the request that he seek clearance from Condon for the few — if any — statements he might make concerning Condon's relationships to the AAAS.

From then on relations deteriorated. After receiving several replies from Meyerhoff, Condon wrote another long letter covering many of the original points and taking a conciliatory tone, but before Meyerhoff received that letter he had written to Condon "...for your guidance I shall say what you undoubtedly have surmised: When anybody says to me 'You will under no circumstances...' my instantaneous rejoinder is 'Go to Hell.' "

Between January 1 and February 10 there were six letters from Condon to Meyerhoff and 14 from Meyerhoff to Condon. Copies of most of those letters were sent to Detlev Bronk, the association's retiring president and chairman, and to Warren Weaver, the president-elect. In addition Condon wrote twice to Bronk and Weaver commenting on Meyerhoff's attitude and his "almost pathological behavior." The letter writing ended with two exchanges. In one Condon asked for factual answers to several questions, not long essays. Meyerhoff's reply consisted mostly of "no," "probably," and "yes." In the other, Meyerhoff invited Condon to review a certain book for the forthcoming book issue of *Science*. Condon refused, saying anything he wrote for *Science* would probably be rejected.

Neither man was one to avoid controversy. Condon had been battling unconfirmed charges by congressional witch hunters who refused to grant him a hearing. And Meyerhoff had been a target of charges and the author of counter-charges in earlier controversies and was at the same time feuding with Weaver over the Arden House statement and the association's annual meeting (see Chapter 3). Bronk's attempt to calm things down was unsuccessful and some members of the board of directors may have not known of the mounting trouble until they received as part of the agenda for the first board meeting of 1953 a 13-page statement by Meyerhoff and Gladys Keener, executive editor of the two magazines. That statement gave their account of the feuds with Condon and Weaver, explained that they felt forced to leave the association, and said they would submit formal resignations after they had made other arrangements (61).

A few days after the next board meeting, which Meyerhoff did not attend, Bronk tried to persuade Meyerhoff and Keener to stay, either until the end of June or indefinitely. Their reply was that they would stay only if Condon and Weaver were to resign from their offices and from the board of directors (62). They did not expect that condition to be met — which it was not — and they left. Raymond Taylor was named acting administrative secretary, Bronk persuaded Bentley Glass to take on the editorial responsibilities on a part-time basis, and the board set out to find a new administrative secretary and a new editor.

A protest movement was quickly started by Richard E. Blackwelder, a member of the council and secretary of the affiliated Society of Systematic Zoology. He organized several meetings of AAAS members in or near Washington to consider the charges and countercharges. Four members of that group, all members of the AAAS council, met with the board to propose that the whole council be informed of what had transpired (63). The board, however, was involved in negotiations with Meyerhoff and Keener concerning retirement rights, salary due, and related matters, and on advice of Warren Magee, the association's legal counsel, did not report to the council until those matters had been settled. Blackwelder, however, felt no such constraints. He and seven cosigners sent all council members Blackwelder's account of Meyerhoff's difficulties with Condon and Weaver and asked those who felt so inclined to write to President Condon to request a special meeting of the council. The response was not sufficient to require a special meeting (64).

Thus council members were not fully informed of the affair until shortly before the 1953 annual meeting, but then they got an overload. At the request of the board Raymond Taylor distributed copies of the minutes of the board meetings of March, May, June, and October, together with a summary "Report of the Board of Directors to the Council of the AAAS." Blackwelder sent each member a 14-page, legal-sized, single-spaced memorandum. It started with a complimentary description of Meyerhoff's services to AAAS. Then followed a one-sided account of the efforts of Condon and Weaver to change AAAS policies and practices without going through the board of directors or other proper channels, along with criticisms of the board for not trying to ameliorate the mounting troubles. When Condon received the Blackwelder memorandum he responded by sending members of the council copies of all 22 letters of the Condon-Meyerhoff exchange. The more-restrained Warren Weaver distributed a two-page statement of his relations with AAAS and Meyerhoff.

When the council met, the first order of business was for Detlev Bronk, as chairman of the board, to report appointment of three new officers: Dael Wolfle as administrative secretary, Duane Roller as editor, and Paul A. Scherer as treasurer. (William E. Wrather who had served as treasurer since 1945 had not been involved in the troubles of the year, but he had been wanting to resign and during 1953 the board had found an excellent replacement.)

Bronk told the council that mutually satisfactory arrangements had been made with Meyerhoff and Keener, that the entire board had recorded their warm appreciation for the devoted and effective service both had given the association, and that the board regretted that both felt it necessary to terminate their services. As for the future, he stated that the board and I had discussed the whole situation and had come to complete agreement concerning the nature of my duties, responsibilities, and opportunities.

He then reported that the board had appointed a special committee to examine the operations of the association and to make such recommendations as seemed desirable. That committee included two members of the board of directors, Wallace R. Brode and Roger Adams, and three council members, Meredith F. Burrill, Clarence E. Davies, and Milton O. Lee. It was no accident that all three members held positions somewhat similar to the one I was entering. They were, respectively, secretary of the Board of Geographic Names of the Department of the Interior, secretary of the American Society of Mechanical Engineers, and executive secretary and managing editor of the American Physiological Society. In addition, Howard Meyerhoff and I were named as advisors and consultants to the committee (65).

Meyerhoff was then a member of the council, for shortly after he left the AAAS office the American Association of Petroleum Geologists had appointed him as their representative. His motion that the report of the board chairman be accepted was unanimously adopted and the council then moved to consider the material they had received from Blackwelder. The whole mood was by then one of conciliation and of wanting to give me an unclouded start in my new position. To reinforce what Bronk had reported concerning my expected relations with the board of directors, Blackwelder introduced a motion stating that the council affirms that "in accordance with the Constitution and Bylaws, the Administrative Secretary shall have the responsibility of conducting the operations of the Association without interference from the general officers of the Association, and that he be responsible for the success of these operations to the Board of Directors as such and not to any officer." The motion then went on to endorse appointment of the special committee to review association operations (66).

Blackwelder's motion was seconded by Warren Weaver and adopted by the council. The board and council had done all they could to make sure there would be no repetition of the troubles of 1953. And there were none. My relations with the next 17 presidents and chairmen and other members of the board of directors were always cordial and cooperative. I could not have wished for better relations than actually existed. But 1953 had not been forgotten: Some years later one member of the board came to my office to tell me that several members were apprehensive that the new president might cause some difficulties, to ask me to let them know immediately if strains did begin to develop, and to assure me they would do all they could to contain any problem. Of course I was grateful for their solicitude, but said I was quite confident that there would be no need to call for help, and there was none.

When he left AAAS Meyerhoff became president of the Scientific Manpower Commission which was just then being established and of which I was a founding director. He held that position until 1963 when he became professor and chairman of the department of geology at the University of Pennsylvania. Within AAAS he continued to be a cordial, constructive, and always well-informed member of the council. Condon served through 1954 as chairman of the board of directors and then largely dropped out of association affairs. Warren Weaver remained on the board through his presidential and chairmanship years and then continued to be a staunch supporter of the association, serving on several major committees, helping to secure financial support for several new activities, and always being available and generous in serving as a wise senior counselor. To paraphrase Will Rogers, I never knew an association president I didn't like, but if I were to pick the one who was most helpful to the association generally and to me personally it would be Warren Weaver (67).

## Relations Between the Executive Officer and the Editor

As the board of directors set out to replace Howard Meyerhoff, one of the decisions made was that the position of editor should be separate from that of administrative secretary (68). Late in the year when the search for an editor was concluded by appointing Duane Roller, I had not yet decided whether to accept the invitation to become administrative secretary. A book-length report I was writing would not be completed for several months (69); I was considering another offer; and, not yet knowing much more than the Meyerhoff side of his difficulties with Condon and Weaver, I hesitated. In the meantime the editorial board had warned that the editorial situation was critical, that Glass could not continue beyond the end of the year, and that "immediate action should be taken to appoint a new editor" (70).

Bronk apologized for appointing an editor without consulting me. But the board really had no alternative; it was necessary to make an appointment and I had not decided whether I would accept the position offered to me. I was soon persuaded, however, by talks with Warren Weaver and other members of the board of directors, plus assessment of AAAS and its potentialities by several friends I consulted.

Thus Roller and I entered our new positions somewhat uncertain about our relations to each other, although the situation may have seemed clearer to him than it did to me, for several months later Wallace Brode, who had been chairman of the search committee, told the other members of the editorial board, including Roller, that "... the Secretarial and Editorial offices were set up by the Board of Directors to be parallel and separate" (71).

Whether that interpretation was shared by other members of the board is not clear. Separate and parallel was the arrangement in the American Chemical Society, of which Brode was a prominent member and a future president, and that arrangement may have influenced his interpretation. Whatever the board's intent, I thought Roller should attend meetings of the board of directors. Thus he was regularly invited, as were Raymond Taylor and John Behnke. They attended board meetings regularly, but Roller did not until October when the board asked him to attend to discuss his resignation. At that meeting, perhaps uncertain of what hidden tension there might be, the board asked to talk in private, first with me and then with Roller. They asked me why Roller had not attended earlier board meetings, as had Taylor and Behnke. I explained that I had invited all three but that Roller had said he was too busy. What he was asked I do not know, but after he left, the board's statement to me was that the two discussions had removed any uncertainties and that to make it evident that the editor was part of the association's single administrative structure they wanted me to submit one or more recommendations for the new appointment (72).

From then on the editor was — and properly so — responsible for magazine content and not subject to instructions from anyone as to what to publish, but was, as Meyerhoff had asked some years before, subject to removal if performance was not satisfactory. However, appointment of the editor, letting printing contracts, filling staff positions, making financial arrangements, and similar matters were not "parallel and separate" but were parts of the regular administrative structure of the association.

To symbolize the relationships involved, my title was changed in two ways. In 1956 when Graham DuShane took over editorial responsibilities, my name went on the masthead of *Science* as "publisher," a title that I asked the board for because I wanted a formal continuing responsibility for the magazine, and one that has been continued by my successors. Some months later Thomas Park told the other members of the board that he thought the title "administrative secretary" no longer seemed appropriate and that they should select a title that more clearly suggested overall responsibility. Debate over several alternatives led to agreement on "executive officer," which in due time was sanctioned by changes in the association's constitution and bylaws.

## The Abelson Years

In 1962 Graham DuShane received an offer he could not resist. Invitations from several other universities had been declined, but Vanderbilt University wanted him as chairman of the Department of Biology and dean of Graduate Sciences. He accepted, but the move turned out tragically. Renting a house from a bird fancier, he became infected, apparently with psittacosis, and had a miserable winter. On a visit to Washington a heart attack sent him to the hospital for several weeks, and on July 26, 1963, while driving across country to visit old friends and attend a meeting of the AAAS Commission on Science Education, a second heart attack was fatal (73).

A year before, when DuShane decided to move to Vanderbilt, the choice of a new editor had been easily and quickly made: Philip H. Abelson, director of the Carnegie Institution of Washington's Geophysical Laboratory, successful editor of the *Journal of Geophysical Research*, chairman of the program committee for AAAS annual meetings, and — perhaps most important of all — one of the most broadly qualified members of the American scientific community. When he had been elected to the National Academy of Sciences he could have chosen to have his membership recorded in any of six of the Academy's scientific sections (74). He chose the section on geology, but when he was elected a fellow of the American Academy of Arts and Sciences it was in the section on molecular biology, and in his biography in *Who's Who in America* he called himself a physical chemist.

The directors agreed that Abelson would make a fine editor. I went to see Caryl Haskins, president of the Carnegie Institution of Washington, to tell him that we wanted Abelson as our editor. Haskins agreed immediately that we had made a fine choice and said that if Abelson decided to accept of course he could, but that the Carnegie Institution had no intention of letting him leave there. I had not discussed with the board of directors the possibility of a part-time appointment, but it did not take many seconds to decide that this part-time appointment would be better than any alternative I could think of.

Happily for AAAS Abelson accepted, and on August 3, 1962, *Science* began to carry his name as editor and soon began to show his mark. One of the first changes was to start using the telephone instead of mail to recruit referees whose advice was wanted on submitted articles. If a first-choice referee was unavailable, one telephone call settled the matter and another possible referee could be called immediately. If the first choice accepted, the manuscript could be put in the mail that day instead of waiting days or weeks for a reply. That speeding up of the review process shortened the average lag between receipt and publication of an accepted paper by about a month. Moreover, it turned out that referees provided more complete and informative comments following a telephone invitation than they did after receiving a letter (75).

As Valentine and other editors had intended or wished, Abelson also wanted to broaden the coverage of *Science* by increasing the number of articles and reports from fields other than biology. He succeeded partly because of his own wide interests and because he gave himself the assignment of knowing the important developments across the broad spectrum of science and partly because history was on his side. The 1960s were years of manned and unmanned satellites and of increased interest in planetary and space science. Abelson made *Science* the journal of quick publication of many reports from that burgeoning field.

## Special Issues

Another of Abelson's contributions was to increase the number of special issues. Earlier there had typically been four a year: one on books, one on instruments, a preconvention issue giving advance information about the annual meeting, and a post-convention issue giving an account of that meeting. He added one-time special issues, each on a timely and rapidly moving scientific field. From a publisher's point of view the most spectacular example was the Lunar Science, or Moon, issue of January 30, 1970.

As Apollo astronauts came home with moon rocks and regolith, those samples were assigned to interested scientists for detailed analysis. In early January of 1970 NASA brought the analysts to Houston for presentation and discussion of their findings. By prior arrangement between AAAS and NASA, *Science* published a special issue containing the papers presented at that conference and did so on a very fast schedule. In 27 days, copies of the manuscripts were received, reviewed by critics, submitted to the authors for their responses to the reviewer's comments, edited for style, and marked for the printer. Simultaneously, illustrations were redrawn and relettered in uniform style and marked for the printer. Authors were given an opportunity to approve or comment on the editing of their papers. Then everything went to the printer; galley proofs were corrected; page dummies were prepared, proofread, and corrected; and revised page proofs were corrected. With all of that done the issue was printed, bound, and mailed. With part of the staff working in Houston and an augmented staff in Washington, D.C., the largest issue of *Science* ever published, including 325 pages of refereed scientific material, was taken from original manuscript to published issue in 27 days, while at the same time the staff and printer were maintaining the regular schedule for the intervening issues of *Science*. That accomplishment must have set some kind of record in the history of publishing scientific journals (76).

The Committee on Publications congratulated the staff for "producing this epic-making issue in an exceptionally brief period of time." Both the content of the issue and the publishing feat received a good deal of attention. Among the reviews and comments the one that amused us most was in the *Journal of the American Medical Association*. After a brief account of what had been accomplished, the *JAMA* editorial concluded, "It will be of little or no interest to physicians. Indeed the sole purpose of this brief editorial is to give notice of an outstanding achievement in scientific journalism. To Philip H. Abelson, editor, and Dael Wolfle, publisher, of *Science*, congratulations on your success in exceeding the speed of light" (77).

Another special annual issue was devised by the advertising staff. In 1961 Earl Scherago, the association's advertising representative and a former equipment salesman, proposed that AAAS publish a guide to scientific equipment useful for research or teaching at the college and university level. AAAS had already had some relevant experience through helping the Council of Chief State School Officers prepare a widely used guide for purchase of equipment for high school laboratories. Scherago's proposal was for a comprehensive but nonevaluative listing of equipment and the sources from which each type could be obtained. He expected that advertising revenue would cover all or most of the cost of preparing and distributing the guide, but he recommended that listings and descriptions be completely divorced from the question of whether or not the supplier bought advertising space. Discussions with a small sample of laboratory scientists indicated that a well-prepared guide would be generally useful (78).

After a trial version the first fully developed *AAAS Guide to Scientific Instruments* appeared in 1963 as a special issue of *Science* and has appeared annually ever since. For laboratory directors and purchasers of equipment it serves as a comprehensive guide to sources of a wide variety of instruments, and the numerical coding system under which instruments are classified has been adopted by a considerable number of laboratories for maintaining records of their equipment.

## Editorials

James McKeen Cattell was never reluctant to use *Science* to express editorial judgments on matters concerning the scientific community. But with AAAS as the owner matters were quite different. True, there was an editorial page, but it was sometimes used for such innocuous items as the description of a newly elected affiliate organization. Editorial statements on more controversial topics might be interpreted as representing the position of the association as a whole or of its officers, and so some thought editorials should be avoided. That was the attitude of Warren Magee, the association's legal counsel, who wanted the association to avoid anything that might be interpreted as libelous or as an attempt to influence future legislation. When Howard Meyerhoff was editor, Fernandus Payne, chairman of the Publications Committee, tried to persuade him to carry true editorials, but, he lamented, "to no avail" (79).

Bentley Glass tried to change matters. However, his attempts to increase editorial comment in the magazine (80) were not always successful. After one editorial he wanted to publish had been approved by Detlev Bronk, chairman of the board of directors, Warren Magee insisted that most of it be cut, so the intended editorial was reduced to a pallid factual statement in the news section.

In the next few years, however, true editorials became more frequent. Graham DuShane, Joseph Turner, the versatile assistant editor during the DuShane period, and I liked to write editorials. It was a taxing chore, however, for one had to select an appropriate topic and then use a fixed amount of space to point a moral, present a challenge, preach a sermon, impart useful knowledge, or in some other fashion try to stimulate readers' thinking.

Greater editorial freedom was partly the result of a changed legal atmosphere. Arthur Hansen, who became the association's legal counsel in 1954, worked from a quite different principle than had Warren Magee. Although not formally expressed Magee's principle seemed to have been "don't let AAAS do anything that might possibly cause trouble." Hansen's, in contrast, could be described as "help AAAS do legally and properly what it needs and wants to do."

One thing the editors wanted to do was to publish editorial discussions of current issues of importance to the magazine's readers. The Publications Committee and the board of directors supported that desire and wanted it to be unambiguously evident that the judgments expressed were those of their individual authors and were not statements of positions adopted by the association as a whole or its officers. Editorials were therefore always signed with the name of the author or the initials of the AAAS staff member who wrote them (81). In 1964 that principle of individual responsibility was made even clearer by placing on the editorial page the statement that the views expressed in editorials, articles, news and comment pieces, and other material represented the views of the individual authors, not those of the association or the author's institution. (That disclaimer still appears in every issue of *Science*).

Late in 1962 a new name began to appear below some of the *Science* editorials, and from then on if readers had been asked which feature of the magazine was most clearly identified with the editor, "the editorial" would probably have been the most frequent answer. Some 500 cogent, well-reasoned, informative, and sometimes fact-filled statements on a wide range of topics came from Abelson's pen. When he retired the association published a selected collection of those editorials under the title *Enough of Pessimism*. In the introduction, Allen L. Hammond, editor of *Science 85*, the association's other magazine, summarized the impact of the Abelson editorials:

> Abelson has been an acute observer and recorder of what was going on around him during his editorial tenure that spanned 22 years of *Science*'s 103-year history. His editorials map the progress of science, the evolution of our society, and the changing interaction between them. You can follow in them the swings and counterswings of opinion on several of the great technological efforts of our times — nuclear power, the space program, industrial competition with other

nations. Often his editorials go further than the popular opinion of the day to anticipate the core of the issue — that which would remain a problem for a long time. It is a record of an inquiring mind, a distinguished intellect, and a great editor (82).

## Accomplishments and Honors

Building upon the accomplishments of earlier editors, and as a result of his own leadership, by 1970 Abelson was editing one of the nation's leading scientific journals. In the scientific literature it was one of the most frequently cited journals and for the public press it was one of the most frequent sources of science news. F. R. Moulton could have felt proud that *Science* had indeed become "at least comparable with *Nature*" (83) and pleased to know that its editor was widely regarded as one of the leaders of thought on scientific issues and matters of science policy. That year was not the end, however, and there were 14 more years before Abelson stepped down from the editorship and then, at the request of his successor, accepted appointment as a deputy editor, a position from which he has continued to add to the growing list of his editorials.

At the time of his retirement the association honored him by establishing the Philip Hauge Abelson Prize, an annual award of $2,500 "to either (a) a public servant, in recognition of sustained exceptional contributions to advancing science, or (b) a scientist whose career has been distinguished both for scientific achievement and for other notable services to the scientific community" (84). Further recognition came from the National Science Foundation which in 1984 awarded him the foundation's Distinguished Public Service Award "for his outstanding contributions as creative scientist, wise counselor, and sterling editor, and for his stalwart pursuit of excellence in science, science policy, and scientific communications" (85).

James McKeen Cattell edited *Science* for half a century. Abelson's 22 years were not half as many as Cattell's 50, but *Science* was larger during the Abelson years, and whether one counts words or pages or articles it brought to the readers more of science itself and more discussions of advancements, problems, and events of scientific interest. Qualitatively and quantitatively he left a record to challenge and inspire any future editor.

*Chapter Five*

# A New Home for the Association

In 1907 Leland O. Howard, the much decorated entomologist who served AAAS as permanent secretary from 1898 to 1920, changed his address for AAAS from "Cosmos Club, Washington, D.C." to "Smithsonian Institution, Washington, D.C." That change ended an era. For its first 59 years the association's office was wherever the secretary lived or worked, but as the scale of operations increased and as it became necessary to employ paid staff to assist the secretary, that arrangement became unsatisfactory. The Smithsonian Institution came to the rescue with a generous offer to provide office space in the Smithsonian's brownstone castle on the Mall. After 39 years that arrangement also became unsatisfactory, for as World War II was drawing to a close, F. R. Moulton, the administrative secretary, knew that AAAS would be taking on new responsibilities and that the already crowded rooms in the Smithsonian building would no longer be adequate.

A new home was obviously necessary and Moulton began planning for it. After discussing matters with the board of directors, he asked AAAS members for help; "one letter," he proudly wrote later, brought in $100,000 even though it "did not even outline plans for location, cost, or character of the proposed building" (1).

Frank Campbell, editor of *The Scientific Monthly*, supported the money raising effort by describing the association's offices in the Smithsonian. Moulton's own office — sometimes called a dungeon — was a converted storeroom on the second floor, reachable only by a narrow back stairway. Three smaller rooms on the third floor served the rest of the staff, except for the editorial office of *The Scientific Monthly*, which was a small, low-ceilinged room on the eighth floor of one of the building's towers. When Howard Meyerhoff joined the staff in 1945 he was given a similar room on the ninth floor. An ancient elevator served the stack of tower rooms, but when Meyerhoff wanted to get to the telephone it was usually faster to use the trapdoor and ladder that connected his office with Campbell's office just below (2).

A few months later when Willard Valentine came to Washington to recruit staff and prepare to assume the editorship of *Science*, there was no more space available in the Smithsonian Institution. The American University generously offered use of a temporary building on its campus at Massachusetts and Nebraska Avenues. That was a welcome offer but the inadequately heated building and its distance of several miles from the rest of the AAAS offices reinforced Moulton's conviction that a new home was essential. Thus Campbell's article, which he described as an "excursion from dungeon to tower," concluded with a request to "tell us what you think of it, preferably by sending us a slip of paper inscribed 'Pay to the order of the AAAS Building Fund' " (3).

By April of 1946, with $100,000 in the building fund, the board of directors appointed a committee on housing needs. One attractive possibility for the new committee to consider was a three-story building at 17th and F streets across the street from the Executive Office Building. Alternatively, a group of Philadelphia supporters were trying to persuade AAAS to move to their city (4). They argued that it would be better for the association to be away from the federal government and the political atmosphere of the Capitol. Although no firm promises were made, there were strong hints that Philadelphia interests would provide some money and an attractive building site without cost, probably on Benjamin Franklin Parkway near the Franklin Institute.

## Purchase of the Massachusetts Avenue Site

Those and other alternatives were quickly abandoned when Moulton suddenly learned of a more attractive opportunity: the whole of one of the small, odd-shaped blocks created by L'Enfant's plan for the city of Washington. It was desirably located on Massachusetts Avenue, a few blocks north of the White House. Within a block or two were the American Chemical Society, the American Council of Learned Societies, the Carnegie Institution of Washington, and the National Geographic Society. And within another couple of blocks were the Brookings Institution and the American Council on Education.

Almost simultaneously the four owners of the five old, red brick houses that occupied the block decided to sell and all four chose the same real estate agent. On Thursday, April 18, 1946, that agent informed Moulton of the availability of the property. Moulton acted rapidly, for he considered the location as good as possibly could be found. The National Academy of Sciences was in session in Washington, and he grabbed the opportunity to consult with several current and former leaders of AAAS. They liked the location and on Sunday, April 24, Moulton convened a meeting of the building committee. Members of that committee examined the site, compared it with other possibilities, met again on Tuesday, and recommended to the board of directors that the association buy the whole block immediately before the separate owners knew that a single purchaser wanted all of their properties. A special meeting of the board of directors on the next day approved that recommendation and by noon Thursday contracts to purchase the four separate properties were drawn up and deposits made with the real estate agent. By Friday evening three of the owners had accepted the association's offering prices. The fourth owner was out of town, but when he returned on the following Monday he too accepted. Ten days after first learning of the opportunity, the association had purchased the entire block for a total of $147,650, a little less than the sum of the asking prices, $153,000 (5).

What the association had purchased was a trapezoidal block with a frontage of 157 feet on Massachusetts Avenue, NW. At the narrower end of the trapezoid, next to Scott Circle, was the largest of the five houses, a red brick mansion of 26 rooms that became AAAS headquarters (see Figure 2). At the wider end of the block on 15th Street, NW, the other four houses were leased to tenants.

On a Monday morning, the 9th of September, 1946, AAAS moved into its new home. Moulton and his staff and *The Scientific Monthly* came from the Smithsonian Institution building, and Valentine and the *Science* staff, together with Theodore Christenson, the advertising manager, moved in from American University. The whole staff was together in ample quarters in a desirable location. In fact they had more space than they needed. In addition to being the editor of *Science*, Valentine was also treasurer of the American Psychological Association (APA), an affiliated society that was looking for space to establish a permanent central office. Valentine suggested that APA might like to rent the unneeded top floor of the AAAS building. Agreement was easily reached, and as AAAS moved into the first two floors and most of the basement, I moved the APA office, which I then organized into the top floor and one room in the basement (6).

The new home was a great improvement over the previous quarters but the board of directors already had more ambitious plans. They immediately asked Moulton to investigate the possibility of erecting a large, new building that would provide space for AAAS and several other scientific societies, one that would include an auditorium for scientific meetings and that could provide archives space for small societies that did not need permanent offices but that did need safe keeping for their records. Harlow Shapley, the president-elect, dreamed of a "Temple of Science" to provide headquarters for a score or so of organizations (7). Others compared the prospect with the great scientific administrative center that George Ellery Hale had in mind for the National Academy of Sciences after World War I.

Special permission would be required to erect such a building, for the AAAS property was in an area zoned for residential use. Some of the houses had been converted to other uses, as had the AAAS building, and the District of Columbia Board of Zoning Adjustment had the authority to grant a variance to permit other kinds of construction. To build a large new office building would require a favorable ruling by the Board of Zoning Adjustment. If that board approved, District of Columbia regulations would permit construction of a building up to eight stories in height and 9,275 square feet per floor. The directors decided to ask for permission to build such a maximum building.

There was not nearly enough money on hand; the $100,000 in the building fund plus a portion of the association's reserves had been used to purchase the land and existing buildings. However, some funds remained, new gifts could be secured, and Moulton favored erecting a building of maximum permissible size. Cost

Figure 2. New AAAS Headquarters, 1946

conscious, however, he wrote to the Committee on Raising Funds for a A.A.A.S. Permanent Home that "The exterior walls are the most costly part of a building structure ... continuous walls are simpler and less costly than those covered with windows, with their sills, exterior and interior casings, and steel lintels to support the loads above the windows" (8). A windowless building was a kind of pet idea with Moulton. Three years earlier he had asked Waldron Faulkner, a Washington architect, to draw a picture of the kind of building he envisaged, a stark building faced with polished stone with no openings except for first-floor entrances and one window on each floor to give access to a fire escape (9). That building was never constructed or even submitted to the Board of Zoning Adjustment, but Moulton liked it and kept the picture on display in his office until he retired in 1949.

## The Building That Was Never Built

Immediately after Moulton's retirement, Howard Meyerhoff returned to AAAS as the new administrative secretary. The need for larger quarters was not yet urgent, and although building plans were never abandoned, the half-dozen years after AAAS occupied its new property must be described as years of uncertainty over building plans and purpose. Moulton's windowless building was not popular, and earlier dreams of a "Temple of Science" were scaled down. In the spring of 1950 the directors decided to settle some of the uncertainties. They agreed that the new building should have windows, that it should be air conditioned, that it should be only about 50 percent larger than was needed for the association alone, and that it should be designed to permit future enlargement (10). Those were positive enough decisions, but they did not last.

A few months later Faulkner submitted plans for a building such as the board of directors had described, but some of the directors were not satisfied and asked that the building project be discussed with Walter Gropius, famous architect from the Berlin Bauhaus and leader of the firm of Architects Collaborative in Cambridge, Massachusetts. Gropius assigned the project to one of his architecture classes as a design exercise and at the end of the year presented the directors with four alternative designs.

The board and council chose one of the alternatives (11), but before any further progress was made, building plans were lured into a dead-end. When land near the National Academy of Sciences came on the market the prospect of a large science center on Constitution Avenue suddenly became attractive. Exploration of that possibility took some months, but by the end of 1951 the directors decided to follow Gropius' advice to build on the property AAAS already owned rather than pursuing the more expensive idea of a Constitution Avenue scientific center. Having made that decision the directors also decided that they preferred Architects Collaborative to the local firm of Faulkner, Kingsbury, and Stenhouse. Soon afterwards, and largely influenced by the well-reasoned arguments of Warren Magee, the association's legal counsel, the directors reversed their earlier decision and decided to seek permission to erect a building of maximum size. Accordingly, they asked Gropius to prepare "several alternative designs in restrained but aesthetic modern style" (12).

When the selected plans were presented to the Board of Zoning Adjustment, that board delayed scheduling a formal hearing and let the association know that they opposed the plan. AAAS persisted, however, and asked for a formal hearing. That hearing was held on October 23, 1952, and resulted in a unanimous decision to reject the association's proposal. There were several reasons for that decision. For one, the case was badly presented. Magee described AAAS as meriting a variance because it was a charitable organization, even though the commissioners knew full well that AAAS was an educational and scientific organization and not, as Magee called it, an "eleemosynary institution." Moreover, they were quite uncertain from the evidence presented that the association either wanted to or would be able to rent all of the extra space to other nonprofit organizations; the zoning board members were not at all certain that enough such tenants could be found. And, finally, they did not like the extreme design of the building. The plan chosen by the board of directors was an eight-story building of five sides with deeply bayed windows and supported by pillars that raised the first floor some distance above the ground-level parking area (13). The building would have been a monument to Architects Collaborative, but it bore no resemblance to the traditional residential architecture of Massachusetts Avenue or the newer buildings that were replacing some of the old homes. The exact words are lost in time, but two of the association's observers remembered one of the commissioners as chiding them for the design and saying that for an organization such as AAAS a building of traditional style would be more appropriate, something, as one commissioner put it, "with columns in front."

After that setback AAAS experienced another of a different kind. In February 1953 Meyerhoff informed the board of directors that he would withdraw from all involvement in AAAS management as of March 31 (see Chapters 3 and 4). That blow to the continuity of AAAS management interrupted, but did not stop, building planning. After giving some thought to possible sites that had become available in southwest Washington the board agreed that in view of the amount already invested in plans for a new building at 1515 Massachusetts Avenue, and because they had not found a more attractive site, they would file an appeal to the adverse ruling of the Board of Zoning Adjustment.

Several weeks later the building committee held a major strategy session to prepare for an appeal. Participants were John Dunning, chairman of the building committee, its other members Wallace Brode, Detlev Bronk, and Paul Klopsteg, plus Raymond Taylor, who was acting administrative secretary following Meyerhoff's withdrawal, Warren Magee, F. P. H. Siddons, chairman of the AAAS Finance Committee and vice president of the association's bank, and Arthur B.

Hanson, a Washington attorney with much experience in handling legal affairs of scientific organizations.

In preparing for the meeting Taylor asked Hanson to consider three questions about AAAS and the relation of those questions to the District of Columbia building regulations: (1) Could the AAAS be classified as an educational organization? If so, (2) did its accounting system permit it to verify that more than 50 percent of its moneys went to the educational processes? And (3) meeting the first two objectives, what type of building would have a chance for approval?

Hanson gave useful answers to all three questions, using for the first two information that Hans Nussbaum, the association's business manager, had collected from a variety of sources on how the association's publications, meetings, and other activities were actually used throughout the United States. The group then reviewed the zoning board's reasons for rejecting the earlier plan and their reasons for not approving the building that had then been proposed. Out of this discussion came the conclusion that a wholly new start was necessary and that the agreement with Architects Collaborative should be canceled and a new architect engaged (14).

The board of directors approved, payment was made to Architects Collaborative, and both parties signed the appropriate documents to indicate that neither had any further claims on the other. When the board next met they agreed that a local architect should be engaged, and they subsequently decided to return to Faulkner, Kingsbury, and Stenhouse, the firm that had been bypassed in favor of Architects Collaborative in 1952.

It was at that time, as building plans were high on the association's priority list, that I became administrative secretary. A problem that needed attention was the question of replacing our legal counsel. Not only had Magee mishandled the association's request for permission to erect a new building but in some other ways he had also been unsatisfactory. In 1953 he had tried to persuade the council that adopting a resolution in support of President Eisenhower's recent "Atoms for Peace" address before the United Nations would jeopardize the nonprofit status of AAAS. And on earlier occasions he had seemed unduly cautious in arguing against activities that he feared might endanger that status even though much precedent to the contrary could be cited.

Arthur Hanson was engaged as Magee's replacement, first to handle the appeal to the Board of Zoning Adjustment, and soon thereafter to serve on a continuing basis as the association's legal counsel. He was the son of Elisha Hanson, a newspaper reporter who had come to Washington years before, earned a law degree while continuing his duties as a correspondent, and later became legal counsel to the American Newspaper Publishers Association. The firm had also become legal counsel to the American Chemical Society and the National Geographic Society and had done legal work for the National Academy of Sciences and several other scientific organizations. Thoroughly familiar with publishing problems and scientific organizations, Hanson was well prepared to serve AAAS and over the years was sometimes able to serve the common interests of several scientific organizations at reduced cost to each.

With a new architect and a new legal counsel in place we went to work to prepare an appeal to the Board of Zoning Adjustment. Hans Nussbaum documented evidence that the majority of the association's activities contributed substantially to the educational programs of schools and colleges throughout the United States. As for the building itself the board decided it should be somewhat smaller than the one of maximum size that had been turned down before. In style, however, and despite the comment of the commissioner who had earlier said a building for AAAS should be "something with columns in front," we did not want to go back to classical Greece for our architectural model.

## A New Home at Last

After considerable discussion of the nature of AAAS and its office needs, Faulkner proposed a modern building of three stories with a basement garage, a penthouse for furnace and air conditioning equipment, and sufficiently sturdy foundations and supporting columns to carry the weight of two additional stories if expansion later seemed desirable. Its most distinctive architectural feature was the use of large vertical louvers or movable sunshades that protected the glass walls of the second and third floors from direct sunlight. Such louvers had been used a few times in the southwest, but the AAAS building was their first usage in the eastern part of the United States.

Those louvers were more than a design feature. Faulkner was interested in planning an energy-efficient building, as we also were. Analysis of the comparative costs of winter heating and summer cooling with various kinds of glass and with and without the external louvers indicated that double-glazed windows with external louvers would be most attractive. The louvers would make it unnecessary to have venetian blinds or interior drapes; the initial cost of air conditioning equipment would be substantially less, enough less in fact to offset the cost of the louvers; and the annual operating costs would be less with that combination than with the others being considered. As an additional contribution to energy efficiency, white gravel instead of black was to be used on the roof (15).

Because of the building's unusual design, Faulkner suggested that we take an actual model to the hearings instead of showing a painting of the proposed building. Thus equipped, with Hanson well informed on AAAS characteristics, activities, and needs, and prepared also by the fact that he had kept the zoning commissioners informed of our plans and wishes, we went to a new hearing on July 30, 1954. As the members of the zoning board entered, the one who had recommended "something with columns in front" walked all around the building model, fingered the tiny vertical louvers, and commented "I like these; they look like columns." Faulkner and I grinned at each other. Hanson ably and accurately described the nature and needs of AAAS, and Faulkner described the proposed building. When the hearing was over I could write to the board of directors that the zoning commissioners had unanimously approved the new building plan (16).

With permission to build granted, the directors authorized the staff to secure bids for construction and to take out a mortgage of up to $300,000 to cover costs in excess of the funds available for the building. The board also authorized a new letter to members requesting contributions to the building fund. In the meantime, the architect's staff prepared detailed construction drawings and secured an estimate of construction costs to use as a basis for evaluating bids from several construction companies that were invited to bid on the job.

On April 4, 1955, the staff vacated the old red brick mansion and moved to temporary quarters in the Stoneleigh Court Building at 1025 Massachusetts Avenue, which served as AAAS headquarters for a little over a year while the five houses on the Massachusetts Avenue property were razed and the new building erected.

The construction contract had been let to the William P. Lipscomb Company for $699,913. While construction was under way 30 design changes were agreed upon and the actual cost turned out to be $700,835, less than $1,000 over the original bid. That amount plus the architect's fee, cost of hearings and licenses, landscaping, and the association's share of the cost of a new sidewalk in front of the lot brought the total cost to about $889,900, plus another $45,000 for new furniture and equipment. Most of the total cost was covered by accumulated surpluses of annual receipts over expenditures, primarily from the years when Howard Meyeroff was in office. Contributions from members helped substantially and the necessary mortgage turned out to be for $175,000 instead of the authorized $300,000.

In June 1956, as the AAAS staff moved into the first two floors of the new building, staff members of the American Chemical Society, the American Geological Institute, the American Geophysical Union, and the American Society of Photogrammetry moved into the offices they had leased on the third floor. For them and for AAAS the only serious problem of the inevitable shakedown period was the fact that the novel louvers simply did not work properly. The company that made them could extrude aluminum well enough, but could not design satisfactory timing controls to position the louvers in time with the earth's daily rotation and its annual change in orientation to the sun. Several months and a new set of controls were necessary to get them to operate as intended, to remain shut at night, and during daylight hours to open and close just far enough to screen the windows from direct sunlight. The louvers not only served that purpose, but also became an identifying symbol; taxi drivers knew where to take a visitor to the "the building with the louvers."

In June 1956 when the board of directors held its first meeting in the new building the discussion, appropriately for the occasion, was largely devoted to a comprehensive review of existing and suggested association activities and a consideration of where AAAS should devote its energies in the future. Part of the time, however, was given to making final plans for a special dedication meeting scheduled for October 12, 1956.

At that time there was much hope for the use of atomic energy to generate electricity, but there was also much uncertainty about the biological effects of atomic radiation. In 1954, the council requested the board of directors to appoint a committee to "collect the facts on the possible biological effects of thermonuclear weapons" (17). Council discussion of that action placed Warren Weaver, AAAS president, in the embarrassing position of knowing that the Rockefeller Foundation had that month asked the National Academy of Sciences to conduct a somewhat broader study of the effects of radiation. However, he did not feel that he should announce those plans, for arrangements with the academy had not been made. Later, when the foundation and the academy had agreed upon the study, Weaver recommended and the council voted to cancel the request for a closely overlapping study by AAAS (18).

The topic was nevertheless interesting, and by October 1956 the NAS study was far enough along to enable some of its participants to discuss their findings. Thus the planning committee arranged a symposium on atomic energy and atomic radiation effects as a timely topic for a dedication meeting. The Carnegie Institution of Washington generously offered use of its auditorium as the site for the symposium. With AAAS President Paul Sears presiding, and following a welcome from Caryl Haskins, president of the Carnegie Institution, papers were presented by Shields Warren, L. C. Dunn, Lawrence R. Hafstad, Detlev W. Bronk, Willard F. Libby, and Laurence H. Snyder. Paul Sears then dedicated the new building and invited the audience to walk down Sixteenth Street a couple of blocks for a reception and an opportunity to explore the new building.

With the dedication ceremonies concluded the staff settled into routine and satisfying use of their new quarters. Later, as expanding activities required, leases for the third-floor rooms were not renewed. And as activities expanded even further the once new and ample building began to get crowded.

## Possibilities and Disappointments

In the 1960s building again became a topic of discussion by the AAAS board as they foresaw the time when we would outgrow our new home. Several times the board decided to erect the two additional stories that were provided for in the original construction and in 1963 the Board of Zoning Adjustment approved that addition (19).

Those plans were never carried out, for before construction was started, or a contract let, two more attractive possibilities arose. Neither turned out to be feasible, but each seemed promising and was thoroughly explored.

Immediately north of the association's property was another trapezoidal block, 25 percent larger than the association's, that was occupied by an old mansion being used as headquarters of the National Paint, Varnish, and Lacquer Association of the United States. Between the two properties was a block-long section of N Street, NW, that provided parking space for several cars, gave access to our basement parking garage, and was used for deliveries to the two buildings, but was not part of any traffic route.

Would the Paint, Varnish, and Lacquer Association sell their property to us? Would the District of Columbia be willing to close the block of N Street between the properties and transfer ownership to AAAS? If those arrangements could be made we would have about 3.5 times as much land, could build a much larger building, and would retain our desirable location. Discussions with our neighbors and with officials of the District of Columbia went on for months. We were encouraged by District of Columbia officials to expect that N Street could be closed if it all came to AAAS, but they were not willing to have it divided with our neighboring trade association. At one stage that organization offered to trade us their larger property and older building for our smaller property and new building. We declined and offered to buy their property. Eventually they declined that offer, but for a time appeared willing to sell if the new building we would erect would include a separate entrance that would be clearly identified as the entrance to their headquarters. After much talk the arrangement that seemed most likely to be achieved was for the District of Columbia to close the block of N Street and transfer all of that land to AAAS; to compensate the Paint, Varnish, and Lacquer Association for their interest in the street we would pay them $200,000; and we would then extend our building upward by two stories and add a five-story, 60-foot extension on the north side of the building.

In January 1966 the commissioners of the District of Columbia held hearings on the plan to close N Street. A favorable verdict seemed likely, but they decided to postpone final action until plans for the new building were approved by the National Capital Planning Commission. Thus another hearing had to be arranged. A staff member of the National Capital Planning Commission inspected the site and the use of the street, reviewed our building plans, and recommended a favorable decision.

Then misfortune struck. On the day of the hearing the staff member who was well acquainted with our plans and who had recommended their approval was ill. Instead of postponing consideration of our request, the staff manager decided to have the proposal presented by another staff member who was not familiar with the plan or the site. Waldron Faulkner, our architect, and I were told that we could attend the hearing but that we would not be allowed to speak. When the substitute presented the case one member of the commission objected because, he said, he did not like to see Washington's vistas closed off. Nobody mentioned that the remaining portion of N Street east of the proposed building addition was a one-way street leading away from the property and that the vista available to motorists was what they could see in their rear-view mirrors, or that the view that would be blocked was a blank wall behind a statue of Samuel Hahnemann, founder of homeopathic medicine. Faulkner and I writhed in our enforced silence. The opposing architect prevailed, and with little discussion the Planning Commission disapproved our request (20).

Informal discussions with the commission's staff told us that they thought the commission had made the wrong decision. That decision was made, however, and the District of Columbia commissioners voted to reject our request to close N Street.

Perhaps that decision could have been set aside on appeal. However, the District was then changing one of its policies concerning street closures. Instead of giving the land to owners of the adjoining property, compensation was being requested (21). We were told that the District might approve the closing if we offered to pay $45 or $50 per square foot for the land involved. However, that payment to the District plus the $200,000 compensation to the National Paint, Varnish, and Lacquer Association would have pushed the cost up above the appraised value. Consequently, the board of directors decided to abandon the N Street possibility and instead go forward with the addition of two floors to the existing building and to purchase land in nearby Virginia on which we could later erect a building to house some of the association's activities that did not need to be as centrally located as was the building on Massachusetts Avenue (22).

The next flurry of excitement about a new building grew out of discussions about the desirability of a new and major center for advanced study in the nation's capital. More than the Moulton and Shapley idea of a center for the headquarters of a number of scientific societies, this dream could be thought of as an evolutionary successor to the wish of George Washington and other early presidents for a national university in the nation's capital. A national university was no longer a realistic possibility, but Washington, D.C., was a mecca for advanced study in many fields, for its resources were rich and varied: the Library of Congress, the Folger Library, the National Library of Medicine; in art, the National, Corcoran, Freer, and Phillips galleries; the informational resources of the Bureau of the Census and other governmental agencies; the Brookings and Carnegie Institutions; and the National Institutes of Health, the Smithsonian Institution, and the agricultural research facilities at nearby Greenbelt, Maryland — the list of agencies that brought scholars to Washington was a long one.

Mature scholars and graduate students came under a variety of arrangements, but for many the local arrangements would be better if there existed a center for advanced study — an institution that could provide visitors with information about local resources, that would have space for seminars and meetings, bedrooms and a dining room for visitors, and, of course, a knowledgeable staff to assist them in finding and making use of appropriate resources.

Early in 1965 the AAAS Committee on Cooperation Among Scientists discussed this idea favorably (23). I discussed it with Frank Bowles of the Ford Foundation, who made no promises but said that he thought the Ford Foundation should do something for the scholars and scientists stationed in or visiting Washington. With that encouragement I arranged a series of dinner meetings for the heads of several of the relevant Washington institutions: Robert Calkins of the Brookings Institution, Caryl Haskins of the Carnegie Institution, S. Dillon Ripley of the Smithsonian Institution, Logan Wilson of the American Council on Education, Louis Wright of the Folger Library, and sometimes others such as Frank Bowles of the Ford Foundation or Robert Horsky, advisor to President Johnson on national capital affairs.

The general idea found a receptive audience. Calkins wrote an article on the history of efforts to establish a national university in Washington (24). Members of the Smithsonian Institution staff wrote several planning papers. Vice President Humphrey suggested to Dillon Ripley, secretary of the Smithsonian Institution, that a congressional resolution calling for the establishment of such a center would help arouse interest. Congressman Frank T. Bow and Senator Clinton Anderson (for himself and Senators Leverett Saltonstall and James W. Fulbright, both of whom were regents of the Smithsonian) introduced a joint resolution proposing establishment of the kind of center we had been planning (25).

By March of 1966 the Woodrow Wilson Memorial Commission, which had been inactive since its appointment several years earlier, was stirred into action and began holding hearings on the kind of memorial that would be appropriate to honor the memory of President Wilson. Because our informal group had by that time decided that the scholarly center we had in mind would probably fare better under the umbrella of the Smithsonian Institution than as a wholly independent organization, we asked Dillon Ripley to present our plans in hearings before the Woodrow Wilson Memorial Commission. When the report of the commission appeared, its section on the program of the proposed Woodrow Wilson Center was but lightly paraphrased from the statement submitted by our group.

One place to which the commission's report was submitted was the Pennsylvania Avenue Commission, which was planning renewal of part of the stretch of Pennsylvania Avenue between the Capitol and White House. Although part of that street of quadrennial inaugural parades was bordered by government buildings, the historic Willard Hotel, and other buildings that should be saved, part had grown rather shabby. Plans called for razing the less-attractive buildings and creating a large opening to be called Market Square at the intersection of 8th Street, NW, and Pennsylvania Avenue.

Around that monumental square new buildings could be erected, one of which was expected to be the Woodrow Wilson Center. Adjoining land would be available for redevelopment by appropriate organizations. The possibility of a new home for AAAS with the Woodrow Wilson Center and perhaps other scientific and scholarly organizations as neighbors seemed a good reason to delay action on expanding the building at 1515 Massachusetts Avenue.

But none of that came to pass. A study commissioned by the Smithsonian Institution concluded that only one block of the proposed Market Square area was so badly deteriorated as to call for immediate renewal. One block would not release enough land for construction of the Woodrow Wilson Center and neighboring buildings for scientific and scholarly organizations. Moreover, land values were so high that razing several blocks to permit more extensive new construction would require financing of the order of half a billion dollars (26). Market Square was not developed, and when Congress did not provide appropriations for a new building for the Woodrow Wilson Center — which was approved and established — one of the Smithsonian's existing buildings became its home.

Once again Moulton's and Shapley's dream of a major center for scientific societies came to naught and as a temporary measure the board of directors decided that the needed additional office space should be rented.

After 1970 renting became the rule, until AAAS offices were scattered across four different buildings. That unsatisfactory condition ended in 1985 when AAAS sold its Massachusetts Avenue building to the government of Tunisia for use as its embassy and brought all of its staff together in leased space in a new office building at 1333 H Street, NW.

Thus the AAAS name was removed from the building that had been designed for its use. Before that was done, however, the *Washington Post* published a series of sketches and verses about distinctive buildings in Washington, D.C. The AAAS building, the building with the louvers, came early in that series (27), where it was described, in part, as

> ... the church of Advancement of Science
> A glass and chrome crypt
> That is fully equipped
> With every last technical modern appliance.

That description included a bit of license on the part of the author. The building was not the "Temple of Science" Moulton and Shapley had dreamed of when the land was first purchased, nor did it ever become the center of a group of scientific societies. A succession of misadventures and disappointments had prevented those possibilities from being realized, but for three decades it served as the association's handsome and distinctive headquarters.

*Chapter Six*

# The Advancement of Science

The board of directors, council, and members of the AAAS staff spent most of their time on financial accounts, publications, meeting arrangements, and the other chores and services expected of a membership society. Sometimes, however, they got closer to the subject matter of science as they were planning and arranging grants to support research; prizes to recognize significant research achievements and to stimulate research in selected areas; publications, lectures, and TV programs to exchange information among scientists; and an annual program of small research conferences, each on a lively and timely topic.

## Gordon Research Conferences

The small conference that brings together scientists with similar research interests is one of the best devices ever invented to exchange information, compare methods, avoid blind alleys, coordinate studies, and thus advance a field of research. In 1931 Neil Gordon, professor of chemical education at the Johns Hopkins University, arranged such a conference for students and faculty members. The idea took hold; another conference was held the next year; and in 1935 the conferences were moved to the Gibson Island Club, a private club on a secluded, scenic, thousand-acre island in Chesapeake Bay.

In 1937 Professor Gordon became secretary of the association's Section on Chemistry and decided that the conferences should be brought under AAAS sponsorship. The board of directors approved, with the cautious reservation that the association would incur no financial responsibility. They need not have worried; conference fees and industrial grants, sometimes supplemented by special grants, have always covered costs.

Early benefactors included some three dozen chemical companies — running alphabetically from American Cyanamid to the Texas Company — each of which contributed $1,000. That fund enabled the conferences to purchase a large house on the island and to remodel and equip it for conference use. That was the conference site for a decade, and the location gave the conferences their name of Gibson Island Conferences of the AAAS (1).

In 1938, with Harold Urey as chairman for the year, conferences on resinous polymers, vitamins, and the relation of structure to physiological action set the permanent pattern for the conferences (2). Each lasted from Monday morning through Friday afternoon. Morning sessions of about two hours were followed by individual or small-group discussions or recreation in the afternoon. An evening session started shortly after dinner and often continued far into the night. Formal papers were discouraged, but at each session a designated member started the discussion with an account of on-going work or problems, with emphasis always on the frontier problems of the particular conference's topic. All discussions were off the record; no proceedings were published; and speakers who later published material they had presented did not identify the publication as being from one of the conferences. Interested scientists applied to attend a conference, but actual attendance was by invitation, with applicants invited only if the conference leaders thought they could make useful contributions to the discussion. To make it easier for the participants to get acquainted with each other and to exchange information, attendance was limited in the early years to 60 and later to perhaps twice that number. Arrangements encouraged informality, free exchange of information, and emphasis on the most current and lively issues of a research field. Conference locations were always far enough away from major cities to discourage part-time attendance.

Each conference was on a quite specific topic, originally some aspect of chemistry, but as their popularity increased their scope broadened to include related fields and later even policy issues.

In 1946 several major changes became necessary. Professor Gordon, who had been ill, and Sumner Twist, his temporary replacement, both resigned, and by then the conferences, which had expanded to 10 a summer, had worn out their welcome at the Gibson Island Club. W. George Parks, professor of chemistry at Rhode Island State University, was selected as the new director and made arrangements to hold the 1947 conferences on the campus of Colby Junior College in New London, New Hampshire (3). With that move the old name of Gibson Island Conferences became obsolete, and in 1948 they were renamed Gordon Research Conferences in honor of their founder (4).

In their New England home the conferences continued to thrive. The 10 conferences of the final year at Gibson Island grew to 16 in 1950 and to 36, on three campuses, in 1960. That number should be the maximum, the management committee thought, having agreed in 1956 that expansion beyond 36 was not desirable. By 1956 the size of the operation and the amount of money handled made it desirable to protect the conference management and the association by incorporating the conferences in New Hampshire, where they were held (5).

In 1956 the management committee had thought 36 conferences a year quite enough, but they soon had to relent. The pressures for new conferences was so strong that by 1970 the number had doubled, with 64 conferences at six campuses in New Hampshire plus eight more in western Washington. By 1988 the number had more than doubled again with 130 in New England and California.

In anticipation of his death, Bentley later transferred to the association stocks with a value approaching $25,000 to support the award in future years or, in the judgment of the association, to be used in some other way consistent with the original purpose of the award. AAAS chose to continue it, but in 1986 changed the name from the AAAS Sociopsychological Prize to the AAAS Prize for Behavioral Science Research.

## Other Prizes

From time to time AAAS has awarded other annual prizes to honor particular kinds of research. For example, the Theobold Smith Award in Medical Sciences supported by the Eli Lilly Company was started in 1936; the AAAS Anna Frankel Rosenthal Memorial Award for Cancer Research was initiated in 1955 and supported by the Rosenthal Foundation; the AAAS Ida B. Gould Memorial Award for Research on Cardiovascular Problems was started a year later and supported by the same foundation; and the Industrial Science Award was begun in 1956 and given to an American industrial firm or other organization for an outstanding achievement in technology. All of those have been terminated. Still others were started and given briefly, considered and then declined, or shifted to some more appropriate organization (11).

The multiplicity of awards led to some unhappiness, for their increasing number seemed to dilute the senior and more prestigious ones and a larger number also led to jealousy over award ceremonies. "Why," a sponsor would ask, "should that award be made at a large general session of AAAS with many scientists in attendance, while mine is given at a sectional meeting with only a small attendance?" Without ever formally deciding upon a proper or acceptable total number of awards or their specific fields, the board's attitude became more resistant; some awards were terminated and some proffered ones were declined.

In 1958, when the association had within the past two years begun two new awards and was being requested to assume responsibility for two more, it seemed time to codify association principles concerning prizes. The board agreed on the following statement (12):

> *Field.* The fields must be appropriate for the AAAS and likely to be helped by the existence of an annual award. Awards administered by the AAAS should be interdisciplinary in character and not restricted to a single field of science.
> *Eligibility.* No restrictions other than those imposed by the purpose of the award.
> *Continuity.* An award will not be offered unless there is a definite plan that it be continued for a reasonable number of years.
> *Name.* The name of the award will start with the name of the association ... as an alternative possibility, it will be acceptable to

name an award the _____ Award of the American Association for Advancement of Science.

*Method of selection.* The method of selecting award winners is subject to the discretion of the association.

*Cost.* The association will be reimbursed for the costs of administering the award.

## Communication Among Scientists

Communication within disciplines and across disciplinary lines is a continuing function of AAAS, and methods other than the publication of journals and the holding of meetings have been used for that purpose.

One method was publication of symposia on scientific topics. That was a special interest of F. R. Moulton, who published the first volume in the association's symposium series in 1938, the year after he assumed responsibility for AAAS activities.

By 1970 a total of 89 symposium volumes had been published. Making money was not the reason for publication, but given a choice, no organization likes to lose money. Thus prices were set at levels the business office and publications committee thought should recapture costs. Overall that goal was achieved; from 1938 through 1970 the direct costs of editing, printing, and distributing the volumes totaled $860,807, while receipts from sales came to $1,176,071. The difference between those two sums was enough to cover reasonable estimates of the never-recorded indirect or overhead costs of the activity, and the average sale of a little over 2,000 copies per volume was enough to give staff and committee members a satisfied feeling that their work was worthwhile.

### Other Publications

In addition to the symposium series other volumes or reports were sometimes produced. One is of interest not only because of its subject matter but also because of the auspices under which the study was carried out.

Around 1960 some members of the council became unsatisfied with that body's role in association affairs. It's large and changing membership, once-a-year meetings, and devotion of so much meeting time to initiatives and recommendations of the AAAS board of directors frustrated some council members who wanted a more active role in association affairs. As one way of meeting that wish, the Committee on Council Affairs was authorized to establish council study committees on almost any topic that was requested by or interesting to a sufficient number of council members to make a study committee seem attractive.

Some of the committees did not produce; the members lost interest or the topic was not one that could be well handled by such a committee. But some were effective, and a particularly good example was the Council Study Committee on the Use of Natural Areas as Research Facilities. In 1962 that committee completed a study done jointly with the Nature Conservancy of how 76 different natural areas owned by colleges or universities were being used. The 76 areas ranged from the four-acre "South Woodlot" of second-growth swamp forest owned by Ohio State University to the University of Florida's 23,000-acre "Welaka Reserve" that included a variety of major biotic communities (13). A year later, also in collaboration with the Nature Conservancy, the committee completed a 347-page report that included a bibliography of research done on natural areas, an inventory of natural areas in the United States, an account of the agencies and organizations holding or administering those areas, and several appendices providing detailed information on those topics (14). The Committee on Council Affairs discharged the committee with a well-deserved vote of commendation (15).

Of the other study committees, some submitted interesting and useful questionnaire studies, arranged symposia for an anuual meeting, or were otherwise active. For example, in 1967 the Council Study Committee on Population Explosion and Birth Control cosponsored a conference on "The Behavioral Sciences and Family Planning" arranged by the National Institutes of Health (16). Yet the idea did not last, and by 1968 the item "Reports of Study Committees" had disappeared from the agenda of the annual meetings of the council.

## Service to the Washington, D.C., Scientific Community

The scientific and technical agencies of the federal government and some other organizations with headquarters in Washington, D.C., included many scientists and engineers and their staffs. In 1958 the National Science Foundation asked AAAS to serve that community by arranging a series of scientific lectures, approximately one a month. An initial grant of $3,000 was provided to cover travel expenses and other costs of the speakers and program administration.

One special subset of the Washington scientific community consisted of the scientific attachés from other countries. Although they came from a variety of scientific disciplines, the nature of their positions meant that they had to be scientific generalists, and all had an interest in science policy and in the plans and activities of the nation's scientific institutions. After AAAS had had several years of collaboration with the Brookings Institution in offering series of seminars for members of Congress (see Chapter 9), some of the foreign attachés told us they would like a similar series. That request was easily met; we simply asked the speakers for the congressional seminars to give a repeat performance the next morning for the foreign science attachés. Arrangements were informal; because the audience was never more than a score or so and strictly defined, there was no need to advertise the program; no supporting grant was requested, for the minor costs involved were simply absorbed or charged off to good will. The result was that for several winters the scientific attachés came together in the association's board room for a series of discussions of current aspects of scientific activity or policy. Another result was that as the attachés returned to their home countries, usually to government positions, AAAS acquired a growing roster of good friends, some of whom were useful in identifying scientists and mathematicians we brought to the United States as lecturers at summer institutes for teachers (see Chapter 8) or for other purposes.

## "Scientists' and Engineers' TV Journal"

A quite different program of communication was one experimenting with television as a means of conveying information to scientists and engineers rather than to the usual TV audience. The program started with an offer by Richard Heffner, director of New York's educational television station, then known as Station WNDT, to provide an hour of late-evening time once a week for a year for an experiment in communication among scientists and engineers. The station offered free use of its production and transmission services if AAAS could find ways of covering such costs as travel expenses of participants, preparation of visual materials, and the salary of a producer for the programs.

Edward G. Sherburne, who had joined the staff in 1961 to lead AAAS programs on public understanding of science, had earlier worked in both commercial and educational television and just before coming to AAAS had been responsible for the University of California's multicampus use of television for educational purposes. He was clearly the staff member best qualified to develop a new TV program and he set to work to take full advantage of the offer.

Eleven other societies cooperated in the venture by developing some of the individual programs. The National Science Foundation provided a grant of $60,000. That grant plus $60,000 worth of contributed time by Station WNDT and small contributions from AAAS and other participating societies made up a total budget of $135,800 for the first year. Twenty-eight programs were broadcast in New York and some were also broadcast in Washington, D.C., Minneapolis, Minnesota, and Lincoln, Nebraska. Most were taped a week ahead of time at the TV station, but a few were broadcast live, for example, as part of a symposium at a AAAS meeting.

It was all quite experimental. Some of the programs did not work out well. Some were criticized as being too technical. Kinescopes (16mm black-and-white sound films made by photographing a high-definition TV image at the studio) were made for use by the sponsoring societies. Efforts to evaluate the programs were initiated by assembling viewing panels or by sending questionnaires to scientists and engineers who might be in the viewing audience. By the end of the first year the representatives of participating societies concluded that the concept of a TV journal was a good one but that several problems needed further work: distributing advance information about programs to members of the appropriate audiences; defining the proper technical level and other characteristics of the most useful programs; and all the other problems of educating the participants and adapting the processes of scientific communication to a new medium (17).

Although no funds were available to continue the "TV Journal" in 1963–1964, reactions to the 1962–1963 broadcasts made both the station and the AAAS Committee on Public Understanding of Science eager to expand the program to a much larger number of stations. The National Science Foundation was willing to help, but did not feel able to provide all of the $290,000 estimated to be required to improve production and provide for broadcast over many stations. NSF did, however, provide $122,000 and the Timken Roller Bearing Company provided an additional $34,000. With that amount available, plans went forward and the 1965 broadcasts were carried over 76 U.S. educational television stations.

With that many stations involved we had to abandon one early idea. Originally we had intended to let the length of each program be determined by the subject matter and not by the clock. That flexibility seemed possible because the programs were always presented after the station's usual sign-off time of 11:00 p.m. However, payrolls, the customs of TV, and the need to accommodate many stations put an end to such flexibility. For the 1965 season all programs were planned for 30 minutes.

Twenty half-hour programs made up the 1965 schedule with one program each prepared by AAAS and 19 other scientific and engineering societies. Some or all of the 20 programs were shown by 76 stations in 31 states. The schedule varied from city to city, but notes in *Science* and elsewhere announced the times and content of the broadcasts. Kinescope recordings were later made available for use by industry, government, or educational organizations (18).

That was the end of the "Scientists' and Engineers' TV Journal." Funds were not available for the following year; Sherburne left AAAS in 1966 to become director of Science Service; and although some 3,000 questionnaire returns from the first year gave substantial support for continuing, there was a real question concerning the initial premise,—namely that TV would be a good medium for conveying technical information to scientists and engineers. The "TV Journal" was never used for first presentation of new material or as an alternate to primary journals. In fact, the AAAS Committee on Public Understanding of Science asked that the programs not be addressed to highly specialized audiences, but instead be of interest to a wide range of scientists and engineers (19). Thus the programs were more like AAAS symposia or articles in *American Scientist* or *Scientific American*. For that kind of material TV had obvious advantages over the print media in portraying motion and some kinds of visual material. But for understanding detailed technical reports that may have to be reviewed, analyzed, and examined in depth, the print media have the advantage.

## Research Grants

For many years AAAS acted as a grant-making foundation. For example, in 1887 Professors Michelson and Morley — of speed-of-light fame — received $175 "to aid them in the establishment of a standard of length" (20) and in 1919 AAAS contributed $500 "to the editorial board of *Botanical Abstracts* [which in 1926 became *Biological Abstracts*] for aid in establishing this new and important periodical" (21).

As research support from federal agencies and other national sources increased, AAAS decentralized its small-grant program by allotting to each affiliated academy of science a certain amount — initially 50 cents and later one dollar — for each member of the academy who was also a member of AAAS. These monies were then used by the academies to make small grants to scientists in their individual states or regions.

AAAS also served as conduit for additional monies to some of the academies, for several of them acquired an "angel," a long-time member of AAAS who also gave the association the largest uncommitted gift it ever received. The benefactor, Charles Matthias Goethe was not a scientist but an enthusiastic supporter who joined AAAS in 1914. Originally from Virginia, he moved to California, contributed to city planning of Sacramento, became a successful businessman with investments in several fields, developed an interest in eugenics, was active in the Save the Redwoods League, and helped start the National Park Service custom of public lectures by park rangers. Within this wide range of interests he and Mrs. Goethe began to patronize state academies of science, initially by contributing directly to the Virginia Academy of Sciences, the academy of their home state. Soon, however, they decided that it would be better to make their contributions through AAAS. Thus on April 17, 1944, Goethe sent Moulton four $25 checks, two of his own and two from Mrs. Goethe, as their quarterly contributions to the academies of Alabama and British Columbia. Later letters also included checks for the North Carolina Academy and for the academies in several western states.

Not appreciating the little gold mine that was opening to AAAS and some affiliated academies, some members of the AAAS staff got a bit irritated at having to handle so many small checks; they wanted Goethe to consolidate his contributions into a single annual check. Fortunately, he was not put off; his contributions to academies continued year after year. By the time of his death at age 91 in 1966 he and Mrs. Goethe had given a total of $8,674 to the academies of Alabama, Arizona, British Columbia, Nevada, and Wyoming, plus their earlier gifts to the Virginia Academy (22).

His purpose was clear. As he wrote to Moulton:

> I know our contributions are tiny. We feel, however, that in this way we are doing our bit to keep alive the flame of pure research in competition with the almost meteoric illumination of commercialized research. I sometimes marvel that the latter blazed so quickly. I think I have told you of my experience, years ago, trying to convince the chief executive of one of the oil companies that his outfit could make money by being somewhat more intelligent in reading the fossil record. I chuckle when I think of his disgust and scorn (23).

Raymond Taylor was the association's principal contact with Goethe, going to see him occasionally when on the West Coast to attend or help plan a meeting of the association's Pacific Division. Other contacts were initiated by Goethe, who sometimes made surprise gifts to AAAS. For example, Taylor or I would receive a check with a letter saying Goethe hoped that we would purchase a supply of a newly issued stamp he thought particularly appropriate for use on AAAS mail.

With Mrs. Goethe already dead and with no children, his will left bequests to over 50 recipients, each with a specified percentage of the total. The largest was to the Save the Redwoods League, with 13 percent. Then came AAAS with 9.7 percent. The California Academy of Sciences followed, and so on down through the long list. Because the estate was largely oil lands in trust, the payment of interest and royalties has continued year after year. In 1984 cumulative payments to AAAS passed the million dollar mark (24).

## Grants to Students

By 1954 research funds from the National Institutes of Health, the National Science Foundation, and other sources had increased so much that the small grants AAAS administered through the affiliated academies of science no longer seemed necessary. At that time there was much interest in improving science education at the high school level, especially for prospective scientists and engineers. Recognition of these conditions led to agreement among the association's Committee on Research Grants, the Academy Conference, and the board of directors that the academy grants should go to talented high school students instead of to mature scientists (25).

South Carolina objected. The academy of that state replied that they preferred to make grants to college students and would give funds to high school students only if there were no suitable requests from college students (26). The board countered by agreeing that college students would be eligible but "with the understanding" that high school students would have priority (27). Accordingly, a new set of rules was drawn up and sent to all of the affiliated academies (28) and the shift of grants from scientists to students was made. That rule was soon modified, however, and the academies were permitted to select recipients as they saw fit, but strongly urged that the grants usually be made to students (29).

Most grants did go to students and soon NSF was also helping to support academy activities. For example, the foundation's 1959 report explained that the state academies of science were uniquely qualified for work intended to strengthen science education for young people because of their involvement with junior academies of science and science fairs and because of the broad ranges of talent in their memberships. Accordingly, in 1959 the foundation had made 30 grants to 23 different academies in 22 states and the District of Columbia (30).

Later one group of talented high school students received an extra bonus that resulted from a much earlier research grant to a famous American scientist.

Early in 1924 AAAS awarded $190 to Robert H. Goddard of Clark University "for work on a high-atitude rocket" (31). At Christmas time of that year Goddard sent the association a progress report, saying that he also had support from the Smithsonian Institution and Clark University and that for a short time had the opportunity to get liquid oxygen for nearly nothing "from a large oxygen concern." Thus he was hoarding the association's $190 for later use, but he wanted to let the AAAS officers know how much he appreciated the grant.

Forty years later, in 1964, I used an editorial in the Christmas Day issue of *Science* to tell the story of that grant and of Goddard's Christmas time progress report. In 1964 the United States was full of excitement about space and the planned trips to the Moon; the Goddard grant seemed a happier topic for a Christmas Day issue of *Science* than some of the heavier topics appropriate for other issues (32).

A few weeks later a letter from Mrs. Goddard brought a check for $500 in appreciation of the 1924 grant. That was a very special $500 and the directors thought it should be used in some special way. After considering several suggestions, we decided to use it to give four-year subscriptions to *Science* to each of the seven top winners of the 1965 Westinghouse Science Talent Search and two-year subscriptions to the other 33 finalists of the talent search. When I wrote to Mrs. Goddard to tell her of that plan she replied "your letter ... has been a source of great pleasure to me. ... I am well acquainted with the Westinghouse Science Talent Search, and am deeply pleased as I know my husband would be, that encouragement from AAAS is going, in my husband's name, to these fine young people" (33).

*Chapter Seven*

# Changes in Governance

Over the years the association's constitution and bylaws have been amended many times and periodically those documents have been more thoroughly overhauled as objectives have been broadened, as the scale of operations increased, or as conditions affecting AAAS activities changed. Those changes and amendments have usually been easily achieved when they constituted clarification of meaning or were primarily adaptations to changes in the association's size or manner of operation. One example already evident from earlier chapters was the change in name of the executive committee to the board of directors. Another was the decision to abolish the office of general secretary and to assign that officer's duties to the permanent secretary and then to change the title of that officer first to administrative secretary and later to executive officer.

In contrast some other changes have been controversial and were debated for years before being finally adopted. Three that were fundamental to the association's formal structure and distribution of responsibility were: clarifying the responsibilities and the relations between the board of directors and the council; reducing the size and changing the composition of the council; and changing the electorate responsible for selecting the association's top officers and the members of its board of directors.

## *The Board vs. the Council*

When it was first used, the title "Executive Committee of the Council" accurately described that committee's role and function. It acted for the council when the council was not in session and it managed the affairs of the association on behalf of the council. Gradually, however, there came to be a clearer distinction between the activities of the two bodies. The council established general policy, elected affiliates, created geographic divisions, approved establishment of new sections, elected officers and board members, and amended the constitution and bylaws. The board of directors handled finances, prepared the annual budget, managed publications and meetings, and appointed and supervised the work of the association's committees.

The difference between the powers and responsibilities of the two bodies was usually clear enough in practice, but it took quite a few years to get the right words down on paper as constitutional statements of the rights and powers of each.

The constitution adopted in 1946 stated the authority of the council in these terms: "Control of all affairs of the Association is vested in the Council, which shall have power to review and to amend or rescind its own actions and all actions taken by the Executive Committee." "There shall be an Executive Committee, which shall execute such commitments as the Council may direct and shall make recommendations to the Council (1)."

Yet even as that wording was being adopted it was out of date, and the executive committee (board of directors) needed more authority. The association had printing contracts for its two journals, was just on the verge of purchasing land on which it would later build its own headquarters building, and in the course of normal operations had to enter into a variety of contracts and other arrangements that would normally be expected to be binding agreements. Efficiency and the normal operating procedures of similar organizations usually left responsibility for making such contracts and agreements to the board of directors or similar body. Yet the existing constitution stated that the council could amend or rescind any agreement made by its executive committee (board of directors). Interpreted literally, that statement seemed to mean that a printer could not feel assured that printing bills for *Science* would be paid unless the council approved the contract, and even then the constitution gave the council "power to amend or rescind its own actions."

Probably nobody expected that kind of behavior, but still the wording was in the constitution. In the constitutional revisions of 1952 the council removed that overly exaggerated statement of its own powers and adopted the limp replacement, "The Council shall perform duties prescribed in the constitution and shall act as an advisory body in matters pertaining to the general policies of the Association." At the same time the powers of the board were stated to be: "The Board of Directors is the legal representative of the Association and as such shall have, hold, and administer all the property, funds, and affairs of the Association" (2).

In six years the statutory relationships had shifted 180 degrees, from having the board of directors be advisory to the council to having the council be advisory to the board of directors. That change, however, was more one of wording than of fact, for it did not alter what the two bodies were doing. The council continued to elect officers and members of the board, make decisions about sections and affiliates, and perform its other traditional duties. The board continued to be responsible for funds, meetings, publications, and other operations.

Nevertheless, the constitutional wording was still unsatisfactory, and in revisions that went into effect in 1961 the statement concerning the council read "Responsibility for the affairs of the Association is vested in the council, which shall have authority to delegate functions to the Board of Directors," and then went on to list a number of specific powers and responsibilities of the council. The comparable statement concerning the board read, "The Board of Directors shall have, hold, and administer all the property, funds, and activities of the Association," and then listed a number of the board's specific responsibilities (3). The details of the two lists were not in conflict, but even as the council was adopting the two statements several members pointed to their apparent contradiction and asked for an explanation or clarification.

A year later Paul Gross reported that the apparent conflict had been discussed with legal counsel; that the new Committee on Council Affairs (of which he was chairman) did not consider the matter urgent; but that a modification of the wording might be submitted later (4). The conflict was later resolved by deleting the troublesome phrase "responsibility for the affairs of the Association is vested in Council" and by enumerating the powers and responsibilities of each body. The formal statement concerning policy has varied a bit in later years, but the division of responsibility has been made clear by such wording as "the Council shall establish the general policies governing all programs of the Association," and "in consonance with the general policies established by the Council, the Board of Directors ... shall conduct the affairs of the Association" with each of those statements followed by a list of specific responsibilities. With such wording the constitution finally caught up with association practice.

## Committee on Council Affairs

The original "Executive Committee of the Council" was in reality a committee on council affairs. However, as the two bodies came to exercise somewhat different sets of responsibilities, there were opportunities for tension or uncertainty as to their relative roles. Toward the end of the 1950s some members of the council came to feel that too much time was devoted to reacting to recommendations submitted by the board or by committees responsible to the board. Thus in 1958 the council created a Special Committee on Council Activities and Organization. A year later that committee submitted an ambitious report recommending that the council take a substantially more active role in association affairs. To that end, the committee recommended:

    1. Codification of the council's rules of procedure and the adoption of some new rules;

    2. Establishment of three standing committees of the council, a permanent "Committee on Council Activities and Organization," a "Committee on Nominations and Elections" (which already existed), and a "Committee on Constitution and Rules";

    3. Establishment of special study committees to review, submit findings, or make recommendations on essentially any topic of interest to the council or a sufficient number of its members; and

4. Appointment of advisory committees or panels which would work by correspondence as advisors to a number of the association's existing committees and which would offer every council member the opportunity to be an advisor to one or another of the association's committees.

That report elicited much discussion at one of the livelier council meetings. Some members supported all of the recommendations, but others argued that a permanent Committee on Council Activities and Organization would overlap the duties of the board of directors and that several of the proposed study committees would overlap existing committees appointed by the board. After a number of trial motions, none of which met with approval, the council decided to approve the report in principle except for the recommendation that the Committee on Council Activities and Organization be made permanent. "In principle" meant that there were a number of discrepancies and disagreements between the recommended actions and the association's constitution and traditions that had to be worked out, and that a year later the council would expect to consider a revised report (5).

Over the next year the council's temporary committee and the board of directors worked together and agreed that the proposed advisory panels and a permanent committee on constitution and rules were not needed, but that a permanent committee on council affairs would be useful. On their joint recommendation the council in 1961 adopted a new bylaw establishing the standing Committee on Council Affairs, which was described as follows:

> The Committee on Council Affairs shall consist of nine members elected by the Council for terms of three years each, the terms of three of whom shall expire on January 14 of each year; the president-elect of the Association, who shall serve as chairman of the committee; and the secretary of the Council, who shall serve as secretary of the committee without vote. The committee shall (a) prepare the agenda for meetings of the Council, (b) receive or initiate, coordinate, and advise on reports of Council committees, resolutions, or actions submitted for consideration by the Council, (c) establish or recommend to the Council the establishment of appropriate study committees to report to the Council on any aspect of Association policy or program or on other matters affecting the advancement of science, (d) recommend to the council appropriate changes in the Constitution and Bylaws (6).

With much the same responsibilities, the Committee on Council Affairs has continued ever since as the executive committee of the council.

With those changes made the organization of the association became as shown in Figure 3.

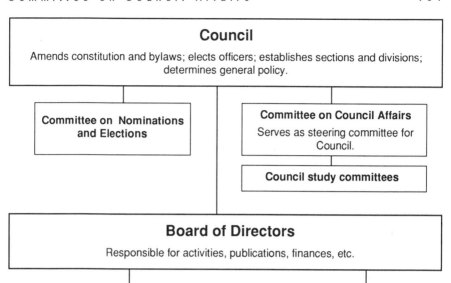

Figure 3. Organization of the AAAS (1945-1970)

## A Leaner, Stronger Council

The Council was a mixed group, including some 50 officers and representatives of AAAS and its sections and divisions, and in 1945 about four times as many representatives of affiliated societies. But never was attendance close to the total membership. In the first four annual meetings after World War II, council attendance ranged from 25 to 85 and averaged 54, about 20 percent of its membership. By 1970 the council included well over 500 members, of whom 185 (35 percent) attended the annual council meeting.

In selecting their representatives some affiliates chose a member living in or near the next meeting city, and thus named a new representative nearly every year. When a named representative did not plan to attend, some affiliates selected an alternate. Thus of the minority of members and alternates who did attend, a fair number were there for the first and perhaps the only time.

As a result, the well-informed and effective council consisted of AAAS officers and board members, section officers, and division representatives, plus a small minority of affiliate representatives. Because the members of that smaller group took their responsibilities seriously, attended regularly, and arrived well informed about association affairs, AAAS was well served and its business usually got done expeditiously.

Nevertheless, there remained nagging problems of low attendance and many inactive members. Whether composition of the council should be changed, and if so how, were questions that were raised periodically throughout the whole quarter century and that were not finally settled until shortly after 1970. In 1950 the committee then debating possible revisions of the constitution concluded that the council included too many nominal members who lacked a sense of responsibility to the association and suggested that a council of 265, which was then its membership, was "too unwieldy to be useful and that a smaller but more responsive Council of about 150 would serve a better purpose as the legislative and policy forming body of A.A.A.S" (7). However, any change in composition of the council would have to be approved by the council itself, for that was the body with power to change the constitution. Consequently, before making specific recommendations, the constitution revision committee polled the council for their advice on reducing the size in order to make the council a better informed and more effective legislative body.

The members were not ready to accept such a change. When the council met a few days later, with 62 of its 265 members signing the attendance record, "the discussion indicated that those present do not consider the size of council ... a serious handicap, and favor preserving the principle of individual society representation" (8).

A few years later the council debated a different method of trying to increase attendance. AAAS routinely reimbursed members of the board of directors and section secretaries for their expenses in attending the annual meetings, including the meeting of the council, but not other officers or the representatives of affiliated societies. As a possible means of securing a larger attendance the council considered assessing each affiliate $25 a year and using that fund to help pay the expenses of the council members who did attend. When that idea was proposed to the affiliated societies, however, most of their replies were negative and the plan was dropped (9).

Uneasiness over the size of the council and the low attendance did not disappear, however, and after the Committee on Council Affairs was established the question of council size and attendance became one of its problems. In 1960, at the committee's request, I wrote to all council members describing several means by which the association could elect a smaller group of members, widely representative by discipline, geography, or both, who would, we hoped, take their responsibilities more seriously and attend more regularly (10). When the council met at the end of the year, Thomas Park, then chairman of the Committee on Council Affairs, reported that the outcome of the discussions and correspondence was "advice that Council go slowly in reorganizing its structure" and that in compliance with that advice the committee was not then submitting any recommendations for change (11).

Time went on and by 1969 the council had grown to include 560 members. Of those, some 400 represented affiliated societies, but the internal association members had also increased to include six or seven representatives of each section, two representatives of each division, and all living past presidents. The Committee on Council Affairs polled the council members, asking several questions about possible changes. Essentially the same questions were also printed in the *AAAS Bulletin* with an invitation to members to reply. Of course that was not, nor was it expected to be, a valid random sample of AAAS members, but it gave members who felt strongly enough about any of the issues an opportunity to express their view's. All replies were passed on to the Committee on Council Affairs.

The major findings of the poll of council members were as follows (12):

1. There was substantial disagreement as to whether the majority of council members should be selected by affiliated societies or by members or fellows of the association.

2. One-third of the council members thought that the council should have no more than 100 members and another third recommended somewhere between 100 and 200.

3. About 75 percent of council members thought that more of the routine business should be handled by mail ballot so that more time at council meetings could be devoted to discussion of some of the larger issues of science.

4. A majority agreed that the customary one meeting a year was appropriate, but there were many votes for two meetings a year.

5. Over 90 percent thought that the terms of office of council members should be at least three years.

As an introduction to discussion of those results and of possible changes in council size and organization, the Committee on Council Affairs reported at the next council meeting: (a) that the committee had followed the expressed wishes of council members by removing a number of items from the agenda, so that they might later be acted upon by mail ballot, and by substituting written reports for some reports that in earlier years had been given verbally; (b) that council [should] inform affiliated societies that the terms of representatives be three years and that in the absence of a representative the president of the affiliated society be accepted as its alternate; and (c) asked council to discuss the other matters, such as the constituency to be represented and the desirable total number of members, so that, if a consensus appeared on any of these matters, the committee might phrase specific recommendations to be acted upon by the council at the 1970 meeting (13).

The second of these three points was the only one calling for immediate action, but the council was not ready to take that action. Nevertheless, the poll results and the council's discussion of association activities and issues (14) were useful to the Committee on Council Affairs in its continued planning.

Recognizing that questions of the size of the council and the method of selecting its members were closely related to the association's sectional structure, relations with affiliates, and election procedure, the board of directors invited the Committee on Council Affairs to help appoint a joint committee to consider restructuring the association. Motivation for that decision came partly from the long history of concern over the size and attendance record of the council, but also came from a decision by the board that it would be desirable to try to increase AAAS membership by an order of magnitude. Nothing but temporary confusion came from that proposed 10-fold increase in membership, but the Committee on Council Affairs did agree to join in establishing a committee on governance that would be asked to consider "such matters as (a) the processes of nominating and electing officers, (b) the makeup of Council, including possible representation of AAAS members who are not members of affiliated societies, and (c) sectional organization" (15).

The new Committee on Governance did not alter the classes or qualifications of membership, the requirements for affiliation, or the association's sectional organization. Instead, it concentrated on changing the method of election and the size and composition of the council.

As the committee discussed its tentative plans with the board of directors and Committee on Council Affairs, agreement was reached on several major policy issues (16).

1. A council of about 150 members would be desirable.

2. Affiliates should have no direct voice in selecting members of the council.

3. Only dues-paying members of the association should participate in the association's governing processes.

4. The president should be elected by popular vote.

5. Any new governing structure should be appropriate for an association of up to perhaps 250,000 members. (The board expected a rapid doubling of membership, but backed off from the proposed 10-fold increase).

At the 1970 council meeting Leonard Rieser, chairman of the Committee on Governance, described the committee's thinking and reported that the committee was attempting to plan a council that would be much smaller than the then-current one; that it was considering recommending that officers be elected by direct vote of the membership instead of by the council; and that the committee hoped for a large attendance at the council meeting a year later when it would have definite recommendations ready for discussion and vote (17).

Thus as 1970 ended the council was twice as large as in 1945 but of essentially the same composition. Change came soon, however, for after further discussion, decision, and the necessary amendments to the constitution, the council voted to eliminate all affiliate representatives and to reduce itself to a body of about 150 that would include all members of the board of directors, all past presidents, representatives of geographic divisions of the association and the Association of Academies of Science, and representatives of the sections of the association, with some 70 percent of the total consisting of section officers and representatives. Thus AAAS achieved the smaller council it had long been seeking and its members have since compiled a better attendance record.

## The Electorate

How should the association's officers be chosen? For one class of officers the arrangements of 1945 were satisfactory and remained unchanged throughout the whole period. That class consisted of the administrative officers, all of whom were selected by the board of directors, subject in some cases to confirmation by the council: the general secretary (as long as there was one); the executive officer; the treasurer; and a secretary for each section. Election of section secretaries normally consisted of ratification by the board of the persons recommended by the section committees, and the system was later simplified by having the section committees elect section secretaries directly.

AAAS also has another class of officers called general officers: the president-elect, the president, the past president, and a vice-president (later called chair) for each section. The president-elect (or before 1946, the President), the eight other elected members of the board of directors, and the vice-presidents were elected by the council, but that method of election was sometimes questioned. The election process started with an invitation to council members to submit nominations. Names of the nominees were then submitted to the general membership for an informal, advisory ballot. Results of the ballot were given to the council, which then conducted the actual election. Under those arrangements in 1945 twenty-two fellows were nominated for the presidency. In the membership ballot the most popular names were Vannevar Bush, with 1,618 votes; James B. Conant, 1,448; Harlow Shapley, 985; and Roger Adams, 476. The council was given that information and also the votes for the other 18 nominees, told that Bush had asked to have his name removed from consideration because he would not be able to serve (18); and asked to cast their votes by mail. Conant received 101 of the 183 votes cast, and, that being a clear majority, was declared president for 1946 (19).

That election was reasonably typical in that council preference agreed with the membership vote and also in that other nominees high in the popular vote were likely to be elected to the presidency soon afterwards, in this case Shapley for 1947 and Adams for 1950.

In one respect, however, that election was not typical, for frequently no nominee received a majority of the votes cast on the mail ballot of council members. When that happened results of the mail ballot were reported to the council at its annual meeting and the council members in attendance then held as many ballots as necessary until one nominee received a majority of the votes.

That process resulted in election of many excellent presidents. Still, not everyone was happy with a system in which the few members of the council actually present determined who would serve as chief officer of an association of nearly 30,000 members and one governed by a representative council of over 250 members. At best, that system seemed rather casual.

During the next quarter century the board of directors made several efforts to have the council consider a broader and more representative method of selecting the president-elect and the other elected members of the board. For example, in 1954 the board asked the council to consider the following scheme (20):

Nominations would be secured by sending a nominating ballot to council members, each of whom would be asked to nominate one person for the position of president-elect. A council committee would examine the list of nominees, select those whose names were suggested most frequently, either send the entire list to all fellows or submit it first to the council for a preliminary vote designed to reduce the list to two or three names, and then send the shortened list to all fellows. A mail ballot by the fellows (by a preferential voting system if more than two names were included or by a simple indication of preference if only two names were submitted) would then constitute the election (21).

The council discussed that scheme in 1954, but the only action taken was to request the temporary Committee on Constitution, Bylaws, and General Operations — which was then considering a number of possible revisions to the constitution and bylaws — to submit one or more possible changes in election procedures that the council could consider at the 1955 meeting (22). After considering several possible methods (23) the committee and the board recommended the following procedure (24):

1. Council members would continue to be invited to submit nominations for the position of president-elect and for election to the board of directors.

2. A tabulation of the resulting nominations would then be given to a new Committee on Nominations and Elections which would prepare a slate. The committee would be guided by—but would not be required to follow exactly—the preference order established by frequency of initial nomination.

3. Names, photographs, and information about the selected candidates would then be published and an opportunity given to add additional nominations endorsed by 30 or more members of the council.

4. Election would then be conducted by mail ballot of the entire council, using a preferential voting system if three or more nominees were presented for any position or a simple indication of preference when there were only two nominees.

The new Committee on Nominations and Elections was to consist of five council members elected by the council for two-year terms, with two or three retiring each year; and two members of the board of directors, named by the board for two-year terms. The senior member from the board was to serve as chairman.

In proposing this arrangement the Committee on Constitution, Bylaws, and General Operations argued that AAAS consisted of many numerically minor groups but no large majority group; that it was desirable to have board membership and the presidency shared by many of those groups; that the proposed Committee on Nominations and Elections could better achieve that objective than had the election procedure then in effect; and that final election by mail ballot was more desirable than election by the particular minority of council members present at one of the annual meetings. That reasoning prevailed and the new procedures were adopted by the council (25).

Under those new arrangements, frequency of nomination by council members was often followed closely by the Committee on Nominations and Elections. But sometimes the committee exercised its discretionary authority in selecting candidates. For example, in 1958 the five top nominees for president-elect were, in order, Chauncey Leake, Margaret Mead, Alan T. Waterman, Thomas Park, and William W. Rubey. Leake was willing to serve if elected, but Mead preferred not to be a candidate that year so her name was not put on the ballot. Waterman was in third place, but the council had just elected Paul Klopsteg, a physicist and associate director of the National Science Foundation, and the committee decided not to nominate as his immediate successor another physicist who was director of the National Science Foundation (26). Instead, the fourth and fifth place nominees, Thomas Park and William W. Rubey, were placed on the ballot. From that slate the council elected Chauncey Leake, but Thomas Park was elected the following year, Alan Waterman three years later, and Margaret Mead (who declined nomination several times) 15 years later.

## The Proper Electorate

Although the new procedures were generally approved they still did not satisfy those officers and members who wanted to change the electorate. There were three realistic options. Election could continue to be by the council, by the fellows, or by all of the members of the association, but only if the council agreed to relinquish electoral responsibility, for any change would require an amendment to the constitution and only the council could amend the constitution.

Some of the officers and board members and also some council members thought that the president and the board members should be elected by AAAS members, or by the fellows, and began to push in that direction. But as they did, another element began to confuse the debates: argument over the use and meaning of the label "democratic."

In a published "Letter from the President" Alan Waterman wrote that AAAS "is not only a very large organization it is a highly democratic one" (27). That statement was soon challenged by Arnold Prostak, who wrote:

> The AAAS has many virtues, but, contrary to the claim made, it is certainly not "highly democratic." The individual members of the AAAS have no voice or vote in the selection of officers or the governing council. Also, the representatives to the council from the affiliated organizations are not generally elected by the members of the affiliated organizations.
>
> Perhaps the government of the AAAS can best be described as a self-perpetuating aristocracy (28).

President Waterman replied that membership was open to any interested person, that any member could at any time express any idea or criticism concerning the association and expect to get a fair hearing, and that government was under control of a council consisting of members selected by many different constituencies, and that he therefore considered AAAS to be "essentially democratic" (29).

He soon backed off a bit from that position, however, and wrote to the board of directors, "Strictly speaking, my use of the term 'democratic' was incorrect. In principle I believe the AAAS has a 'representative' type of administration rather than a 'democratic' one" (30).

Waterman had by that time become chairman of the board of directors and at the next meeting he asked the board to consider whether the president-elect and the members of the board should be elected by the council or by the fellows of the association. That question had been raised before without achieving any change. Nevertheless, the board agreed with its chairman's views and voted to recommend to the council that ballots for president-elect and board members be sent to all AAAS fellows instead of to council members (31).

The council was by no means of one mind on that issue. Some members defended the proposal. Others pointed out that election by fellows would be more expensive; suggested that a sample of fellows might be used instead of the whole list; suggested that there might be better ways of giving members a feeling of participation in association affairs; or worried that the nominees would not be known to all the fellows and that the vote might therefore be low and sometimes not well informed. The council debate also illustrated a clouding of the main issue by disagreement over the meaning of the word "democracy." The agenda item before the council had tried to avoid that confusion by describing the proposed change in electoral process without applying adjectives to it or to the current method. But that did not prevent some members from labeling election by the council as "undemocratic" whereas election by fellows would be "democratic." At about that point in the discussion, as I recall, Don Price, who was then a member of the board and was shortly to become the first political scientist to serve as AAAS president, rose to remind the council of the distinction between a "direct democracy" and a "representative democracy." He pointed out that AAAS was organized as a representative democracy, as was the United States, and recommended that the debate be on the form of government the council wanted, not on a label.

The end result of the debate that year was to ask the board to prepare a specific amendment to the constitution to be voted on a year later (32). The board did as requested, but a year later the council defeated the proposed amendment that would have provided for direct election by fellows of the association (33). That action left electoral responsibility with the council but did not end the argument over which method was "democratic" or which was preferable. In a letter to *Science* Lawrence Cranberg regretted the council's decision to continue an "undemocratic" procedure (34). Soon came a reply from Ellis Yochelson, who took credit for defeating the proposed amendment, because, he said, he had been the first and last speaker to protest the proposed amendment and then went on to explain his reason: "... a division of the membership into aristocrats (qualified to vote) and proletarians (disenfranchised) is meaningless in the context of present day science." The issue of membership classes should be addressed before the question of voting is considered. As for voting, he added, "I do not see that voting for officers by representatives is any less democratic than direct voting" (35).

Two years later, when Alfred Romer was chairman of the board, he told the council that he hoped the association would change to have the officers elected by AAAS fellows instead of by the council (36). As the Committee on Council Affairs discussed that recommendation, they reviewed the reasons for its earlier rejection. There seemed to be three: it was easier and cheaper to handle 200 to 300 ballots from the council than the 10,000 to 15,000 that might be cast by fellows; existing election procedures were consistent with the fact that in other respects AAAS had a representative form of government; and the historical record indicated that there was little reason to expect that the results of an election would be much different if fellows instead of council members were voting (37). The Committee on Council Affairs therefore tabled the board's proposal for those three reasons and because they thought it better to defer action on election procedures until after decisions had been reached on the composition of the council and the possible restructuring of AAAS sections (38).

As plans for those changes were being considered, the Committee on Council Affairs decided to poll council members on several matters including the question of the electorate (39). Results of the poll showed the council to be quite divided in their judgment concerning election procedures. More members voted for continuing election by the council than for any of several alternatives, but the vote was less than a majority and was not strong enough to sway the Committee on Council Affairs or the association's temporary Committee on Governance. Both committees gave greater weight to what they considered appropriate and desirable, and both were moving increasingly in the direction of recommending giving members a greater voice in elections, of reducing the size of the council, and of eliminating direct representation of affiliated societies (40).

As the board of directors considered those changes they once again agreed that the president-elect should be elected by popular vote (41). Discussions continued through 1971 and 1972 and finally ended with agreement. The council gave up its role as the association's electoral body, and all AAAS members soon began receiving ballots for election of the president-elect, members of the board of directors, members of the Committee on Nominations and Elections, and officers of the sections (42). A major change had been accomplished, but no one could accuse the council of having acted hastily.

## Presidents and Board Members

In one very important respect the method of election made little difference: Year after year AAAS enjoyed the generous, constructive, and thoughtful services of an excellent succession of presidents and board members. The names of all past presidents and of some of the other members of the board of directors are annually published in the association's *Handbook*, but for the quarter century under review their names should be recorded here, for they were the guides and directors of the association's development during the whole expansive quarter century (43).

The presidents are listed in Table 3 under the year of their presidency, although beginning with President Sinnott, election was to the office of president-elect, a year earlier than the year each served as president.

Twenty of the 28 presidents listed came from universities; three from the federal government; two from industry; and one each from three other kinds of institutions. In their scientific specialties 11 came from the life sciences, pure and applied; nine from physical sciences and engineering; five from the earth, atmospheric, and space sciences; two from mathematics; and one from the social sciences.

Mina Rees was the only woman, but I believe Margaret Mead, who was one of several women later elected to the presidency, could have been elected in the

## Table 3. Presidents of AAAS, 1944–1971

1944, Anton J. Carlson, University of Chicago
1945, Charles F. Kettering, engineering, General Motors Corporation
1946, James B. Conant, chemistry, Harvard University
1947, Harlow Shapley, astronomy, Harvard University
1948, Edmund W. Sinnott, botany, Yale University
1949, Elvin C. Stakman, plant pathology, University of Minnesota
1950, Roger Adams, chemistry, University of Illinois
1951, Kirtley F. Mather, geology, Harvard University
1952, Detlev W. Bronk, physiology, Johns Hopkins University
1953, Edward U. Condon, physics, Corning Glass Works
1954, Warren Weaver, mathematics, Rockefeller Foundation
1955, George W. Beadle, genetics, California Institute of Technology
1956, Paul B. Sears, plant ecology, Yale University
1957, Laurence H. Snyder, genetics, University of Oklahoma
1958, Wallace R. Brode, chemistry, Department of State
1959, Paul E. Klopsteg, physics, National Science Foundation
1960, Chauncey D. Leake, pharmacology, Ohio State University
1961, Thomas Park, animal ecology, University of Chicago
1962, Paul M. Gross, chemistry, Duke University
1963, Alan T. Waterman, physics, National Science Foundation
1964, Laurence M. Gould, geology, University of Arizona
1965, Henry Eyring, chemistry, University of Utah
1966, Alfred S. Romer, paleontology, Harvard University
1967, Don K. Price, political science, Harvard University
1968, Walter Orr Roberts, astronomy and meteorology, National Center for Atmospheric Research
1969, H. Bentley Glass, genetics, State University of New York at Stony Brook
1970, Athelstan Spilhaus, meteorology and oceanography, Franklin Institute
1971, Mina Rees, mathematics, City University of New York

---

1960s. In several years she was high on the list of nominees submitted by council members and was invited by the Committee on Nominations and Elections to be a candidate. However, one year she was chairman of the committee and refused to have her name on the ballot. In several other years she had other reasons for declining. How the elections would have turned out had she been a candidate is but speculation, but my knowledge of the nominating ballots and feeling for council votes led me to think that she would have been elected. That did not happen, however, and when Mina Rees was elected Mead told me she considered

Rees a more appropriate first woman president than she would have been, for Rees came from mathematics and she from anthropology.

In age the presidents ranged from 49 to 72 at the time of their presidency, with the median age of 61. Charles F. Kettering was the youngest of the series, with Edward U. Condon and George Beadle as the next younger ones. Most had joined AAAS fairly early in their careers; Edmund W. Sinnott, the youngest, joined when only 20 and Paul Klopsteg and Kirtley Mather at 23.

Whatever their age, specialty field, or institutional affiliation, those presidents served AAAS well. Most had had previous experience as members of the board of directors and many had served on major committees. Illness, bad weather, or other obligations sometimes prevented their attendance at board meetings, but only one on the list had a poor attendance record. The exception was Charles F. Kettering, vice-president and director of research of General Motors, who was AAAS president in 1945 when GM was heavily engaged in war production and then in conversion to the postwar period. He missed several board meetings but contributed substantially to AAAS in other ways. At least during the years from 1954 through 1970 when I worked directly with them, the presidents came to board meetings well informed about the issues to be considered, contributed continuously and effectively to the well-being of the association, and were always ready and willing when I needed their help.

## Elected Board Members

Each year in addition to choosing a president-elect, the council elected two members of the board of directors. With four-year terms, sometimes subject to re-election, and with some members later being chosen as president-elect, there was considerable continuity in board membership. But with the past president retiring each year and with limitations on re-election of other members, there was also constant induction of new members. Members elected or appointed to the board of directors from 1945 through 1970 are named in Table 4.

The 63 persons on this list served on the board for an average of six to seven years, including the time some of them spent in the presidential succession. Slightly over one-third came from the biological and medical sciences, about the same number from engineering and the physical sciences, and slightly less than a third from the rest of the scientific spectrum. When a term on the board was left incomplete,—as when a member was elected president-elect,—the board itself made an appointment to fill out the remainder of the term. Using that authority the board frequently, but not always, chose from one of the less well-represented areas. Thus it picked three mathematicians and three social scientists, all but one of whom were later re-elected for full terms through the usual council procedures.

## AAAS Treasurers

The board also chose the association's treasurers, of whom there were four who served during the 1945–1970 period. William W. Wrather was treasurer from 1945 through 1953, but because of changes in the composition of the board of directors he was a member of that body in only two years, his first and last as treasurer. Paul A. Scherer became treasurer in 1954 and resigned at the end of 1962 to avoid a conflict of interest, for he then became assistant director of the National Science Foundation, which was supporting a number of AAAS activities. Paul E. Klopsteg, who had earlier been a member of the board and had served through the three years of the presidential succession, was drafted to come back as treasurer, and served in that office from 1963 to 1969. William T. Golden became treasurer in 1969 and as of 1988 was still in office.

Except for the 1946–1952 period the treasurer served as a member of the board of directors. In all years the treasurer, as the title implies, was the association's chief financial officer, responsible with the executive officer for the annual budget and with an advisory committee for investment policy and management. Working with both the treasurer and the executive officer, the staff member primarily responsible for the association's financial health was Hans Nussbaum. After earning a doctorate in economic sciences form the University of Cologne, he came to the United States in 1938, gained several years of commercial experience in New York, and at the end of 1945 came to AAAS as business manger, a post he held until 1975 when he reached the association's nominal retirement age. But he was not allowed to retire. His services had become so valuable that he was asked to continue as the staff officer responsible for the association's investments and collections, a part-time position he still held in 1988, nearly 43 years after he joined the AAAS staff.

Those four treasurers, the executive officer, and Nussbaum shared a problem common to all investors: In making investment decisions how much weight should be given to safety and preservation of principal? How much to income? And how much to expected capital appreciation? In making those policy choices, and in selecting specific investments, they had the advice of the Committee on Investment and Finance, which in 1940 was reconstituted to consist primarily of officers of banks and investment firms. Chairman of that committee from 1940 to 1961 was Frederick P. H. Siddons, vice-president of the American Security and Trust Company, one of the major banks of the District of Columbia. For those 21 years AAAS was the beneficiary of his generous counsel and services. To the treasurer and that committee the board delegated responsibility for investing and reinvesting the association's funds other than the bank accounts used for current expenditures.

As Table 5 shows, the association was operating at quite different financial levels when the four treasurers entered office. They served under different conditions and brought to the association different policies.

**Table 4. Members of the Board of Directors, 1945–1970**

Election to the board was for four-year terms, indicated in the following list by the first and last of those four years, although the actual term did not begin until the middle of January of the first year shown and ended in the middle of January of the year following the last one shown. Terms shorter than four years occurred because the member was appointed by the board of directors to complete an unexpired term vacated by some other member, or because the member was elected president-elect and therefore vacated the earlier board position, or, rarely, because the member resigned. Terms without further designation were for service as one of the eight elected board members. Terms indicated by (P) were for the three years in the presidential succession. Other terms are identified by the name of the position held.

The record shown here begins with 1945, although members Otis W. Caldwell, Anton J. Carlson, Arthur H. Compton, Burton E. Livingston, Kirtley F. Mather, Forest R. Moulton, Elvin C. Stakman, and William E. Wrather were already on the board at the beginning of 1945.

Roger Adams, chemistry, 1945, 1948, 1949–1951(P)
George Baitsell, zoology, 1947–1950
George Beadle, genetics, 1954–1956 (P)
William Bevan, psychology, 1970–1974 (executive officer)
Richard H. Bolt, physics, 1969–1972, 1973–1976
David Blackwell, mathematics, 1970–1972
Lewis M. Branscomb, physics, 1970–1973
Wallace Brode, chemistry, 1953, 1954–1956, 1957–1959 (P)
Detlev W. Bronk, physiology, 1951–1953 (P)
Harrison Brown, geochemistry, 1960–1962
Otis W. Caldwell, botany, 1935–1947 (general secretary)
Anton J. Carlson, physiology, 1945–1948 (president in 1944)
Barry Commoner, botany, 1967–1970, 1971–1974
Arthur H. Compton, physics, 1945–1947
James B. Conant, chemistry, 1946–47 (president and past president)
Edward U. Condon, physics, 1952–1954 (P)
John F. Dunning, physics, 1950–1951, 1952–1955
Henry Eyring, chemistry, 1961, 1962–1963, 1964–1966 (P)
Edwin B. Fred, bacteriology, 1948–1951
John Gardner, psychology, 1963, 1964–1965
H. Bentley Glass, genetics, 1959–1962, 1963–1966, 1968–1970 (P)
David R. Goddard, botany, 1964–1967
William T. Golden, corporate director and trustee, 1969— (treasurer)
Paul M. Gross, chemistry, 1956–1959, 1961–1963 (P)
George R. Harrison, physics, 1956–1959
Hudson Hoagland, physiology, 1966–1969

Gerald Holton, physics, history of science, 1967–1970
Walter S. Hunter, psychology, 1951–1954
Mark Ingraham, mathematics, 1952, 1953–1956
Charles F. Kettering, engineering, 1945 (president), 1946–1949
Paul E. Klopsteg, physics, 1949–1952, 1953–1956, 1957, 1958–1960 (P), 1963–1969 (treasurer)
Karl Lark-Horovitz, physics, 1947–1950 (general secretary)
Chauncey D. Leake, pharmacology, 1955–1958, 1959–1961 (P)
Burton E. Livingston, plant physiology, 1920–1930 (permanent secretary), 1945–1946
Kirtley F. Mather, geology, 1945–1946, 1947–1949, 1950–1952 (P)
Margaret Mead, anthropology, 1955–1958, 1959–1962, 1974–1976 (P)
Howard A. Meyerhoff, geology, 1949–1953 (administrative secretary)
Walter Miles, psychology, 1945–1948
Forest R. Moulton, astronomy and mathematics, 1937–1948 (administrative secretary)
Thomas Park, animal ecology, 1954–1957, 1958–1959, 1960–1962 (P)
Phyllis H. Parkins, biology, 1970–1973
Fernandus Payne, zoology, 1946–1949, 1950–1953
Don K. Price, political science, 1959, 1960, 1961–1964, 1966–1968 (P)
Mina S. Rees, mathematics, 1958–1960, 1962–1964, 1965–1968, 1969, 1970–1972 (P)
Leonard M. Rieser, physics, 1968–1971, 1972–1974 (P)
Walter O. Roberts, astronomy and meteorology, 1963–1966, 1967–1969 (P)
Alfred S. Romer, paleontology, 1960–1963, 1965–1967 (P)
William W. Rubey, geology, 1957, 1958–1961
Paul A. Scherer, engineering, 1954–1962 (treasurer)
Paul B. Sears, plant ecology, 1950, 1951–1954, 1955–1957 (P)
Harlow Shapley, astronomy, 1947–1948 (president and past president)
Edmund W. Sinnott, botany, 1947–1949 (P)
Malcolm Soule, biochemistry, 1949–1951
Athelstan F. Spilhaus, meterology and oceanography, 1964–1967, 1969–1971 (P)
Laurence H. Snyder, genetics, 1952–1955, 1956–1958 (P)
Elvin C. Stakman, plant pathology, 1945–1947, 1948–1950 (P)
Burr Steinbach, zoology, 1964–1965, 1966–1969
Kenneth V. Thimann, plant physiology, 1968–1971
Alan T. Waterman, physics, 1957–1960, 1961, 1962–1964 (P)
Warren Weaver, mathematics, 1950–1952, 1953–1955 (P)
John A. Wheeler, physics, 1965–1968
Dael Wolfle, psychology, 1954–1970 (executive officer)
William W. Wrather, geology, 1945 (treasurer), 1953 (treasurer)

Table 5. Selected Financial Records for 1945, 1954, 1963, and 1970

|  | 1945 | 1954 | 1963 | 1970 |
|---|---|---|---|---|
| Receipts: | | | | |
| Dues payments | $129,943 | $303,664 | $664,510 | $1,426,715 |
| Subscriptions to *Science* and *Sci. Mo.* | 41,247 | 62,087 | 114,831 | 358,681 |
| Advertising in *Science* | 71,605 | 144,347 | 1,375,174 | 2,195,910 |
| All other income | 20,387 | 100,295 | 235,624 | 745,140 |
| TOTAL | 263,182 | 610,393 | 2,390,138 | 4,726,446 |
| Expenditures | 222,668 | 563,301 | 2,259,736 | 4,851,247 |
| Net revenue | 40,515 | 47,092 | 130,402 | (124,801) |
| Grant expenditures (not included in above figures) | 00 | 4,167 | 834,546 | 466,171 |
| Fund balances of net worth | 324,795 | 639,186 | 2,154,121 | 3,509,032 |

Wrather came at a time when frugal management was necessary so the association's funds were invested very conservatively. When he arrived in 1945, 81 percent of the invested money was in bonds, almost all U.S. bonds and treasury certificates. Safety of principal was surely given greater weight than amount of income. A few years later, however, Wrather and the board agreed to withdraw four savings accounts from banks paying 0.99 percent a year in interest. The $30,000 involved was transferred to six savings and loan associations that were paying 3.0 percent a year (44).

During Scherer's tenure the division of investments between stocks and bonds changed substantially to increase the equity portion to over half. He also introduced two major changes in investment practice. One was to consolidate 15 separate funds, several of which had their own individual portfolios of stocks and bonds, into a single investment portfolio from which each separate "fund" would draw earnings in proportion to its size as adjusted annually by additions and withdrawals. Internal record keeping was not materially changed by that action, but it enabled the Investment and Finance Committee to focus more broadly on investment strategy instead of having to deal separately with many small funds (45).

Scherer's other change was to bring in external advice. Most members of the Committee on Investment and Finance were professionally engaged in money management. Nevertheless, Scherer argued, it would be desirable to engage the services of an investment counselor whose recommendations would then be weighed by the committee in making its investment decisions. The board agreed (46) and the firm of Scudder, Stevens, and Clark was engaged for that purpose.

When Klopsteg became treasurer he continued the two changes Scherer had introduced, and during his tenure the investment balance was further shifted so that when he resigned less than a quarter of the portfolio was in bonds and treasury bills and more than three-quarters in equities.

When Golden became treasurer in 1969 the association had as its treasurer an experienced and successful professional money manager. He made no immediate changes in the bond-stock balance of the portfolio, but soon introduced a major change in investment practice: The treasurer, with the concurrence of the Committee on Investment and Finance, stopped investing in a relatively large number of individually selected stocks and, instead, began to rely more on closed-end investment trusts selling at substantial discounts from liquidating values and on mutual funds with good performance records.

## The Board as a Whole

All of the board members who served between 1945 and 1970 provided valuable leadership for the association, but one name calls for special mention: Paul E. Klopsteg. In addition to serving on a variety of AAAS committees, he was a member of the board from 1949 through 1957, then moved immediately into three years as president-elect, president, and retiring president. AAAS got along without the benefit of his counsel for a couple of years, but in 1963 brought him back as treasurer, a post he held until 1969. From 1949 to 1969, except for a two-year sabbatical, he was a member of the board. He brought to AAAS the scientific experience and wisdom gained through 26 years with Leeds and Northrop and Central Scientific Company, 11 years as professor of applied science and director of research at Northwestern University's Institute of Technology, seven years as assistant and associate director of the National Science Foundation, and service to numerous scientific, governmental, and educational institutions. He gave the association much and AAAS kept him in harness longer than any other officer in the period under review (47).

Second prize for longevity during the period went to Mina Rees. The board of directors first selected her to fill out the remaining three years of the term vacated by Klopsteg when he became president-elect. She was then off the board for a year but was brought back to complete the three years of the term vacated by Paul Gross when he became president-elect. The council in 1965 then elected her to a four-year term and four years later to another, which she vacated after one year when she became president-elect. With only one year out, she served on the board from 1958 through 1972.

Let thanks to those two long-termers symbolize the appreciation the members owe to all their colleagues who helped govern the association.

*Chapter Eight*

# Science Education

Improvement of education in science has always been an interest of AAAS, earliest and most naturally at the college level, for most of the founding members were professors. Yet interest has also extended to younger age levels and some of the association's earliest records showed "a continuous tendency for members to worry about what was being stuffed into young heads in the ... Little Red School House" (1).

In 1925 a grant from the Commonwealth Foundation helped AAAS to conduct "a thorough study of the role played and to be played by science in general education" (2). Twenty years later, in 1945, AAAS was participating in the work of the National Educational Association for Study of Science in Schools and had three educational committees of its own: one on improvement of education at the college level, another on general education, and, most significantly, the new Cooperative Committee on the Teaching of Science and Mathematics.

## The Cooperative Committee

The Cooperative Committee on the Teaching of Science and Mathematics was born in 1941 when five societies, not including AAAS, decided to work together on problems of science and mathematics that no single organization could handle alone. The five founding societies were the American Association of Physics Teachers, the American Chemical Society, the Mathematical Association of America, the National Association for Research in Science Teaching, and the Union of Biological Societies. They called their new joint committee the Cooperative Committee on Science Teaching and, with the help of a grant of $3,000 from the Carnegie Foundation for the Advancement of Teaching, set out to try to improve the school curricula in science, the education then being given to prospective teachers of science and mathematics, and the requirements and procedures for licensing elementary and secondary school teachers (3).

In 1945 the committee asked to become part of the AAAS structure while still consisting primarily of representatives of other societies (4). The board of directors accepted that proposal, took the committee into AAAS, changed its name to the Cooperative Committee on the Teaching of Science and Mathematics, and adopted essentially intact the enlarged composition, working procedures, and financial arrangements proposed by the committee.

By 1949 the committee had achieved a bibliography of 23 articles, notes, and reports in such journals as *School Science and Mathematics, The Science Teacher, The American Journal of Physics,* and the *American Mathematical Monthly.* In cooperation with the National Science Teachers Association, the committee prepared for UNESCO a report on the courses, materials, and apparatus used for science in schools and colleges of the United States, a report UNESCO distributed widely to help in rebuilding the educational systems of countries devastated by World War II.

The committee's largest task up to that time was carried out in response to a request from John R. Steelman, director of President Truman's Scientific Research Board. Steelman asked the committee to assess "the effectiveness of science training to increase the quality as well as the supply of scientists for government, industry, and university research, and for high school and college teaching." The committee's 100-page answer filled most of one volume of the five-volume "Steelman Report," *Science and Public Policy* (5).

Through its own studies the committee helped stimulate interest and activities on educational problems that were later picked up by the National Science Foundation and other scientific organizations. As those programs developed and as NSF increased its support for course content improvement programs, institutes for science teachers, and other efforts to improve science education, the relative importance of the Cooperative Committee understandably declined. In 1952 Howard Meyerhoff told the board of directors that the committee "is rapidly losing effectiveness since the resignation of Dr. Lark-Horovitz" (6).

As things turned out Meyerhoff was too pessimistic about the committee but quite right in identifying Karl Lark-Horovitz as the committee's most important member. Born in Vienna and educated as a physicist, in 1925 he received the second International Research Fellowship awarded by the Rockefeller board. He came to the University of Toronto for a year of research and also spent brief periods at Chicago, Stanford, and the Rockefeller Institute. In 1928 he was invited to give a series of lectures at Purdue University. His reception there was so favorable that a year later he was persuaded to return permanently to lead and develop Purdue's Department of Physics. At Purdue, he continued his research, built up the department, and gave much attention to teaching, not only in teaching prospective scientists and engineers but also students who were not planning on scientific or technical careers. It was he who first proposed the original Cooperative Committee on Science Teaching. When AAAS took over that committee in 1945, he was appointed chairman, a role he continued for the next five years. He also served AAAS as a member of the board of directors, as the association's last general secretary from 1947 until that position was abolished in 1952, and as a member of the editorial board from 1949 until his death in 1958 (7).

In the early 1950s there was increasing national recognition of the need for improvements in science education. It was also becoming evident that the downward trend in the number of college graduates qualified to become teachers of science or mathematics was quite out of phase with the number of such teachers who would be needed when the children of the postwar baby boom began moving through school. Members of the Cooperative Committee thought that AAAS should tackle some of the problems involved and that more than their own committee effort was needed. To that end the committee met in Washington in October 1954 to formulate what they called an "Action Program." In order to broaden the range of views considered and to increase interest on the part of other organizations, they invited representatives of the National Association of Manufacturers, the National Education Association, the National Research Council, the National Science Foundation, Science Service, the U.S. Chamber of Commerce, the U.S. Department of Labor, the U.S. Office of Education, and the Washington, D.C., school system.

The conclusions and recommendations of the three-day meeting of the Cooperative Committee and that group of consultants were immediately reported to the AAAS board of directors, which commended the Cooperative Committee "for its forward looking program, received the recommendations with enthusiasm, and voted to express the board's intent of supporting the principles of the committee's report and recommendations" (8).

The timing was right, for the recommendations were part of a much wider recognition of the need to improve science education. Scientists and engineers had been in short supply during World War II and continued to be during the Cold War that followed. So were qualified teachers of science and mathematics (9). In response to those problems the Research and Development Board of the Department of Defense had established a Panel on Human Resources; the Conference Board of Associated Research Councils had appointed its first Commission on Human Resources; the President's Committee on Scientists and Engineers had been appointed; and scientists and mathematicians were establishing special task forces to develop improved texts, apparatus, and other teaching materials for high school students. In 1955 the whole problem was given added urgency by publication of Nicholas DeWitt's *Soviet Professional Manpower: Its Education, Training, and Supply*, a book that reported the surprisingly large number of engineers being trained in the Soviet Union.

The National Science Foundation quickly became a leader in the effort to improve science education by financing the course content improvement projects. Taking a cue from General Electric and Westinghouse—which had supported a few summer institutes for high school teachers—NSF began to support and encourage universities and colleges to offer summer institutes for high school teachers. A single summer institute offered in 1954 by the University of Washington for teachers of high school mathematics expanded to six institutes for science and mathematics teachers in 1955, to 18 in 1956, skyrocketed to 90 in 1957, and then went on increasing. Those programs were popular with Congress, and the NSF budget for science education grew from $7,200 in 1952 to over $1.4 million in 1956 and then jumped to $14 million a year later.

AAAS was already involved in this effort. In addition to sponsoring the Cooperative Committee, AAAS was the home of the Academy Conference — which later adopted the more formal name of National Association of Academies of Science. The Academy Conference met annually at the AAAS meetings and served as a loose coalition of the academies of science that existed in most states. Most of those academies sponsored junior academies of science, which typically met annually to give their members a platform for reporting on research studies they had carried out. Starting in 1946 representatives of the junior academies came together each year at the Junior Scientists' Assembly, a part of the AAAS annual meeting. As one other association activity, AAAS in 1951 began cooperating with the National Science Foundation in sponsoring an annual conference on scientific manpower. Held in conjunction with the association's annual meeting and later published by NSF, those conferences provided an annual forum for discussing problems, prospects, and possible remedies.

I had a personal and professional interest in this whole set of problems and thus in helping the association move toward a greater involvement, for in 1954 I completed the report of the first national study of the needs for and the prospective and potential supply of scientists, engineers, lawyers, physicians, teachers, and members of the other professions based on higher education and specialized knowledge (10).

## *The Science Teaching Improvement Program*

It was in that climate of widespread concern over the shortcomings of education and the nation's need for scientists, engineers, and teachers that AAAS in 1955 decided to adopt a completely different style of work. Instead of relying on the voluntary efforts of the Cooperative Committee and similar bodies, AAAS began to build a professional staff to devote full time to efforts to improve science and mathematics education at the precollege level.

As described by the Cooperative Committee in its Action Program, a strong national effort was needed to secure more and better educated teachers of science and mathematics and to reward them with higher salaries, better working conditions, and greater prestige (11).

All of the more specific objectives of the Action Program seemed desirable, but some were not within the competence of AAAS or any other scientific society. For example, AAAS might advocate increasing teachers' salaries and making the teaching profession comparable in prestige and rewards with those of other professions requiring comparable education, but effective action on that issue was for society as a whole, working through local and other levels of government, not for an organization such as AAAS.

Nevertheless, the Action Program pointed toward some highly desirable goals toward which AAAS could contribute, and when the board of directors received the report of the Cooperative Committee it resolved to adopt the proposals as an AAAS program, to authorize me to seek funds of at least $25,000 a year for that purpose, and to encourage the Cooperative Committee to continue its studies and plans of ways to meet the expected shortage of teachers (12).

The sum of $25,000 a year did not seem sufficient to support a program as ambitious as the Cooperative Committee was recommending, but the board had authorized an attempt to get $25,000 "or more." I discussed the proposed program with John Gardner, the new president of the Carnegie Corporation of New York, and then sent him a formal proposal asking for $300,000 to be used over a three-year period. The Carnegie Corporation approved and that amount was granted.

The chairman of the Cooperative Committee at that time was John R. Mayor, who was professor of mathematics and also professor and chairman of the Department of Education at the University of Wisconsin in Madison. That year he was also acting dean of the School of Education and some of his colleagues were recommending him for permanent appointment to that position. However, Mayor was a mathematician, and while it might be acceptable to have a mathematician serve as the elected chairman of the Department of Education, apparently some thought he did not have enough "education" in his record to hold the position of dean of education.

But for AAAS, Mayor was the first choice to direct the new program (13). As chairman of the Cooperative Committee he had helped develop the plans the board had just approved. As professor of mathematics in one of the nation's major research universities he clearly belonged to the scientific community. As professor and chairman of the Department of Education at the same university he also had standing among educationists. The necessity of cooperation between scientific and educational organizations was a basic premise in the whole development of the association's educational program, and Mayor's dual citizenship in science and education always gave him a good basis for securing that cooperation.

Thus we offered the post of director of the education program to Mayor. He accepted, left the University of Wisconsin to come to AAAS, resigned the chairmanship of the Cooperative Committee, and started to recruit staff for what soon came to be called the Science Teaching Improvement Program, STIP for short. One of the earliest and longest continuing members of that staff was Orin McCarley who served the association's educational programs in many ways, including editorship of *Science Education News*, until her retirement in 1985.

The activities that developed under STIP were of three, somewhat overlapping types:

1. Actions intended to stimulate or assist other organizations that could contribute to improvement of science education;

2. Actions designed to strengthen the qualifications of teachers or to give them improved materials for teaching; and

3. Actions intended to increase students' interest in and understanding of science.

While some of the specific activities might equally well be described under two of those headings, the three can serve as a basis for organizing the discussion of the Science Teaching Improvement Program and the association's other educational activities.

## Conference and Advice

Interest in improving education was widespread and the AAAS provided a prominent platform. As a consequence, there were many calls upon Mayor and his staff and much of their time went into a kind of missionary work. In 1957 he told the board of directors that in the preceding 12 months he and the assistant director of the program, Eugene Wallen, had visited 24 colleges and universities, appeared on the programs of nine academies of science, 11 National Science Foundation institutes, and the meetings of some 20 scientific societies. In addition he had appointed 20 regional consultants, all scientists interested in the improvement of education for students and for prospective science teachers. With their travel expenses covered by a grant from the General Electric Educational and Charitable Fund, these regional consultants had visited many colleges and school systems to give information, advice, or other help on problems of science education (14).

Another activity that started early and continued for years exemplified the convening function of AAAS. A few months after his arrival, and in cooperation with the National Science Foundation, Mayor arranged a meeting of the chief staff officers and the chairmen of the educational committees of a number of scientific organizations to exchange information and advice about their efforts to improve science teaching, the training of future science teachers, and the recruitment and education of exceptionally talented students. Another early conference brought together representatives of several mathematics groups to encourage cooperative planning among them. Later those groups worked together in organizing the School Mathematics Study Group which developed materials for teaching what was then called "the new math."

Results of this missionary work and convening of conferences were quite impossible to evaluate in any precise fashion, and we made no effort to do so. Interest was widespread, the association's advice and leadership were frequently sought, and it seemed worthwhile to try to respond to as many of the requests as we could.

## University Consultants for Science and Mathematics

One of the major objectives of STIP was to encourage college and university departments of science and education to work together on problems of teacher education and to encourage scientists to accept greater responsibility for teacher education. An early example was an experimental program designed to assess the effectiveness of having well-informed consultants available to any teacher in the consultant's region who wanted help on problems of teaching science or mathematics.

The idea of providing teacher consultants was part of the original Action Program recommended by the Cooperative Committee. It was based on four lines of thinking:

1. The shortage of well-qualified teachers would continue as school enrollment continued to rise.

2. In a region in which there were, say, 50 teachers of high school science, an additional well-qualified teacher would be useful in the classroom, but more useful as a counselor, tutor, or guide to as many of the other 50 as wanted help on their teaching problems.

3. Having the counselors operate out of a state university would help build bridges between the science and mathematics faculties of the university and the surrounding school districts. As employees of the university instead of the local school district or the state department of education, the counselors would clearly be seen as helpers, not administrators, and were therefore not a threat to the individual teachers they were helping or to the concept of local control of education.

4. The position of "teacher counselor" constituted a higher rung on the professional ladder and thus provided an opportunity to recognize and reward some particularly capable teachers.

With those ideas in mind, arrangements were made with the state universities of Nebraska, Oregon, Texas, and Pennsylvania for each to engage one science counselor and one mathematics counselor to work in their regions.

Most of the costs were borne by the Carnegie Corporation grant to AAAS, but in each state the university was sufficiently interested to bear part of the cost. In each case endorsement was received from the state department of education and from the local school districts that were to be served. Each counselor was expected to be available to about 100 teachers within a geographic range of about 100 miles. Their functions were left quite open-ended. They were available for help on problems of subject matter, teaching procedures, equipment, examinations, science fairs or other special projects, the use of educational resources outside the school system, or anything else on which a teacher needed advice, information, or help.

In the first year, 1956–1957, the eight counselors visited 128 different schools, where they counseled or helped a total of 739 teachers, with an average of four or five conferences with each teacher, and in the process traveled some 136,000 miles (15).

The project continued in essentially the same fashion through 1957–1958, but there was no intention to continue it indefinitely under AAAS auspices or Carnegie Corporation funding. Two years, we thought, should provide a reasonable opportunity for trial and demonstration. Throughout those two years letters from high school teachers and principals, from representatives of state departments of education, and from university scientists and educators showed genuine enthusiasm for the program. Teachers wanted more counselors. They said that visits by the counselors had reduced their feelings of isolation and provided intellectual stimulation. Administrators said that teacher morale had been improved. An interesting side effect was that many of the teachers joined scientific or science-teacher organizations. From the universities came letters reporting that fruitful relationships had developed between university scientists and educators and between university faculty members and state and local school administrators and personnel (16). It seemed that the teacher counselor idea was one that universities could well continue, and we hoped that funds would be available to extend the counselor plan.

Immediately, however, it was continued only in Texas, where the University of Texas paid the counselors' salaries and the local school districts paid for their traveling and other expenses. Nebraska and Oregon did not have funds to continue in 1958–1959, but hoped to restore the plan later. Pennsylvania sought a grant from the U.S. Office of Education to extend the counselor services statewide, but the Office of Education did not make the requested grant. Nevertheless, for a time it seemed that the idea might be adopted widely. In 1958 two comprehensive aid to education bills were debated in Congress. One bill, sponsored by Senator Lister Hill and Representative Carl Elliott, picked up the teacher counselor idea from AAAS and offered matching funds to any state that decided to adopt that plan. The other bill, sponsored by the Eisenhower administration, included a comparable provision, but changed the title from "counselor" to "supervisor" and changed the base of operations from the state university to the state department of education.

When Laurence Snyder, then chairman of the board of AAAS, testified before the Senate Committee on Labor and Public Welfare on those two bills he explained why the use of "counselors" rather than "supervisors" was desirable and urged the Senate to choose those provisions of the Hill–Elliott bill rather than the supervisor provisions of the administration bill (17). A few weeks later when I testified on the two bills before the House Committee on Education and Labor I made the same recommendation (18).

We were not persuasive enough. The compromise between the two bills that became the National Defense Education Act of 1958 followed the administration bill in offering matching support to states that engaged science or mathematics supervisors in the state department of education.

From the standpoint of the original objectives of the counselor program that decision was a mistake. The supervisors who were appointed under provisions of the National Defense Education Act did not become consultants or educational extension agents from the university, as were the original counselors. Instead they became part of the educational bureaucracies of their states and their time quickly came to be so filled with supervisory, administrative, and reporting duties that they had little time to visit schools and counsel teachers.

Yet the idea did not die. The science and mathematics counselors in Texas became professors of science and mathematics education at Michigan State University, the Pennsylvania science counselor became professor and director of the science teaching program at the University of Maryland, and the mathematics counselor in Nebraska became professor of mathematics at Purdue. They and others in similar positions or serving as science or mathematics supervisors have carried on the concept of the teacher counselor. When in the 1970s the next flurry of national interest in improving science education arose there were two school systems in the United States that had adopted the teacher counselor idea and built upon it to the point of having well-staffed and flexible teaching resource centers available to help teachers with the kinds of problems that led to the original 1956–1957 teacher counselor program. Those two were the Science Center for Instructional Materials and Processing that served 17 small school districts in the Genesee Valley of New York and the Science Materials Center of Fairfax County, Virginia. Both had gone well beyond the AAAS demonstration by providing not only individual consultation and help but also kits of demonstration and experimental materials and in-service training programs (19). Those two centers provided good examples of what could be accomplished by systematically making well-qualified consultants and their materials available to support and supplement the capabilities of many classroom teachers and the resources of the schools in which they taught. That example has since been followed by several other school systems and also by some of the science centers or museums that provide prepared kits of teaching materials for teachers to use, send vans of equipment with special teachers to neighboring schools, or in similar ways supplement the resources of nearby schools.

## The National Defense Education Act of 1958

The National Defense Education Act was one of the nation's reactions to the shock of *Sputnik*. As indicated earlier, that act was a compromise between an administration bill and one sponsored primarily by Senator Lister Hill and Representative Carl Elliott. The provisions of those two bills and a number of other ideas were debated in congressional hearings and committee meetings. To supplement those formal hearings the board room of the AAAS building provided a congenial site for Stewart McClure, a staff member of the Senate Committee on Labor and Public

Welfare who was handling the legislation, to meet with scientists and educators for informal discussions of ideas and options.

The National Defense Education Act failed to include the provisions for science counselors that AAAS had recommended, but the association soon became involved in planning the implementation of other provisions of that bill. At the request of the President's Committee on the Development of Scientists and Engineers, AAAS convened a panel of scientists and school counselors to prepare recommendations for a kit of guidance materials for school use. A few weeks later, Mayor served as director of a conference called by the Council of Chief State School Officers to prepare recommendations on standards for equipment in science, mathematics, and modern foreign languages that would be used by the state departments of education in applying for aid under the act (20). And, as probably its largest contribution to the educational activities supported by that act, AAAS developed the lists of books that served as the standard guides for schools all over the country in purchasing books on science and mathematics for library and general use (see later section on "Traveling Science Libraries," this chapter, or see pages 163–169).

## Use of Special Teachers in Elementary Schools

In a typical elementary school class one teacher is responsible for teaching English, arithmetic, science, social studies, and perhaps other subjects, while art, music, and physical education are frequently taught by special teachers of those fields. To many scientists it seemed reasonable to expect that pupils would learn more about science and mathematics from teachers who specialized in those subjects, and the AAAS staff decided to make an experimental test of that proposition. The study sought to answer two questions:

1. Can science and mathematics be more effectively taught by special teachers in those subjects than by the regular classroom teacher?

2. How do the effects differ for children of different ability?

Four rather different school systems participated in the study: Cedar Rapids, Iowa; Lansing, Michigan; Woodford County, Kentucky; and Washington, D.C. Each of the four school systems selected four fifth-grade and four sixth-grade classes of children with similar socioeconomic backgrounds. Each school system also selected what they considered a particularly well-qualified teacher in science and one in mathematics who taught those subjects in two of the four classes, while the other two matched classes were taught science and mathematics by the regular classroom teacher. At the beginning of the year and again at the end pupils in all of the classes were given five standardized tests that were designed to measure interest, general ability, and knowledge in science, mathematics, and social studies (21).

As things turned out it did not make much difference whether the pupils were taught by the regular classroom teacher or by the specialists. There was some weak evidence of greater achievement in science by pupils who had been taught by special teachers of science, but no evidence of any difference in learning of mathematics. As for social studies, there was no loss, and even some slight evidence of greater gain, by pupils who had had special teachers for science and mathematics. To some extent those findings may have been due to the fact that the whole study was weaker than it should have been, for the special teachers were not much better qualified in science and mathematics than were the regular classroom teachers. However, if the pupils did not gain much from the demonstration study, the special teachers did; most of them were promoted to supervisory positions in their own schools or received more attractive offers from other school systems (22).

## The Education of Science Teachers

A recurring theme in discussions of educational issues has been the topic of education for prospective teachers. How much of the curriculum should be devoted to learning the subject matter to be taught and how can that subject matter most effectively be presented? How much time should be devoted to gaining a better understanding of child development? How much to instruction on teaching methods and other educational topics? Most of the interest of scientists has been focused on the nature and amount of education in the subject matter. From its beginning the Cooperative Committee took up those questions (23) and later returned to them on several occasions (24). The topic also engaged much attention from the association's science education staff, partly in the development of guidelines for the preparation of elementary and secondary school teachers, as will be described later, and partly in a variety of services, conferences, and experimental studies.

During the academic year 1960–1961, AAAS made small grants to several colleges to allow them to try out innovative ideas on the education of prospective science teachers. Three examples can illustrate the range of ideas involved. San Francisco State University maintained a laboratory of teaching materials that was used, apparently with enthusiasm, by students who expected to become science teachers. Emory University and the University of Tennessee provided opportunities for prospective science teachers to work as research assistants to members of the university faculty. And Hunter College compared an experimental problem-solving course in science with a conventional science course offered for prospective teachers (25).

As is often true of experimental programs, the results of these novel methods were generally reported with enthusiasm, and at least one was later continued and expanded with a grant from the U.S. Office of Education. Yet there is little evidence that they had any major or permanent effect on college and university curricula for prospective teachers. Arguments over the proper content of those curricula still continue.

## The Commission on Science Education

In the mid-1950s when there was much concern over the need to improve education in science and mathematics, it was the grant of $300,000 from the Carnegie Corporation of New York, and its later renewal, that enabled AAAS to employ professional staff and initiate a major program on science education. Simultaneously, however, the National Science Foundation's budget had increased to a level at which it was making substantial grants for work on science education, including support of two AAAS programs that were not considered part of the Science Teaching Improvement Program: the Visiting Foreign Lecturers program serving summer institutes for science and mathematics teachers and the Traveling Science Libraries for elementary and secondary school students, both of which will be described later.

As those AAAS programs were developing, NSF was also supporting the School Mathematics Study Group, the Biological Sciences Curriculum Study, the Physical Sciences Study Committee, and other groups that were developing new teaching and learning materials for high school science and mathematics courses. As those materials began to be available there arose a substantial recognition that improved materials for the elementary and junior high school levels were at least as necessary as for the senior high school years. The National Science Foundation explained:

> By 1960 it seemed clear that the time had come for a careful and thorough examination of the problems of science in elementary and junior high schools. A number of questions had to be answered: Should science be an important part of the curriculum in these grades? If so, what kind of science? What content? How should it be presented? What sort of effort would be required to create model materials and programs? To what extent is this an appropriate area for substantial Foundation support? By fortunate coincidence, the American Association for the Advancement of Science itself was already seriously interested in the general question of the kind of science that should be taught at this level. A grant from the Foundation enabled the Association to carry out a so-called "Feasibility Study" on science and mathematics for elementary and junior high schools. Nearly 200 leading scientists, school administrators, and teachers participated in three independent regional conferences to consider the problem and develop recommendations. The three groups reached remarkably similar conclusions: science should be a basic part of general education for all students in grades kindergarten through nine; the instruction should deal in an organized way with science as a whole; there must be a greater progression in science study from year to year; there should be no single national curriculum in science; the teaching should stress the spirit of discovery characteristic of science itself; there is urgent need to prepare improved materials for science in these grades — an effort requiring the

combined talents of scientists, classroom teachers, and specialists in learning and teacher education; there is also great need for new materials for pre-service and in-service education of teachers. The final report of the American Association for the Advancement of Science study also recommended the establishment of a national correlating and stimulating body to guide these efforts, the establishment of several major study groups to prepare instructional materials, support for independent and experimental groups of smaller scope, encouragement of fundamental studies on learning processes, and expansion of science and mathematics preparation for teachers (26).

The AAAS board of directors responded promptly to one of those recommendations by establishing the AAAS Commission on Science Instruction in Elementary and Junior High Schools, a commission that soon asked to have its name changed to Commission on Science Education in order to indicate more accurately the wide range of its activities. From its formation in 1962 until its termination in 1974 that commission provided guidance and direction to most of the association's educational activities (27).

## The End of the Cooperative Committee

The new commission quickly became a constructive and productive body. Yet its effectiveness had the sad consequence of making the Cooperative Committee on the Teaching of Science and Mathematics less necessary and less useful. The Cooperative Committee had served as a useful advisory body for the Science Teaching Improvement Program. It had continued to provide a forum for representatives of a number of science, mathematics, and teaching organizations. And it conducted some studies of its own. Nevertheless, when the Commission on Science Education was established and was expected to play a coordinating role in work to improve science education at the elementary and junior high school levels, several members of the Cooperative Committee began to ask whether it should disband. Meeting twice a year with limited funds and with an unincorporated organization that made it ineligible to handle substantial grants, it could not engage in the kind of large programs being conducted by AAAS and other scientific organizations working on educational problems. Although it had recently published a highly regarded report on the desirable qualifications for science and mathematics teachers (24), the committee devoted most of its October 1962 meeting to a soul-searching discussion of whether it should restructure its activities to serve some fewer selected functions more effectively or should go out of business (28). It chose to continue, but in 1970 the committee proposed to the AAAS board that it merge with or become the AAAS Commission on Science Education. Sadly, that arrangement did not seem feasible to the board, for the members of the Cooperative Committee were appointed by too many different organizations for too many different reasons to constitute the well-focused body the Commission had to be in order to carry out its responsibilities (29).

A year later the Cooperative Committee decided it had served out its usefulness and decided to disband (30). During its 30 years the 157 individuals who had served as members had been selected by 27 different organizations. They and the organizations they represented could take much satisfaction in the fact that other agencies and institutions had assumed responsibility for working on most of the problems that had earlier engaged the attention of the Cooperative Committee. It could go out of business not because it had failed but because it had succeeded.

## The New Commission Starts To Work

As the Commission on Science Education began to develop its own programs it also continued some of the activities of STIP. For example, because new materials for teaching science and mathematics at the high school level were being developed by several of the course content improvement study groups and because thousands of teachers were attending summer institutes or taking advantage of in-service opportunities to learn more about science and mathematics, it seemed desirable to give school administrators opportunities to learn about the improved teaching materials, some of the developments in science, and the reasons why science teachers needed better grounding in their teaching fields. During 1962 NSF provided funds that enabled AAAS to hold conferences in nine cities around the country for a total of 1,200 high school principals and city school superintendents. Those conferences were jointly planned by AAAS, NSF, and the U.S. Office of Education (31). At each, teams of representatives of the course content improvement projects demonstrated and discussed their materials, and scientists and science educators addressed the needs for improvement in science education. A report of the conferences (32) was distributed to 50,000 school administrators (33).

As a follow-up to those regional conferences, in 1964 AAAS and the American Association of School Administrators arranged a seminar in Washington, D.C., at which five interesting and articulate scientists discussed with an audience of school administrators some of the new and exciting developments in their fields of science. That meeting received such a favorable response from those in attendance that NFS agreed to continue the program with two seminars a year, one in Washington, D.C., and the other at the Pacific Science Center in Seattle. The program continued, on a yearly basis, until the 1970s (34).

After beginning its support of summer institutes for teachers of science and mathematics in secondary schools, NSF in 1959 added a program of fellowships to assist selected applicants to attend some of those institutes. In that year there were 1,578 applicants, far more than could be awarded fellowship. AAAS convened 26 three-member panels to review applications and to select the 628 to whom NSF fellowships were awarded (35). AAAS continued to convene selection panels for this purpose through 1964.

The Commission on Science Education also started several new activities, such as development of "Science — A Process Approach," a whole new set of materials for teaching science in the elementary grades. In guiding that and other activities the Commission's primary responsibility was to AAAS, but it also played a wider role, as the Cooperative Committee had done. By inviting representatives of other groups working on improvements in science and mathematics education to attend some of its meetings, by regularly having representatives of other groups working on improvements in science and mathematics education attend some of its meetings, by regularly having representatives of the National Science Foundation and other government agencies at those meetings, and by publishing *Science Education News*, it provided a forum for considering interests shared by many of the groups working on the improvement of education. All of that gave the commission a busy schedule. It met about four times a year, usually for two-day meetings, thus devoting as much meeting time a year to educational affairs as the board of directors devoted to its responsibilities. In fact it was the board of directors for the association's educational activities, for although it reported regularly to the AAAS board and although the AAAS board named its members and approved its budgets, it was the commission that planned and guided educational activities (36).

The commission had a series of able and devoted chairmen: Paul B. Sears, Leonard Rieser, Robert B. Livingston, John A. Moore, and Albert V. Baez. John Mayor, who had directed the Science Teaching Improvement Program, continued as director of the commission's programs. To help him handle the expanding program of activities, in 1963 Arthur Livermore, a chemist on the faculty of Reed College, came to AAAS as deputy director of the Commission on Science Education's program. In 1971 and 1972 Livermore took leave of absence for 18 months while he and Joseph Dasbach, who is still a member of the education staff, worked on a science education project in Malaysia. When Mayor retired in 1974, Livermore succeeded him as director of the association's educational activities, a position he held until his own retirement a few years later (37).

The major new programs initiated by the commission will be reviewed later, but first some other educational programs that developed in parallel with STIP and the commission on Science Education will be described.

## *Traveling Science Libraries*

One day in 1955 I had lunch with Harry Kelly, the National Science Foundation's assistant director for scientific personnel and education; Bowen Dees, his deputy; Samuel Brownell, United States commissioner of education; and Ralph Rackley, the deputy commissioner. The reason: to consider the state of school libraries in the United States. A typical school library might have multiple copies of *Ivanhoe* and other classics of British and American literature, but in only a few school libraries could more than six or seven percent of the holdings be classified as scientific and some of those were quite out of date.

Kelly and Dees thought students should have access to more science books than their libraries typically provided and suggested a plan: If the commissioner of education approved the idea, and if AAAS would manage the program, NSF would provide funds to send traveling libraries of books on science and mathematics to a number of the nation's high schools. Brownell and Rackley approved, I promised AAAS cooperation, and the traveling science libraries program was born.

It would be satisfying now to be able to give credit to whoever first proposed the idea. It probably originated within the NSF staff, but no confirming file memorandum or correspondence has been found in the NSF archives (38). There were some precedents: County libraries, books by mail, and bookmobiles have all been used to serve rural and remote areas, and farther back in history, in 1839, the American Society for the Diffusion of Useful Knowledge prepared a traveling library called "The American School Library" and sent sets of those books to schools in frontier regions (39). Whether based on such precedents or not, someone had the bright idea of the traveling science libraries and started a chain of activities that gave hundreds of thousands of young people access to collections of books that they would not otherwise have seen.

The goals of the traveling science libraries were four: to develop among students the habit of reading good books in science and mathematics, to stimulate the choice of a career in science by those with scientific aptitude, to afford science and mathematics teachers an opportunity to broaden their reading and enliven their courses with up-to-date materials, and to facilitate the acquisition of well-chosen science books by school and public libraries. (40).

We were fortunate to find an excellent leader for the program. Hilary Deason, an ichthyologist who had spent most of his career in the U.S. Fish and Wildlife Service, had retired a few years earlier to become a consultant on population problems to the Diocese of Washington. Deason went to work immediately to select books on a variety of scientific subjects that seemed likely to be interesting and informative to high school students. With the help of catalogues, representatives of trade book publishers, and especially Margaret Scoggins, coordinator of young adult services of the New York Library System, a tentative list was prepared and sent to all 421 members of the AAAS council for their evaluation, criticism, and additions.

By those means 150 books were selected, and in the try-out year of 1955–1956 those 150 were sent to 66 high schools (41). Circulation records and questionnaires filled out by students in the program schools indicated that 38 of the 150 titles had been used but little and should be replaced. They were, and 50 additional titles were added, bringing the total to 200 books for the second and all subsequent years. That year also saw adoption of what became the standard routing practice. Each school received 50 books at a time and after about two months sent those 50 to another school and received a new set of 50. Thus in any year all of the schools received the same 200 books but in different orders.

The traveling libraries received much publicity, partly through exhibits at meetings of the National Science Teachers Association, the American Library Association, and other organizations. The Library of Congress featured them in a special exhibit. Many of the summer institutes for science and mathematics teachers had the books on display for examination and use.

As more schools learned of the opportunity, more wanted to participate, and the number of high schools served grew from 66 the first year to 104 the second, 216 the third, and 1,301 the fourth. Then for the next three years the number remained at about 1,650 a year (42). Although some schools were included in the distribution for more than one year, in general there was a new list of recipients each year.

All told, in its seven years of operation the traveling high school libraries were sent to 5,713 public and private secondary schools in all 50 states, the District of Columbia, Puerto Rico, and the Panama Canal Zone. Sets of the books were also sent to 73 Armed Forces dependent schools at various foreign bases, six schools operated in foreign metropolitan areas for children of U.S. nationals stationed abroad, 43 public libraries in the United States, and 301 summer and academic year institutes for science and mathematics teachers.

Enthusiastic letters of appreciation came from teachers, principals, librarians, and sometimes students. Yet in most high schools most students did not use the library. On the average one book was withdrawn for every two students in the high school. Some students, however, made much use of the library. In one representative year 61 percent of the responding students read none of the books; 23 percent read one or two; 11 percent read three or four; 3 percent read six to ten; and 2 percent read more than ten (43). Student use of the library was closely related to the amount of teacher interest, with more students using the library at schools where teachers themselves checked out a number of books or were reported by students to have made helpful suggestions of what to read. Every student in one school read at least one of the books, and at half a dozen other schools where the teachers were particularly active in arousing interest in the traveling libraries between 59 and 87 percent of the students read one or more of the books. From other schools, in contrast, came letters saying the students had too many activities and little or no time for reading.

Each year as new books became available or as circulation records indicated that a few books were rarely withdrawn it was necessary to make some changes in the list. Thus the annotated catalogue of books in the library had to be brought up to date each year. Those catalogues became popular far beyond the schools that actually received the traveling library. Thousands of copies were requested by school librarians, colleges, and community libraries. Originally published in 1955 in an edition of 3,000 copies, the 1956 edition numbered 9,000, and the 1957 edition, 22,000.

Reviewing new books, selecting prospective replacements, and some of the other work could be spread throughout the year, but each summer there was a high peak load. Some of the books needed physical repair. Some had to be removed and new ones added. New sets had to be assembled. New circulation cards had to be prepared for each volume. The wooden cases that did double duty as display cases while the books were in a school and as packing cases while in transit sometimes needed repair. Routing schedules had to be made out for the several hundred sets involved. Each summer AAAS needed an annex — an empty store building or some similar space near the AAAS office—and each summer there were jobs for several dozen high school students. In the peak year of 1959 there were 75 on the annex staff. As the years passed many of those summer helpers became college students and a few of them were hired as foremen for the summer. Deason was a good mentor and supervisor for those summer helpers and the annex was a lively place to visit, for one found an industrious and generally happy group of teenagers getting the traveling library ready for the coming year.

By 1959 results of the high school traveling libraries were so encouraging that NSF and AAAS decided to extend the program to the elementary school level. For that purpose 160 new books were chosen and starting in 1959–1960 were sent each year to about 800 elementary schools, with 80 in each school for the first half of the year and the other 80 in the second half (44).

Requests for sets of those books came from 3,000 schools and student interest turned out to be greater in elementary schools than in high schools. In high schools, as mentioned before, one book was checked out for every two students; in the elementary schools the ratio was three books for every two students (45). Part of the difference may have been due to differences in the way the schools were selected. The high schools were broadly representative of the whole range of high schools in the United States, large and small, city and country. Elementary schools were required to have an organized library so there would be a central location for display and a librarian responsible for making the books available.

Although the traveling libraries usually served schools for only one year, the effects were often more lasting. At the beginning of the 1958–1959 year schools on that year's schedule reported that on average they already owned 30 of the 200 traveling libraries titles. Two years later the same schools owned an average of 51 of the same 200. Several schools had bought all on the list that they did not already own, and in a couple of cases a local benefactor bought the entire set for the school. The reports from schools were supported by reports from publishers of suddenly increased sales of titles in the library. In fact we put a number of good books out of print.

Influence of the traveling libraries also spread far beyond the schools that were directly involved. The annotated *Books of the Traveling High School Library* and the comparable catalogue of the elementary school libraries were widely popular, but those lists included only a fraction of all of the science and mathematics books that could be recommended for school libraries. Accordingly, with the advice and help of many scientists who reviewed, criticized, or extended preliminary lists in their fields, some 900 titles recommended for high school and public libraries were chosen for inclusion in *The AAAS Science Book List* (46). From an edition of 75,000, free copies were sent to every superintendent of schools and high school principal in the United States, thanks to the cooperation of the Council of Chief State School Officers and the state departments of education.

In 1960 a similar annotated list of over 1,100 titles recommended for elementary school libraries and for children's divisions of public libraries was published as the *AAAS Science Book List for Children* (47) in an edition of 65,000 copies, of which about 50,000 were distributed without charge to elementary schools, again with the cooperation of the Council of Chief State School Officers and the state departments of education.

Endorsement of both of these lists by the Council of Chief State School Officers and the Library Services Branch of the U.S. Office of Education quickly made them the standard guide for the purchase of science and mathematics books (other than textbooks) under the provisions of the National Defense Education Act of 1958. Both lists continued to be popular even long after the traveling libraries themselves were terminated, and new editions were published in the 1970s and 1980s (48).

Popularity of the lists of books in the high school libraries plus the cost of hardcover books led one of the summer helpers to suggest that AAAS publish a catalogue of recommended paperbacks in science. The resulting *An Inexpensive Science Library* was first published in September 1957. Within several months all 22,000 copies were gone. A new edition of 50,000 lasted for awhile but new editions were needed every year until 1961. A single unsolicited mention of the list in *U.S. News and World Report* brought some 7,000 requests for copies and other news reports of its availability brought thousands of other requests. Annual editions of 50,000 were all sold out.

In 1961 Deason made arrangements with the New American Library of World Literature — a publisher of many popular paperbacks — to publish the association's list of paperbacks. He compiled an annotated list of about 1,100 science and mathematics books, added Warren Weaver's AAAS presidential address of 1955, and essays by Margaret Mead and Rhoda Metreaux, H. Bentley Glass, and Joseph Gallant, all of which showed the need for greater general science literacy. The result was *A Guide to Science Reading* (49). In reviewing that guide Isaac Asimov wrote, "To anyone interested in science education for himself or for others (and I presume that includes all of us), the book is as essential as a knife and fork at a steak dinner" (50).

In passing it might be noted that the AAAS book lists were not the association's first effort to provide guidance for reading about science. In the early 1930s, with a grant from the Carnegie Corporation of New York, AAAS compiled short bibliographies in 27 different fields of scientific interest and sold them for five cents a copy.

Of a different nature from the book lists was a 12-page leaflet entitled *A Selected List of Career Guidance Publications* that had originally appeared as an appendix to the annotated catalogue of the traveling high school libraries. AAAS used it to answer letters from students and teachers. The first edition of 12,000 copies ran out and a new edition was needed within a couple of years (51).

The sales and distribution figures indicate that the AAAS book lists filled a widely recognized need, and other evidence showed that their use went well beyond elementary and secondary schools. The Veterans Administration adopted those lists as acquisition guides for libraries in VA hospitals and rehabilitation centers. The U.S. Navy sent copies to some 300 Navy stations for use in acquiring books for Ships' Service units. They were used by the departments of education of several states and by several TV stations as recommended lists.

The Asia Foundation bought several sets of the traveling libraries to donate to teacher training institutions in other countries. UNESCO placed a set in the International Pedagogical Center at Sevres (52). At the 1962 World's Fair in Seattle the United States Science Exhibit was far and away the most popular attraction on the fairgrounds. Its souvenir shop sold only one kind of item: science books selected by Hilary Deason and his staff. The University of Washington's student-owned bookstore managed sales and donated the profits to the University's scholarship fund (53).

Despite all of these evidences of usefulness, in 1959 NSF and AAAS agreed that the traveling high school libraries should be brought to an orderly termination in 1962. As NSF later reported, "Widespread distribution of the traveling libraries has so encouraged the acquisition of science books by school libraries that further Foundation support of this phase of the activity is no longer necessary" (54). Two years later, in 1964, the elementary school libraries were also terminated.

Over the nine years of the program—seven for the high school libraries and five for the elementary school libraries—the NSF supporting grants totaled $1,898,300. As a result school libraries added several hundred thousand volumes to their previously meager holdings in science and mathematics; some two million students read books that they would not otherwise have seen; and nobody knows how many additional books were read or purchased by the hundreds of thousands of teachers, librarians, high school and college students, and others who bought or received one or more of the book lists published by the libraries program. There was no way to know the full effect of the program, but the available evidence left both AAAS and NSF with the satisfied feeling that the program had been altogether worthwhile.

Termination left AAAS with well over 100,000 books to dispose of. Retrieving them from the field, warehousing them, and handling sales would have cost a considerable amount, and neither NSF nor AAAS wanted to go into the secondhand book business on a large scale. Instead, we decided to give the books away. Sets went to NSF, state and territorial departments of education, public libraries in small communities, UNESCO, Armed Forces and other schools abroad, and to universities and colleges that wanted sets for teacher training institutes and curriculum material laboratories. But most went to the schools where they were last on display; we simply notified many schools that they might keep that portion of the total set.

Termination of the traveling libraries and their associated book lists created a problem for schools. The many efforts to improve instruction in science and mathematics had led to an increased use of collateral reading in science courses and school librarians were interested in securing more and better books for general reading. However, many schools were not permitted to purchase books that had not been approved by some recognized authority. There were other book selection aids, but none was as complete, up to date, and highly regarded in science and mathematics as the AAAS lists. All over the country those lists had become the standard source of information about books appropriate for purchase. Deason thought that AAAS should continue to meet the need for that information. The Committee on Publications and the board of directors agreed, and so a new quarterly publication *Science Books* was born (55). Intended primarily to give librarians and teachers brief descriptions, evaluations, and indications of the appropriate age level of new books on science and mathematics, that publication also turned out to be useful to some other subscribers. Later retitled *Science Books & Films* when reviews of science films were added, it now appears five times a year and continues the heritage of Hilary Deason and the traveling science libraries.

## *The Certification of Teachers*

One of the early interests of the Cooperative Committee on the Teaching of Science and Mathematics was the education of prospective teachers of those subjects and the qualifications required for their certification (56). The Science Teaching Improvement Program continued that interest and collaborated with the American Association of Colleges of Teacher Education in forming a Joint Commission on the Training of Teachers of Science and Mathematics. That commission included four members from each of its two parent organizations plus one member each from the Association of American Colleges, the American Council of Learned Societies, the National Council of Teachers of Mathematics, the National Science Teachers Association, the U.S. Office of Education, and the National Science Foundation, which financed the work of the Joint Commission.

That membership illustrated a major premise of the whole AAAS educational effort: the problems of improving science education could not be solved by scientists working alone. Everything that went on in the school room was the business of school teachers and professional educators; cooperation with those groups was therefore essential. The Joint Commission adopted that premise and admonished scientists and mathematicians, members of the education profession, and classroom teachers all to become more cooperative, for

> Scientists have sometimes been ... skeptical of the need for much, if any, professional preparation for teachers. Educationists have often complained that scientists are neither informed about nor really interested in the problems of secondary school teaching, and not realistic in course requirements; and classroom teachers have often criticized both groups as being ignorant of the problems with which teachers must cope (57).

Having agreed on the need for three-way cooperation the Joint Commission went on to propose a program to examine experimentally a variety of modifications of the ways in which teachers were being trained. The AAAS board approved the ambitious plan and authorized me to find the necessary half-million dollars. Although I did not find such a grant, the spirit of the proposal remained alive and some of the suggestions were later carried out as parts of the AAAS educational program.

As was often the case the Cooperative Committee was working along a parallel track. Early in 1959 that committee released a set of recommendations for improving the preparation of high school teachers of science and mathematics (58). They recommended that about half of the undergraduate credit hours of a prospective teacher of science or mathematics be in those disciplines, leaving the other half for courses in the humanities, social sciences, and professional education. The American Chemical Society sent copies of the report to all its local branches; the 1959 National Conference on Teacher Education and Professional Standards used the report as a major basis for its discussions; and the *New York Times* gave it favorable review (59).

Both professional and public interest in improving education and the preparation of teachers surged upward as part of the post-*Sputnik* worry about the state of science and technology in the United States compared with the Soviet Union. As part of that wave of interest the National Association of State Directors of Teacher Education and Certification (NASDTEC), in part based on recommendations made by John Mayor in addressing their annual meeting, decided that they needed an authoritative definition of the subject-matter content of curricula appropriate for preparing future teachers of science and mathematics. NASDTEC was a tiny organization, yet from the standpoint of strengthening the preparation of prospective teachers it was perhaps the most influential single organization in the country, for its membership consisted of the one person in each state who was most directly responsible for overseeing compliance with the state's requirements for the education and certification of school teachers. Because those directors had that responsibility, they also had much influence in establishing the certification requirements. Unlike AAAS or the Cooperative Committee—which could only recommend what someone else should do in setting certification requirements,— NASDTEC was asking its own members what they should do to improve the education of prospective teachers of science and mathematics.

NASDTEC had already established cordial working relationships with Mayor and his staff and used the AAAS board room to prepare a grant proposal for a national study of certification requirements and how those requirements should be strengthened. They sent their proposal to the Carnegie Corporation of New York and named as the study's prospective director William P. Viall, chief certification officer for the state of New York and president of NASDTEC.

The Carnegie Corporation was sympathetic, but had a problem. NASDTEC was obviously an appropriate organization to conduct the study, but Carnegie could not make a grant to an unincorporated organization. In that situation, Frederick Jackson of the Carnegie staff asked Mayor and me if AAAS would be willing to sponsor the study. If we were willing, he suggested that Mayor be named director of the study with Viall as associate director. Both agreed; the AAAS board approved; the Carnegie Corporation made a grant of $81,000 to AAAS; Viall moved to Washington; and what gradually developed into a series of three national studies of teacher certification began on December 1, 1959 (60).

The objective of the first study was to describe the science and mathematics courses that should be included in the undergraduate programs of prospective high school teachers of those subjects. The state accreditation officers could then use those standards to judge college programs for the education of those teachers. When a college had an "approved program," individual students who had complied with the requirements of that program could be recommended for accreditation or licensure.

To help Viall carry out the work, NASDTEC appointed an advisory committee consisting of five NASDTEC members, one representative of the Council of Chief State School Officers, and two scientists recommended by AAAS: Edward G. Begle, professor of mathematics at Yale University and director of the School Mathematics Study Group, and H. Bentley Glass, professor of biology at the Johns Hopkins University and chairman of the Biological Sciences Curriculum Study. Working with the advice of that committee and members of the association's education staff, Viall reviewed previous recommendations made by the Cooperative Committee and other bodies; arranged conferences in different parts of the country; and conducted many consultations with scientists and mathematicians, school teachers and administrators, and college and university teachers of education. On that basis he developed guidelines that were sent to many critics for review and then, as amended, were approved by NASDTEC and recommended to the states for adoption (61). The guidelines were widely publicized by scientific and educational organizations and government agencies and 17,000 copies were soon distributed in response to requests from government offices, colleges, other organizations, and individuals. Pennsylvania, the first state to adopt the guidelines, was quickly followed by others. Eighteen months after publication the guidelines had been adopted or approved, wholly or in part, by 26 states and Puerto Rico and were in process toward probable adoption in most of the other states. In addition 170 colleges and universities were known to have adopted the guidelines for their own programs (62).

As that study was nearing completion, NASDTEC and AAAS decided that it would be desirable to develop a comparable set of guidelines describing the education in science and mathematics that should be acquired by prospective teachers at the elementary level. The Carnegie Corporation agreed and in the spring of 1961 made a second grant to enable Viall to remain at AAAS for two more years while the new guidelines were being developed. In doing that he followed the same procedure he had used before of reviewing existing recommendations, scheduling conferences, and holding consultations with scientists, mathematicians, teachers, school officials, and educators. The resulting guidelines for the education of elementary school teachers were published in 1963 (63).

A third activity of the NASDTEC–AAAS program was an NSF–supported study of the actual, as distinct from the desirable, qualifications of teachers of high school science and mathematics. For that purpose a questionnaire was sent to a stratified random sample of nearly 4,000 teachers who were actually teaching high school science or mathematics classes in the spring of 1963. A 76 percent return provided a good basis for concluding that many of the teachers did not have the qualifications called for by the guidelines; that most of the teachers of science, but not of mathematics, were teaching two or more different subjects; and that most of the people teaching physics and chemistry were primarily teachers of some other subject (64).

With that survey completed and the two sets of guidelines being widely adopted, Viall wrote an account of the work for all members of AAAS (65) and left to accept a new position as professor of education at Western Michigan University. He soon learned, however, that he was not finished with the development of educational guidelines. Motivated by the NASDTEC–AAAS collaboration, teachers of English and of modern foreign languages decided that their fields also needed sets of guidelines for teacher preparation. AAAS was not the right sponsor for those studies, but William Viall was surely the most experienced director they could find, and after leaving AAAS he conducted both of those studies.

Nor was AAAS finished with preparing guidelines for the preparation of elementary and secondary school teachers. During the 1960s new materials for teaching science and mathematics were developed by the course content improvement projects. Concurrently the National Council for the Accreditation of Teacher Education was changing some of its standards. There was also a significant shift in the thinking of many educators as to what the guidelines should attempt to do. Emphasis of the earlier NASDTEC–AAAS guidelines had been on what teachers should know. By the end of the 1960s there was a substantial body of opinion that new guidelines should be phrased in terms of what teachers should be able to do. Accordingly, AAAS with David Ost of the University of Iowa leading the work, prepared a new set of guidelines for the preparation of elementary school teachers (66). A little later AAAS and NASDTEC, with Dudley Herron of Purdue University responsible for leadership, prepared a new set for high school teachers of science and mathematics (67). Both studies were supported by the National Science Foundation and both were carried out under the guidance of the association's Commission on Science Education.

## *Visiting Foreign Scientists*

Visits by foreign scientists have often been a useful and pleasant means of exchanging information and ideas and of cementing the international character of scientific work. It was within that tradition that the National Science Foundation began in the late 1950s to aid several organizations to bring foreign scientists to the United States for brief visits to some of the summer institutes for science and mathematics teachers that NSF was supporting.

NSF liked the program and so did many of the directors of the institutes that were visited, but the whole program was having troubles. Looking toward future improvements, in 1960 NSF invited that year's visitors, some of the institute directors, and several consultants to gather at Estes Park, Colorado, at the end of the summer to review the program. I was one of the invited consultants and it did not take long to find out why: NSF wanted AAAS to select all of the future visitors, orient them to the program, make arrangements with the institutes to be visited, and handle the details of the visitors' travel, schedules, and reimbursement.

Under the earlier arrangements one organization had selected the chemists, another the biologists, and so on. There was little coordination among the sponsoring organizations, so one campus might have two or more visitors at the same time while most campuses had none. Other criticisms were that the visitors were not adequately informed about the nature of their audiences or how they could be most useful, while the institute directors did not get timely information about the visitors and their special interests and so could not make best use of the short time the visitor was available.

Yet NSF considered the whole program valuable and wanted to continue it. The summer institutes were popular and hundreds were being offered each year. A few days' visit by a scientist from far away was an added attraction to the regular program. The visitors could give interesting lectures on scientific topics and were sometimes asked to describe the educational systems of their own countries, a topic of special interest to many teachers. Most institute directors who had received a visitor wanted to have another the next year. As for the visitors, with great unanimity they reported that while they sometimes had doubts about how valuable they had been to the institute participants they were certain that they themselves had had a profitable and enjoyable summer.

So NSF turned to AAAS, hoping we would consolidate the several programs and continue and improve the whole. When the board of directors first considered this request they regretfully decided to decline it. The staff was already busy enough. Foreign relations was more the business of the National Academy of Sciences than of AAAS and the academy was responsible for other programs of scientific exchange.

Overnight, however, there were some second thoughts. The next morning the board reconsidered the request, discussed the fit between the new program and the rest of the association's efforts to improve science education, decided that the association should become more active on international problems of science education, and agreed that AAAS would accept the request if adequate help could be secured to handle the additional work load (68).

With the summer of 1961 only eight months away the first task was to decide about the kind of people that we wanted to recruit. Discussions with NSF staff members, institute directors, and some of the visitors led to formulation of four criteria to be used in selecting future visitors:

1. Sufficient knowledge of some field of science or mathematics to be able to lecture with confidence and authority;

2. Sufficient knowledge of secondary education to be able to select topics and prepare lectures of interest to secondary school teachers;

3. An adequate English vocabulary and sufficient freedom from accent to permit easy communication with an audience largely unaccustomed to anything but American speech; and

4. Ability to get acquainted easily and quickly with the members of a succession of new groups of teachers.

Those criteria were emphasized rather than the kind of institution in which a visitor was employed. Some came from universities, some from teacher training institutions, and some were science masters at secondary schools. Visitors who attended the Estes Park meeting strongly recommended that scientists engaged in training teachers or who were exceptionally well-qualified teachers in secondary schools would often be better adapted to the requirements of the program than would many university professors. That proved to be good advice.

For help in identifying good prospects I called upon colleagues with whom I had worked on matters of science education under the auspices of UNESCO or the Organization for Economic Cooperation and Development (OECD): for example, Hans Lowbeer, deputy minister of Educational and Ecclesiastical Affairs in Sweden; Robert Major, director of the Institute of Scientific and Industrial Research in Norway; Joseph Lauwerys of the University of London; and Francis Parkinson of the Pan-American Union in South America. To each of those advisors and others in several other countries we sent a description of the program, the criteria we planned to use in selecting visitors, and a plea for help. Help was generously given and in each of several countries the colleagues on whom I had called enlisted the aid of other scientist-educators in identifying good prospects.

I then took advantage of attending a meeting in London of an OECD committee to travel to several Western European cities to interview prospective visitors and found enough to staff the program for both 1961 and 1962.

Thereafter biennial visits to Western Europe (usually with William Kabisch, who assumed much responsibility for this program after he came to AAAS at the end of 1961), correspondence, contacts made at international meetings, long-distance telephone calls to test ability in English, and — very importantly — the generous help of several former visitors whose enthusiasm for the program made them effective recruiters in their own countries enabled us to identify a good supply of well-qualified visitors of diverse interests. Most came from Western Europe but others came from India, Pakistan, Peru, Turkey, and elsewhere. One of the most distinguished came from Australia.

Management of the program rather quickly settled into a regular pattern. At the beginning of the summer the visitors came to Washington, D.C., for an orientation conference where they were given information about the American educational system, the preparation of secondary school teachers, and the kind of institutes they would visit. Two or three of the visitors gave the lecture they had planned to use at the institutes that they would visit. Those "guinea-pig" lectures were then critiqued by the other visitors and by the institute directors and other Americans at the orientation conference. As the conference came to an end each visitor was given a detailed schedule of the six to 10 institutes to be visited, information about how to get to each, the necessary railroad, airplane, or other tickets, an insurance policy, and whatever special information that each wanted or we thought would be useful.

The visitors varied in age and prominence and of course in salary and earnings. We did not try to take these differences into account in setting honoraria but instead established a standard fee. We did vary travel arrangements, however, to try to make them as pleasant as we could. If a visitor wanted to travel by sea that was fine, for many passenger liners were still crossing the Atlantic. If airplane travel was preferred, that was arranged. Once here, priority had to go to matching a visitor's field and special competence with the requests from institute directors, but in so far as we could we arranged itineraries to suit visitors' preferences. Some who were fairly familiar with the East Coast were sent to Western states. One who had dreamed of a train trip through the Colorado Rockies was routed that way. Some wanted to visit a particular institution or former colleague. Side trips were always at the visitor's expense, but we helped make the necessary arrangements to make the experience a pleasant one for the visitors. From what they told us and said to later visitors, we usually succeeded.

The institute directors also tried to make the visits interesting and useful, sometimes at considerable inconvenience. One visitor with a free weekend at the University of Idaho was asked by the institute director what he would like to do on Sunday. Not yet really acquainted with American distances, he said that he had long heard of Crater Lake and would enjoy a drive to see it. Swallowing hard, the director warned that they would have to get a very early start. They did, saw Crater Lake, and long after dark got home from a round-trip drive of well over a thousand miles.

There were other surprises. Some years later, R. V. Jones, professor of physics at Aberdeen University whose honors entitled him to write CB, CBE, FRS after his name, wrote to me that at the first college visited he was asked to speak at a college assembly as well as to the institute. Obviously not knowing that CBE stood for Companion of the Order of the British Empire the college president flamboyantly introduced him as "the champion of the British Empire" (69).

In preparing for each season's tours the directors of the several hundred institutes scheduled for that summer were all given an opportunity to let us know whether they wanted a visitor, and if so with what special interests. Most directors did not, but upwards of 200 usually did. It was never possible to fill all of the requests, but visitors were sent to about 135 institutes each summer. Each visit lasted from two to five days, depending upon the director's wishes, and in the six-week period a visitor usually got to about eight different institutes.

Evaluation of the program was only by judgment and impression. Most institute directors did not try it. Of those who did a large majority expressed enthusiasm and asked for another visitor for the following year; a minority expressed more restrained approval; and a few were disappointed, either with the particular visitor or for some other reason.

The visitors were enthusiastic about their experience. They admired the summer institute program. They hoped that they had added something useful to the institutes they had visited. They knew they had had an interesting summer. Some wanted to come back, and a few for whom we had received particularly favorable accounts were brought back for a second summer. Some returned later under other arrangements and as late as 1984 one was still making periodic visits to the United States to continue a collaborative research project that he had started with the director of one of the institutes he visited (70).

The first year that AAAS managed this program, 1961, was a trial year for both NSF and the association. Responses were so favorable that agreement on continuation was quickly reached, and the program was continued until into the 1970s. Some activities leave a good feeling all around and are fun to manage. This was one.

## Science for Elementary School Students

In the 1950s, in the midst of much concern over the nation's schools, many scientists and mathematicians believed that the most rapid and effective contribution they could make to improving precollege education would be to develop new, more accurate, and more interesting texts and other teaching materials. To carry out that idea groups of scientists and teachers began working together to develop new materials that would be available to any high school teacher of science or mathematics who wanted to use them.

By 1960 that interest had spread from the senior high school and college level to the earlier grades. As described earlier in this chapter, AAAS and the National Science Foundation's interest in this movement merged in the three regional feasibility conferences that led AAAS to appoint the association's Commission on Science Education and led NSF to support several programs designed to develop new materials for teaching science in the early school grades.

As those several programs developed they followed most of the recommendations of the AAAS "Feasibility Study." But one recommendation could not be wholly complied with: the recommendation that a national correlating and stimulating body be established to guide the several programs. NSF, the financial supporter of all of those programs, could not delegate that responsibility to some ad hoc body or to one of the program groups. Nevertheless, informal coordination did develop. Communication among participants in the several groups and meetings with the AAAS Commission on Science Education helped insure that different ideas, materials, and procedures of the several groups were all directed toward the objectives that had been emphasized in the three regional conferences.

One of the groups developing new materials for teaching science in the elementary grades was AAAS itself. With financial support from NSF, and under the continuing guidance of the Commission on Science Education, the association started its elementary science program by convening two working conferences in the summer of 1962, one at Cornell University and one at the University of Wisconsin. The scientists, teachers, science educators, and school administrators at those conferences were joined by representatives of several of the already experienced groups working at the high school level. Together they reviewed the work of a number of elementary and secondary school curriculum projects and the results of work on science education and then went on to make specific recommendations concerning objectives, scope and sequence of instruction, and laboratory or other experiences appropriate for elementary and junior high school students (71).

Both conferences agreed that the most desirable outcome of science instruction at the elementary level was not knowledge of the facts of any particular science, but rather an understanding and feeling for the processes of science. Children should learn how to observe, compare, contrast, measure, draw inferences, and make and test predictions. Through such experiences they would not only learn how scientists work but would also learn valuable lessons about how any intelligent person should approach and gain an understanding of natural phenomena.

Thus was conceived "Science — A Process Approach," the major AAAS contribution to instructional material for children in kindergarten and the first six grades of school. "Science — A Process Approach" (or SAPA, for the initials soon became the familiar acronym) consisted of a kit of material for children to use and written material to guide the teachers in using those materials and appraising the outcomes of their use.

As the first step the commission drew up several working papers to define in more detail the basic philosophy of the effort (72) and the kinds of material that were desired, namely, an orderly progression of experiences and experiments that would build upon earlier ones, that would be aimed toward specific and measurable objectives (73), and that would develop skill or competence in observing, measuring, communicating, inferring, and predicting (74).

In the summer of 1963 some 25 scientists, elementary school teachers, educational specialists, and child psychologists spent eight weeks at Stanford University preparing the first version of the new materials for kindergarten and grades 1–3. In the following school year those materials were tried out by 106 teachers in 12 school systems, with each tryout school advised and monitored by one member of the commission or of the 1963 writing team. In the summer of 1964 the materials were revised to take account of the first year's experience with their use. Although never considered completely finished and always subject to revision and improvement, the usual cycle was write, tryout, revise, tryout again, revise, and then maybe another tryout and revision. Under this schedule the 1965 summer writing conference at Michigan State University constituted a peak of activity, for the K–3 materials were being given their second revision and materials for grades 4–6 were being developed or revised. During that summer 39 participants worked together for eight weeks and another 50 were there for part of the time.

As the materials moved through the tryout and revision process they reached a stage when commercial production and distribution became desirable. Following instructions from the National Science Foundation, AAAS informed publishers of the availability of the new materials; provided detailed information to those that showed interest; reviewed proposals from several publishers; and, with NSF approval, selected the Xerox Corporation as the commercial publisher (75). By 1970 all parts of "Science — A Process Approach" were available from the Xerox Corporation: kits of material for student use and guides and other instructional materials for the teachers (76).

Also available were the "competency measures," for SAPA included a built-in evaluation program consisting of tests of competency in using the processes at the level appropriate for each grade. In developing those measures the program managers insisted that achievement be demonstrated by some form of observable behavior. Words such as "understand" or "appreciate" were deliberately, and sometimes painfully, ruled out. Instead, the pupils had to name, to construct, to describe, to state a rule, or do something else that other people could observe and evaluate. Using such measures, the objective was that 90 percent of the pupils in a class should be 90 percent correct on those competency measures. In the tryout years, when AAAS was following classes most closely, that standard was reached, and because the program was one of learning by doing instead of by reading, that standard was reached in inner city schools as well as in schools of more affluent neighborhoods.

SAPA was one of three major new programs for teaching science in the elementary grades, the other two being the Elementary Science Study (ESS) of the Educational Development Center of Newton, Massachusetts, and the Science Curriculum Improvement Study (SCIS) developed at the Lawrence Hall of Science of the University of California at Berkeley. All three emphasized activity and discovery rather than reading. All three were supported by the National Science Foundation, and all three became available for general adoption at about the same time. In order to provide school administrators an opportunity to learn about those three programs and two others that were not as as popular, the AAAS Commission on Science Education and the American Association of School Administrators held seven regional conferences. Materials from all five programs were displayed and discussed, and the more than 700 elementary school administrators who attended had opportunities to examine and compare the new programs (77).

"Science — A Process Approach" was adopted by a number of school systems throughout the United States and by all schools in Europe for U.S. dependents. International interest in the program sent Henry Walbesser, a member of the education staff, to Montevideo, Uruguay, for two weeks of discussion with interested South American scientists and teachers. Fulbright funds supported that trip, and a few months later the Organization of American States took Arthur Livermore to Chile for three weeks of discussion of "Science — A Process Approach." Thanks to a Spanish translation of the materials, SAPA was soon adopted by some schools in Latin America. Later in 1968 Robert Livingston, then chairman of the Commission on Science Education, represented U.S. activities in science education at a conference on that topic sponsored by the Commission Interunions de l'Enseignement des Sciences held in Bulgaria (78).

The other two major K–6 programs were also adopted by numerous school systems, but none of the three ever captured a truly large share of the market. In 1979 a national survey of education in science, mathematics, and social studies (79) found the record of current and prior use of the three major elementary programs that is shown in Table 6.

All three of these programs were subjected to many comparisons by many investigators to determine whether they were more effective than the usual textbook and the read–recite–test–discuss method of teaching science that they

Table 6. Percentage of Teachers Using or Having Used Each of Three New Programs for Teaching Science in Kindergarten and Grades 1-6

| New Science Programs | Percentage of Teachers Using Selected Programs | | | |
|---|---|---|---|---|
| | Grades K–3 | | Grades 4–6 | |
| | Using in 1976-77 | Used prior to 1976-77 | Using in 1976-77 | Used Prior to 1976-77 |
| Elementary Science Study (EES) | 5% | 7% | 9% | 14% |
| Science — A Process Approach (SAPA) | 4 | 10 | 9 | 13 |
| Science Curriculum Improvement Study (SCIS) | 11 | 16 | 12 | 16 |

Sources: I. R. Weiss, note 79, p. 305.

were intended to replace. For both elementary and high school programs the collective evidence of 105 evaluative studies demonstrated that the new science curricula did indeed improve science education. As measured by a variety of criteria the average performance of students using one of the three new elementary curricula "... exceeded the performance of 63% of students in traditional science courses" (80).

All three of the programs showed advantages over more traditional methods of instruction, but the amount of improvement should not be overemphasized, for many of the differences were too small to be significant. Of 400 comparisons on different kinds of measures, about a third showed significantly greater achievement (at the .05 level of confidence or better) by pupils using the new curricula over those using traditional materials but in nearly two-thirds of the comparisons the differences were not statistically significant (81).

As its name was meant to indicate, "Science — A Process Approach" emphasized a particular kind of achievement: understanding and using the processes of science—observation, comparison, inferring, and so on. SAPA achieved that objective better than the other new curricula and much better than the traditional methods of instruction as evidenced by the measures of achievement summarized in Table 7 (82). The measure of effectiveness used in the table is the number of percentile points on a scale from 1 to 100 by which students using one of the new curricula exceeded those being taught by more traditional methods. Most of the achievement areas in the table are obvious. "Attitude" included attitude toward science generally as well as toward the particular course; "related skills" meant competence in reading, arithmetic computation, and communication.

Despite all the evidence that SAPA and the other two programs really were quite superior to the kinds of instruction they were intended to replace, the amount

Table 7. Percentile Gain in Performance by Students in Classroom Using SAPA, ESS, or SCIS Compared with Students in Traditional Classrooms

|  | Percentile Points Gained | | |
|---|---|---|---|
|  | SAPA | ESS | SCIS |
| Achievement | 7 | 4 | 34 |
| Attitude | 15 | 20 | 3 |
| Process skills | 36 | 18 | 21 |
| Related skills | 4 | * | 8 |
| Creativity | 7 | 26 | 34 |
| Piagetian tasks | 12 | 2 | 5 |

*No studies reported the use of these measures for ESS.
*Source:* James A. Shymansky, William C. Kyle, and Jennifer M. Alport, note 82, p. 305.

of use has dropped substantially since the 1970s. Some newer texts have adopted some of the ideas of SAPA, ESS, and SCIS and a few schools still use each of these programs. Perhaps one percent of the nation's elementary schools now use SAPA. But these programs cost more than textbooks, initially for kits of materials and consumable supplies and in later years for maintenance, repair, and replacement. They require more time for preparation and are more demanding of teachers' knowledge, understanding, and time. In many schools the teachers using SAPA or one of the other programs were left to their own resources to understand and teach the learning modules and to maintain, repair, and resupply the needed materials. Those disadvantages seem to have outweighed the fact that pupils learned more than did pupils using the traditional text and recitation methods. Thus we have reached a discouraging disparity: Studies and surveys show that children enjoy science centers and museums, science programs on TV, and science stories, yet there is widespread avoidance of science courses in junior and senior high school. The pupils who do take those courses are often faced with textbooks of the traditional type that contain only a faint echo of the interesting activities of the course content improvement programs of the 1950s and 1960s and instead require seemingly endless memorization; the most widely used seventh grade life science text, for example, contains 2,500 technical and unfamiliar terms (83).

However, something better may lie ahead for teachers and school systems willing to take advantage of the opportunity. SAPA is still given attention in some courses for prospective elementary school teachers — how the program was developed, its rationale, and its evaluation methods. Moreover, early in 1987 the National Science Foundation announced the award of three grants totalling $6.6 million in a new program supporting three multi-year programs to develop new materials for teaching science in the elementary grades. Ironically some of those new programs promise to revive some of the ideas, methods, and materials of SAPA, ESS, and SCIS (84).

In the meantime interested teachers or schools can still secure "Science — A Process Approach." After the second commercial edition was published, Ginn and Company, then a subsidiary of Xerox Corporation, abandoned SAPA. Delta Education, Inc., of Nashua, New Hampshire, obtained the rights to publish the program and assemble and sell the necessary kits and also secured the rights to the materials for the Elementary Science Study and the Science Curriculum Improvement study programs, so all three of those carefully designed and experimentally proven sets of materials are still available.

## Science Education for Junior High School Students

After work on "Science — A Process Approach" was well under way the Commission on Science Education turned to the topic of junior high school science. As a starter for the work, in 1965 the Commission held a three-day conference at Michigan State University at which representatives of all groups working on the development of course materials for junior high school students came together to discuss the problems of education at that level. One of the recommendations of the group was that the AAAS Commission on Science Education develop a new curriculum for teaching science to junior high school students (85). The Commission promptly took up that task and developed some promising plans, but those plans never materialized (86). Because staff members of the U.S. Office of Education had expressed interest, AAAS submitted a request for nearly $2 million for a five-year project to develop the proposed materials. Although all of the referee reports were favorable, the Office of Education had to reply that they did not have the funds in their 1967 budget to make the grant. Informally, members of the staff added that they hoped to support it in 1968 and therefore hoped we could find some other source of funds to get the project started.

Over the next couple of years the project continued to be discussed; plans were revised somewhat to make a better fit with some of the newly developed science courses for senior high school students; and in 1968 we asked the National Science Foundation to support the first year of the project. By then, however, the period of rapidly increasing federal support for research and education in science was coming to an end, and by 1969, as part of a general reduction of expenditures, the NSF had to give each of its grantees, including AAAS, an expenditure ceiling somewhat less than the sum of amounts already granted for currently active projects. With no support being obtained, the junior high school curriculum disappeared from the commission's agenda (87).

## *Holiday Science Lectures*

In the London of 1829 Michael Faraday started the Royal Institution's famous "Christmas Course of Lectures Adapted to a Juvenile Auditory." In 1959 Detlev Bronk, then president of Rockefeller University, transplanted that idea to New York by arranging a series of "Christmas Lectures" for an invited audience of New York high school students. With René Dubos and Paul Weiss, two excellent lecturers, giving the 1959 and 1960 series the program was an immediate success and that success led the National Science Foundation and AAAS to expand it to a national scale with a new name — Holiday Science Lectures — to indicate that lecture series would be given at Thanksgiving and Easter time as well as during the Christmas holidays.

From the beginning the Holiday Science Lectures differed from other AAAS programs on science education in one important respect. Other programs were intended to improve science education for all students. In contrast the Holiday Science Lectures were intended for high school students who had already demonstrated interest and superior achievement in science courses. Principals of the high schools in or near a city in which a series of lectures was to be given were therefore invited to nominate a specified quota of their best and most interested students, with the number depending upon the size of the school and the total who could be accommodated in the auditorium in which the lectures were to be given. Each school was also entitled to send a small number of science teachers. Edward G. Sherburne, Jr., who carried much of the responsibility for this program, then sent individual letters from AAAS to the home addresses of the selected students telling them of the program and inviting them to attend.

With attendance on an invitation-only basis, the speakers had to be carefully chosen, both for their scholarship and for their ability to communicate clearly and enthusiastically. They also had to be willing to spend a good deal of time in preparing the content and the supporting demonstrations and materials for a series of four or five lectures designed specifically for the audience. Because they did spend much time in preparation they were always invited months in advance and if the first series went well were asked to repeat their lectures later in another city.

Reaction to the very first of the AAAS series, five lectures in San Francisco by Paul Weiss, was typical of the responses to later series by other lecturers. Weiss wrote: "The audience kept growing, rather than falling off ... till there were hardly any spare seats. The last day, the kids gave me a touching extra ovation. Many came to thank me in person. Some told me they had changed their minds about college preference and were now going into biology ... On the whole it was very worthwhile, but I am pretty well fagged out from the strain" (88).

After every series we received enthusiastic letters from sponsors, school administrators, teachers, and sometimes students. The speakers reported that attendance held up well and sometimes increased, that questions and discussion after each lecture continued for as long as twice the scheduled hour, and that the hand-picked high school students constituted more alert and interested audiences than they usually had in college classrooms. Teachers wrote of projects started and enthusiasm generated among their students. And students wrote to the lecturers asking for reprints or other information and to AAAS to praise the lecturer or to say they hoped they would have a few teachers of comparable quality when they got to college.

Starting with two series of lectures in the first year the program quickly expanded to 10 or 12 each year. In each city a university, a scientific academy or society, or a science center or museum provided the necessary auditorium and other supporting facilities and services. The audience consisted of 300 to 600 students and 30 to 60 teachers. For most series the students and teachers all came from within commuting distance. In 1962, however, a residential series was tried out on an experimental basis. Paul Weiss, who had lectured in San Francisco the year before, went to Seattle for a series held in the new Pacific Science Center. His audience included 250 high school students from in and near Seattle and another 250 from elsewhere in the state or from Alaska, Idaho, Oregon, or Montana. The visitors lived in University of Washington dormitories and when they were not attending lectures were invited to visit several of the university's laboratories.

That residential series was so well received that there were usually two a year thereafter in order to extend the opportunity to students from rural areas or cities too small to provide an entire audience. The idea of extending the opportunity to hear such lectures beyond students in big-city schools was particularly attractive to some members of the board of directors, but led to a rare event, a conflict with NSF. In 1966 NSF told us that they no longer wanted to pay for the subsistence costs of the students attending residential series. The board protested strongly and NSF relented (89).

In the first 10 years, from 1961–1962 through 1970–1971 some 50 busy scientists had the extra and often wearing duty of preparing and delivering Holiday Science Lectures, but also the satisfaction of knowing that well over 30,000 bright and interested students had had the privilege of attending one of those series of lectures. The program was intended to add to the education and motivation of students who were already interested and competent in science and to provide a reward for their academic achievement. The frequent "outstanding," "enlightening," "excellent," "wonderful," or similar expressions in letters from teachers, students, and sponsors gave ample evidence that the program was a very welcome one (90). The lectures continued to be popular but were terminated in the economic downturn of the early 1970s.

## *Publications*

The primary responsibility of the Commission on Science Education was the association's own programs. Beyond that, however, it served all the other course content and, indeed, the whole national interest in improving education in science and mathematics by collecting and disseminating information. In addition to publishing a periodic newsletter describing its own activities the commission was responsible for several other publications. *Science Education News*, covering new or continuing science and mathematics programs, was sent quarterly and without charge to over 5,000 recipients (91). From 1963 to 1977 an annual *Report of the International Clearinghouse on Science and Mathematics Curriculum Developments* was published jointly by the commission and by the Science Teaching Center of the University of Maryland, giving annual (later, biennial) reports on the status of as many as 300 course content improvement projects and related activities. John Mayor, while continuing to direct the association's educational programs, also wrote *Accreditation in Teacher Education: Its Influence on Higher Education* for the National Commission on Accrediting (92).

In 1969–1970 the Commission on Science Education began exploring the field of environmental education and John A. Moore, a member of the commission, began collecting bibliographic information on the relations of science and society, with three editions supported by NSF and with further support from the E. I. du Pont de Nemours Company, the Xerox Corporation, and the General Motors Corporation. The bibliographies quickly became very popular. The Battelle Memorial Institute of Columbus, Ohio, which was already publishing its own quarterly review of publications on science policy (93), printed and distributed 65,000 copies of the first edition (94), sending it to school administrators and science teachers. Rand McNally distributed another 10,000 copies to teachers on its mailing list. AAAS sent out many copies in response to requests from Brazil, Canada, England, Japan, Laos, and other countries, and from every state. The bibliography was useful not only to high school teachers but also to college and university students and faculty members at a time when courses with such titles as "Science and Society" were beginning to appear on many campuses (95).

## *Achievements and Disappointments*

By 1970 it had been 15 years since the Science Teaching Improvement Program had been started, and eight since the Commission on Science Education was established. Looking back, the Action Program of the Cooperative Committee on the Teaching of Science and Mathematics can serve as a starting point from which to evaluate what AAAS had accomplished, for it was that program of the Cooperative Committee that led directly to the association's Science Teaching Improvement Program.

Although all the goals stated in the Action Program seemed desirable, some were for governments and voters rather than for AAAS or any other scientific society to accomplish, for education in the United States is the responsibility of thousands of individual and independent states, cities, districts, and other units.

The Cooperative Committee and AAAS could hope their advocacy and studies would influence some of those many school systems to provide their teachers with higher salaries, greater prestige, and better working conditions, but that was all they could hope for on those goals. However, some of the other objectives of the Action Program were within the competence of AAAS. Vigorous programs were undertaken, and some successes were achieved.

1. One objective was to promote the selection and utilization of consultants in mathematics and science in representative schools. The value of such consultants was demonstrated in the four states: Nebraska, Oregon, Pennsylvania, and Texas. Although the consultant idea was not adopted in the National Defense Education Act of 1958 it still survives and is being implemented in some locations.

2. A second objective was to encourage departments of science and mathematics in colleges and universities to accept responsibility for educating future teachers of those fields. The association's teacher-counselor program, studies of the education of prospective teachers, and work on teacher education and certification all helped bring about increased acceptance of responsibility for teacher education by some university departments of science and mathematics. Overall, change has continued to the extent that by 1986 there was a strong move by several of the nation's large and influential universities to insist that all prospective high school teachers should earn undergraduate degrees in the fields in which they expected to teach and, of course, that departments of science and mathematics should provide appropriate courses for future teachers of those fields (96).

3. Another objective was to provide for the recognition of exceptionally able teachers. The record was a mixed one. Although many teachers benefited individually, there was little success in overcoming the organized educational community's objections to merit systems for rewarding individual teachers.

4. The AAAS education staff also hoped to assist in interesting appropriately qualified high school students to prepare for teaching or other careers in science. "Science — A Process Approach" and the elementary and high school traveling libraries of science surely fostered in some students both greater interest and better understanding. The NASDTEC–AAAS certification standards gave many students better prepared teachers than they would otherwise have had, and the Holiday Science Lectures provided additional knowledge and inspiration to thousands of high school students who were already doing well in science. No effort was made to find out how many of the students involved followed one career path rather than another, however, and of course other factors were also influencing the choice of college majors and later careers.

Some successes resulted from the work of the AAAS, the other course content improvement projects, the National Science Foundation, the National Defense Education Act of 1958, and other forces working toward the improvement of elementary and secondary school education, yet when the nation's public school system is looked at as a whole its problems were—and are—far from solved.

Neither AAAS nor the other actors involved had the funds, the authority, or the influence to produce fundamental changes in the ways in which schools are organized and managed or teachers are educated, selected, recognized, and rewarded. Since the 1960s peak of efforts to improve education, social pressures, local control, traditionalism of the education profession, disruptive influences of several kinds, and the increasing rewards of other fields of work have so diminished the attractiveness of a teaching career and so exacerbated some of the problems of the nation's schools that the calls for improvement are as strong in the 1980s as they were in the 1950s (97).

Yet looking at the educational system as a whole may not be the appropriate way to evaluate the effectiveness of attempts to improve science and mathematics education. The educational system of the country is greatly decentralized. Variety is one of its hallmarks. Thousands of independent school districts, hundreds of thousands of teachers, and millions of pupils are individually responsible for their own decisions. What AAAS did was to create a variety of educational opportunities: SAPA, traveling libraries, conferences, Holiday Science Lectures, interesting foreign visitors to summer institutes, standards for the education of prospective teachers, lively and informative seminars for school administrators. Use could not be compelled, but those opportunities were made available, to states, to schools, to teachers and administrators, and to students. Most did not take advantage of those opportunities, but many did. The evidence — sometimes statistical, sometimes anecdotal, sometimes in the form of careful evaluations — showed that many schools, teachers, and students did seize upon and value those opportunities, enough to give AAAS staff, the members of the Commission on Science Education, the many teachers and scientists who worked with them, and the funding agencies that supported their work the satisfaction of knowing that their efforts had been worthwhile.

## Chapter Nine

# Public Understanding of Science

"To increase public understanding and appreciation of the importance and promise of the methods of science in human progress." Those words described the objective that the AAAS board of directors had in mind in 1945 when they agreed with representatives of the Westinghouse Company to initiate a series of awards to honor excellent science writing for the public. A year later AAAS began awarding two annual AAAS–Westinghouse Science Writing Awards of $1,000 each, one for an outstanding example in newspapers of popular writing on the natural sciences or their engineering or technological applications and the other for similar writing in a general–circulation magazine(1).

That was the association's first successful postwar venture into the field of improving the public understanding of science. It was only a start, however, and the board hoped to develop a larger and more varied program of improving public understanding of science (2). A larger program did gradually develop, but not until after the initial AAAS–Westinghouse prizes had collapsed. In 1953 the Westinghouse Educational Foundation announced that they had decided not to support the prizes any longer. Informal talks elicited the explanation that Westinghouse no longer thought the prizes necessary, for other awards for excellent writing in individual fields of science were then also being offered. Moreover, they thought that science writers were not sufficiently interested or appreciative and that Westinghouse was not getting its money's worth in publicity (3).

A few years later they reversed that decision and in 1959 the awards were resumed (4). In 1966 a third prize was added for writing in newspapers of less than 10,000 circulation and in 1971 a fourth and fifth prize for radio and television, respectively. Except for the 1954–1958 gap the AAAS-Westinghouse awards have been continued for over 40 years. They constitute the association's second oldest prize (after the Newcomb Cleveland Prize) and its senior effort to encourage and reward outstanding contributions to the public understanding of science.

### Popular Books and Lectures

If AAAS could recognize and reward good writing for the public, why not also produce it? An idea that never panned out but that seemed so plausible and natural for AAAS that it came up again and again was the proposal that the association publish a series of popular books on scientific topics. Committees met and made recommendations about such books. So did outsiders; in 1945 the association was offered $200,000 to start a revolving fund for the publication of such a series (5). Staff members conferred with commercial publishers. A committee was appointed to choose a publisher and develop procedures for selecting manuscripts (6). Basic Books actually published *New Roads to Yesterday*, a book based on articles on archaeology that had appeared in *Science*. The disappointing sale of only 4,400 copies in the first 10 months brought into question the ability of AAAS to judge the market, and publishing popular books on science did not develop into one of the association's successful ventures.

As it turned out, AAAS's science lectures were no more successful. The Lancaster, Pennsylvania, Branch of the association handled that responsibility well at the local level (7), and the National Geographic Society lectures at the AAAS annual meetings were always popular. Suggestions were made from time to time that AAAS should put good lecturers on the public circuit. So after studying the Sigma Xi and American Chemical Society lecture series and after polling some colleges, members of the council, and a sample of association members, the board in 1953 decided to start a public lecture series, initially in the southern and southeastern sections of the country (8). The series never developed; that effort was no more successful than were the popular books.

## AAAS as a Source of Science News

AAAS itself was not successful as a popularizer of science, but its annual meetings and *Science* provided good source material for professional journalists. Each week as *Science* came off the press, copies were rushed to professional science writers for their use in preparing newspaper and magazine articles. As a result stories credited to *Science* appeared each year in hundreds of newspapers and magazines. In a 12-month period in 1978–1979, for example, material drawn from *Science* was published in over 400 newspapers and magazines with a combined—and of course overlapping—circulation of over 230 million. Stories from *Science* were published somewhere in the United States on 238 of the 365 days of that period (9).

Each year one of the standard parts of the annual meeting was the press room where newspaper and magazine reporters and free-lance writers could secure copies of manuscripts, interview program participants, write their stories, and phone copy to their publishers. The number of reporters varied depending upon the location and program interest, but usually ranged around two hundred.

Sidney Negus was the reporters' friend who made the press room the valuable aid and support to science writers that it was. For most of each year Negus was professor of biochemistry at the Medical College of Virginia, but each fall and extending to a little past New Year's Day, preparing for and managing the AAAS press room was high on his agenda. Year after year writers for individual publications and for the wire services filed many stories from the AAAS meetings.

Magazines sometimes gave the meetings special coverage. When the 1949 meeting had the largest registration yet recorded, *Life* magazine headlined its eight-page account "U.S. Science Holds its Biggest Powwow and Finds it Has a New Einstein Theory to Ponder." The article included a group photograph of 41 eminent scientists from many fields; a picture of M. T. Cook of Louisiana State University, who was attending his 50th AAAS meeting; an account of the talk the reporter found most exciting, one by Donald Menzel of Harvard on solar prominences; and several other items including a piece on Albert Einstein (10).

Five years later *Life* featured the AAAS council's endorsement of the board of directors' statement on shortcomings of the nation's security system (see Chapter 2). Prose and pictures described the applause and support for retiring president E. U. Condon, the award of the Newcomb Cleveland Prize to Daniel Alpert for the most nearly perfect vacuum ever achieved on earth (described as "one millionth of a billionth" of the natural atmosphere), a rapt student listening to a lecture, and the youngest speaker on the program (18-year-old Ray L. Harris of Enterprise, Oregon) (11).

After a quarter of a century helping reporters secure those and hundreds of other stories, Negus died while in New York City to attend a joint meeting of the association's Committee on the Public Understanding of Science and the Council for the Advancement of Science Writing (12). Years before, Daniel Wilkes, director of public information for the University of California at Berkeley, had written that American science

> is most fortunate in having the services of Dr. Sidney Negus. I have no idea what Sidney receives for doing this job, but I must say that I feel sure the AAAS could not possibly pay him according to the worth of the job he does .... He has a place in the esteem of the science writers that is unique (13).

Two years later, when I visited the press room at a meeting of the British association for the Advancement of Science, several of the reporters there told me how helpful Negus had been in responding to their requests for information concerning scientific activities in the United States. Tributes to his services were many, and after his death his press room traditions were carried on by Edward G. Sherburne, Jr., who had joined the staff in 1961 to head the AAAS program on public understanding of science; his assistant Kneeland Godfrey (14); and Thelma Heatwole, a member of the professional staff of Philip Morris Company in Richmond, who for years had served as Negus's assistant in the press room.

## *Radio and Television*

Prior to 1956 AAAS was occasionally involved in broadcasting programs on science. Examples included the radio and television coverage of the 1948 centennial meeting (see Chapter 3) and a radio program developed a few years later with the help of *The Scientific Monthly*. But it was not until 1956 that a sustained effort in the electronic media was started, and that came about as a direct outgrowth of the activities of the annual press room.

In 1956 John Behnke suggested to Columbia Broadcasting System officers that the next annual meeting, which would be in New York City, would provide an opportunity for a program reviewing recent developments in science. CBS agreed and on Sunday afternoon of December 30, as the 1956 meeting was drawing to a close, CBS devoted an hour to a review of recent developments in science. The favorable public reaction prompted Frank Stanton, the CBS president, to plan to broadcast similar reviews on an annual or perhaps even a quarterly basis. CBS did broadcast another annual roundup at the end of 1957, but before that program was aired, CBS, the Monsanto Company, and AAAS had entered into a three-way partnership to produce a regular series of science programs called "Conquest." Monsanto gave financial support; CBS provided the producer, Michael Sklar, and the necessary technical support; and AAAS, initially in cooperation with the National Academy of Sciences, provided ideas and program material (15).

To collect, winnow, and develop ideas and plans, we engaged the services of Willard Bascom, an imaginative and ingenious scientist-engineer-oceanographer turned explorer, entrepreneur, and filmmaker. As a member of the staff of the National Academy of Sciences he had directed the first, and successful, phase of the Mohole Project (16). When the National Science Foundation was unable to proceed immediately with further stages of the Mohole Project, Bascom became available for something else. His own ability as a film producer was evident from the film he made of the Mohole drilling, and that and other films he made later earned awards for excellence. John Coleman, then NAS executive officer, thought Bascom would be the ideal representative of AAAS and NAS in developing the Conquest programs. I trusted Coleman's judgment, and a cable to the South Seas brought Bascom back to Washington to start developing our new TV series, which was to begin on December 1, 1957 (17).

At the same time that the "Conquest" series was being developed the association received a request from Broadcast Music, Inc., to help select scientific topics for a series of 15-minute radio programs. Earlier that organization had presented a series on the history of the United States that Alan Nevins had written and that were later published as *The American Story*. To follow that successful series Broadcast Music wanted AAAS to cooperate with the American Council of Learned Societies and its chairman, Howard Mumford Jones, in selecting materials for programs on science and the humanities. The AAAS board approved, and in March of 1958 "The World of the Mind" began alternating 15-minute weekly programs on scientific topics with similar programs in the humanities, all carried on some 525 radio stations throughout the United States (18). The series received a special citation from the Thomas Alva Edison Foundation "in recognition of distinguished public service in the interests of science and the humanities."

Concurrently, the "Conquest" programs were honored as the TV program that had made the greatest contribution to science for youth in 1958. Those hour-long programs were aimed at adults as well as youth and drew an audience of about 15 million viewers. To our amused pleasure, on one evening in the 1958–1959 series, "Conquest" outdrew the then widely popular "Lone Ranger" program. In 1959 Monsanto and CBS agreed to continue the series for a third year, but with a format consisting of 20 half-hour programs to be shown at 5:00 p.m. on Sundays instead of the smaller number of hour-long programs of the two previous years (19). At that time also the National Academy of Sciences withdrew, leaving AAAS as the sole scientific sponsor.

That turned out to be the final year. The program continued to win awards, but Monsanto authorities decided that they had put enough money into "public-image" advertising. CBS wanted to continue the program, but "Conquest" had been so closely identified with Monsanto that it did not appear to be a good advertising vehicle for a new sponsor and no other supporter was found. Thus the program ended leaving no nationally televised science program for the general public (20).

For three years, however, thanks to Willard Bascom and the scientists he interested in participating, to Michael Sklar and the technical facilities of CBS, and to financial support from Monsanto, AAAS had a highly regarded and widely viewed series of television programs on the air.

## Searching for a Role

The 1950s were years of widespread interest in the nation's scientific and technological status. There were worries about impending shortages of scientists and engineers and about the nation's technological capabilities in comparison with those of the Soviet Union. When the decade opened the Cold War was on, and before it ended the Soviet Union had beaten the United States into space with much heavier satellites than the United States could launch.

One response to those worries was an upsurge of efforts to improve education in science and mathematics and an increased interest in improving the general public understanding of science, its achievements, its methods, and its relations to technology and national welfare. That was a congenial response for AAAS, for improving public understanding of science had long been one of the association's objectives and one that had been emphasized in the Arden House statement of 1951 (see Chapter 3).

AAAS meetings and *Science* were widely used as sources by science writers. Opportunities to contribute to the "Conquest" TV programs and "World of the Mind" radio broadcasts were accepted with enthusiasm and the results were reviewed with pleasure and satisfaction. But those were responses to targets of opportunity and not a total program.

There was no dearth of other opportunities. Among those declined—because the association did not have the necessary resources, because the plans did not seem right for AAAS, or because the AAAS name would be associated with an activity over which we would not have had adequate control— were an offer of $200,000 to publish science books for laymen (21); a request to join in the establishment of the "American Foundation" to grant honors and large awards to scientists and to produce documentary films dramatizing the accomplishments of its own and Nobel laureates (22); an invitation to help produce motion picture films of the lives of 10 Nobel laureates (23); a proposal to join Science Service in sponsoring a national amateur science program (24); a staff proposal for a national science information center through which AAAS would provide information to science writers, newspapers, magazines, book publishers, radio and television producers, and anyone else involved in presenting information about science to the public (25); and possibilities discussed with representatives of the National Association of Science Writers that were so large and expansive as to overwhelm the association's resources (26).

Although the board decided against all of those opportunities, their discussion led to a policy decision and several procedural decisions. The policy decision was that AAAS could be more useful by helping agencies and individuals already involved in providing science information to the public than by attempting to reach the public directly. That decision was not meant to exclude continuing the "Conquest" television programs, had that been possible, or to become involved in similar programs. In general, however, we believed AAAS should focus on helping science writers, publishers, producers, and others to increase or improve their wares. On the procedural side the board appointed a committee to study the best options for AAAS, asked the staff to bring together information on what was being done by AAAS and by other organizations to provide a basis for defining AAAS's role, and later authorized hiring someone full time to develop and monitor the association's public understanding of science programs (27).

The anticipated program was expected to cost more than AAAS could afford; the board therefore asked the Ford Foundation for $500,000 for programmatic support of a new AAAS program for five years (28). When Ford declined (29) smaller requests were submitted to the Carnegie Corporation of New York and the Rockefeller Foundation (30), but those requests were also declined (31).

Nevertheless, the board gave the effort high priority and decided to try to proceed without foundation support. Several possible appointees to lead the work were discussed and the board authorized me to find a program director (32). The result was appointment of Edward G. Sherburne, Jr., a 1941 graduate of the Massachusetts Institute of Technology with a major in mathematics. Sherburne had finished World War II as Head of Engineer Technical Intelligence in the

European Theater. A couple of years in an advertising agency had been followed by seven years of television experience, first for a commercial network, then for the Navy, and then for educational television. He had also had statewide responsibility for educational television for the seven campuses of the University of California. That range of experience commended him to the AAAS board and in the spring of 1961 he joined the staff (33).

To provide policy guidance for the public understanding of science program the board converted the temporary Committee on Public Information and Science into a permanent Committee on Public Understanding of Science. When Sherburne arrived, that committee consisted of Warren Weaver, chairman, and Willard Bascom, Victor Cohn (science writer for the *Minneapolis Star and Tribune*, Laurence M. Gould, Richard D. Heffner (then of the National Educational Television and Radio Center and soon to become director of New York's educational television station, Channel 13), and Thomas Park, plus Sherburne himself, who had been appointed to the committee before he joined the association staff.

When Sherburne came to AAAS he began visiting other organizations to see what they were doing to improve public understanding of science, but as he was getting started on that survey we received an urgent request to help a quite different effort to improve public understanding of science and decided that the best thing Sherburne could do for a few weeks was to respond to that request.

Three years earlier a visitor from Seattle had told me that a group of state and civic leaders in western Washington were planning to hold a world's fair, "Century 21," emphasizing future developments and possibilities. Would scientists help plan such a fair? he asked. He did not know it, but the plans for the popular science exhibits at the World's Fair then about to open in Brussels had gotten quite out of hand. An international committee had planned to have exhibits about four themes, had identified exhibits to illustrate each of those themes, and had allocated the development and construction of those exhibits among a number of cooperating countries. But when it came to construction the prospective public audience had somehow been forgotten. Most of the actual exhibits had grown too complex for their original purpose. Knowing the dissatisfaction that had developed with that situation, I told my visitor that I thought scientists would be glad to help and that I could provide an exceptionally good opportunity to find out. One week from the following Saturday, I explained, AAAS would be holding a special meeting (the Parliament of Science discussed in Chapter 2). No session was scheduled for Saturday evening, so if representatives of the Seattle group were willing to come to Washington, D.C. for dinner that evening I would arrange for them to meet most of the American scientists who had had leading responsibilities for the American exhibits at the Brussels Fair; scientists from the National Science Foundation, the Department of State, and other relevant government agencies; and some others who had a special interest in improving the public understanding of science.

The Seattle people decided to risk another trip across the country. At the appointed time Edward Carlson, president of the organization planning the fair and also president of Western International (later Westin) Hotels, Ewen Dingwall, the director of planning and later general manager of the fair, and Senator Warren Magnusen had an opportunity to explain their plans and hopes to a dozen-and-a-half selected representatives of the scientific community. They posed their question: Will scientists help us? It did not take long to find that the answer was "Yes" if they meant science and not gadgetry and if they meant an exhibit of interest and value to the general public.

That, they said, was what they did mean and that agreement quickly led to a further agreement. Leonard Carmichael, representing the National Academy of Sciences, James Mitchell, representing the National Science Foundation, and I, from the AAAS, would select a "national advisory committee" to help plan what became the U.S. exhibit at the fair (34). Senator Magnusen got a bill through Congress to finance the U.S. exhibit and the national advisory committee started to plan the kinds of exhibits that seemed most desirable.

Then came a change of administrations. President Eisenhower was succeeded by President Kennedy. The new Secretary of Commerce, Luther P. Hodges, wanted to appoint his own commissioner for the fair that was scheduled to open in a little over a year. When Athelstan Spilhaus was appointed to that position he desperately needed help. A complex of five beautiful buildings was being erected as the U.S. Science Pavilion, but no one yet knew what exhibits would fill those buildings. The most useful thing Sherburne could do for the next few weeks was to become "Interim Science Coordinator of the United States Science Exhibit of the Century 21 World's Fair in Seattle," to quote the long title he held while AAAS loaned him to the Department of Commerce. He went to work immediately; soon Spilhaus and his deputy, Craig Colgate, were on board; and when the fair opened on April 21, 1962, the U. S. Science Pavilion quickly became by a large margin the most popular exhibit there. The science exhibit was not an achievement of AAAS, but the association could take some of the credit, for Sherburne had filled a critical gap; Spilhaus, a future AAAS president, was the director; and Hilary Deason, director of the AAAS Traveling Libraries of Science, selected the books on science and mathematics that were sold in the gift shop of the U. S. Science Pavilion.

Louise Campbell, then editor of the *AAAS Bulletin*, produced a special issue of that periodical entirely devoted to a largely pictorial account of the U. S. science exhibit, an account that concluded with the observation that although visitors sometimes disagreed as to what they liked best, "On one thing most visitors are likely to agree: this imaginative view of the scientific adventure is too good to dismantle" (35). All summer long the Seattle sponsors also heard that message, and on the day after the fair closed they reopened the science exhibit under its new name, The Pacific Science Center, a permanent asset dedicated to science education for the citizens and school children of the region.

## Analyzing the Problems

With his temporary responsibilities to the U. S. science exhibit completed Sherburne returned to the task of analyzing the association's public education opportunities and problems. In order to have a solid basis for planning a long-range program, he planned a series of studies of what the public actually knew about science and technology and a formulation by scientists of what the public should know. Lack of time and resources prevented doing all he hoped to do, but Wilbur Schramm, director of the Institute for Communication Research at Stanford University, made a start by analyzing the literature on the public understanding of science in a report entitled "Science and the Public Mind" and one of his graduate students, Phillip J. Tichenor, wrote a doctoral dissertation entitled "What the Public Understands about Science," an analysis of responses to about 100 questions on science and technology that had been asked in public opinion polls conducted by George Gallup, Elmo Roper, or other pollsters (36).

Those studies were made as the Apollo program to send astronauts to the Moon was getting underway. Several years earlier the United States had planned to launch a data-gathering satellite into orbit as part of the nation's contributions to the studies under the International Geophysical Year of 1957–1958. Led by Margaret Mead, and in anticipation of that launching, we tried to gather a group of natural and social scientists to plan studies of public attitudes toward going into space and of the impact of that ability on popular attitudes and social customs. The proposed meeting, set for March 14–15, 1957, had to be postponed because most of the invitees could not come at that time. And thereby the opportunity was lost, for on October 4 of that year America was shocked to find that the Soviet Union had successfully launched the first small sputnik, a success that was followed six weeks later by a larger sputnik with the dog Laika aboard. With the space age thus already opened, the AAAS meeting was never rescheduled.

Fortunately, the National Association of Science Writers had better timing. Aided by a grant from the Rockefeller Foundation, the science writers and the Survey Research Center of the University of Michigan planned a broad, nationwide study of the audience for science news. In March and April of 1957 interviews of upwards of an hour and a half were conducted with a sample of the U.S. adult population. With that time available for each interview, a large amount of information was collected about popular information and attitudes toward science. On the specific topic of space more than half of the people interviewed had never heard of space satellites and only about one in eight could state specific reasons for their launching. More generally useful to AAAS staff members and to anyone else interested in the public understanding of science was the finding that science information — not just about satellites but about other topics as well — was quite unrelated to the amount of time spent with radio or television. Information about science and related topics was almost always learned from reading newspapers, magazines, and books (37).

Complementing that study of what the adult population knew and thought about science was a study conducted by Margaret Mead and Rhoda Metreaux on the images of scientists held by high school students in the United States. To secure that information the authors designed a set of open-ended introductions for essays that high school students were asked to complete. For example, one read "If I were going to be a scientist, I should like [or in an alternative form "not like"] to be the kind who ...." Another, to girls, read "If I were going to marry a scientist I should like [or "not like"] to marry the kind who ...."

The Rockefeller Foundation provided $9,000 to cover costs. Hilary Deason arranged with 145 high schools around the country that were receiving the association's traveling science libraries to have their students write the requested essays. From among the 35,000 returns the two authors selected a sample for detailed content analysis (38).

The results were clear. The students respected scientists for their contributions to medicine and their other values to society, but most of them neither wanted to become scientists nor to marry scientists. Their general image of scientific work was that it was difficult and sometimes dangerous; that the requirements were too exacting; and that scientists were lonely and isolated, neglected their families for their work, were not well paid, and did not have enjoyable lives. It seemed that the students had been watching too many portrayals of the lonely, white-coated scientist working in a secret laboratory toward some sinister purpose. Mead and Metreaux concluded that both schools and the mass media should present a more accurate and attractive picture of scientists and their work to help ensure an adequate supply for the nation's future needs.

## Help to Science Writers

Many science writers regularly attended the association's annual meetings, and the National Association of Science Writers held their annual meetings in conjunction with those of AAAS. Some professional science writers served on AAAS committees, the association regularly selected some of the trustees of Science Service, and overall there was a cooperative relationship between AAAS and the professional science writers of the United States. In addition, many science writers had developed close working relationships with individual scientists to whom they could turn for background information on a newly breaking story or for judgment as to whether a news report was important or trivial. Those relationships were so useful that in 1959 the National Association of Science Writers asked AAAS to establish a national panel of scientist advisors to whom they could turn for counsel and information, with the understanding that the science writers would treat the information obtained as confidential and not for attribution. With board of directors' approval, the AAAS staff developed a list of scientists willing to work in the proposed fashion and with the understanding that they were not formally representing AAAS but were giving personal advice to the writers who called upon them (39).

Concurrently, science writers were planning to establish an independent, nonprofit Council for the Advancement of Science Writing, a council that could establish programs for training new science writers, develop new techniques for presenting science news in the media, provide fellowships, or in other ways advance their professional field. The group planning the council expected it to be governed by a board of directors consisting of members representing several different constituencies. In response to their invitation the AAAS board agreed to nominate two of those board members and for three years contributed a thousand dollars a year to help the new council get started (40).

While AAAS relationships with science writers were close and frequent, that was not true of relationships with editors and publishers of the periodicals for which science writers wrote. In the main, editors and publishers were neither scientists nor especially interested in science. Their interests were more like those of other intelligent nonscientists, whereas interests of their science writers resembled more closely those of professional scientists. Nevertheless, some editors were becoming increasingly aware of the upsurge of interest in science, of reports of the recently completed International Geophysical Year, the developing space race with the Soviet Union, and the new curricula being developed for teaching science and mathematics in the schools. Reflecting that growing interest the *Minneapolis Star and Tribune* — with co-sponsorship by Carleton College, the University of Minnesota, and the Mayo Clinic — invited a number of newspaper editors and publishers and scientists to come together for a few days at a rustic hunting lodge in the spectacular autumn colors of northern Minnesota. The meeting was neither planned nor conducted by AAAS, but it had a distinctly AAAS cast. Of the three dozen participants, three were science writers who were all regulars of the annual AAAS press room: Alton Blakeslee, Victor Cohn, and Earl Ubell. The others were about evenly divided between scientists and senior editors of the *Wall Street Journal,* the *Washington Post, Minneapolis Star and Tribune*, leading dailies from other cities, and the Hearst empire. Of the scientists seven had been or were to be presidents of AAAS: Laurence Gould, Roger Revelle, Paul Sears, Harlow Shapley, Athelstan Spilhaus, Alan Waterman, and Warren Weaver. I was there, and some of the other participants had served the association in various ways.

From arrival on Friday to departure on Monday there were no interruptions; the site was too isolated for distracting side visits or part-time participants. The hosts assigned one editor and one scientist to each room to help the two groups become better acquainted. Talks by scientists on developments in their fields, discussions by everybody, and free-for-all exchanges filled the days. After the meeting was over Alfred Friendly, managing editor of The *Washington Post* and president of the American Society of Newspaper Editors, wrote:

> The consequences of the meeting was a realization by the newspaper types of the fascination of the new world of science, of what terrific copy they are for the daily press, and that this material is as important to cover as the courts, the Congress, or the county board — and much more interesting and readable (41).

With that enthusiastic memory of the Minnesota meeting, Friendly wanted to give a larger number of editors an opportunity to hear some of the same kind of talk he had heard. He came to me to ask if AAAS would plan a one-day version for the next annual meeting of the American Society of Newspaper Editors, which was to be held in Washington, D.C. We agreed; the National Science Foundation provided a grant of $3,900 to cover expenses; and the first day of the editors' annual meeting of 1962 consisted of talks by five articulate spokesmen for their fields of science: Frank Brown, Frank Press, Francis O. Schmidt, Harlow Shapley, and Warren Weaver. Ted Sherburne came back from the meeting reporting that the hundred-plus editors who heard those talks seemed to enjoy the day and that some of them were already considering what might be done the next year (42). After we had paid the bills we were able to return about half of the grant to NSF.

## Understanding

The opportunity to meet with editors who established policy for major daily newspapers was welcome, but most of the association's relations were with the science writers. As a continuing contribution to them Sherburne and Pierre Fraley of the Council for the Advancement of Science Writing (CASW) soon began to discuss the possibility of publishing a news periodical to report activities directed toward improving the public understanding of science. They agreed that Fraley and the CASW would have primary responsibility for preparing the editorial material, with help from AAAS, and that the association would be responsible for printing and distribution. The resulting quarterly was named *Understanding* and was sent without charge to a list of approximately 3,000 persons from newspapers, magazines, book publishing houses, museums, radio and television stations, adult education organizations, government offices, scientific societies, and individual scientists with a special interest in public education (43).

Assistance to the National Association of Science Writers and the CASW, a special session for members of the American Society of Newspaper Editors, publication of *Understanding*, and seminars for television script writers may all have helped, but the problem of improving the public understanding of science is not one that can be "solved" and then dropped. Science and technology change and so do their relations to contemporary social issues. Whether the motivation for efforts to improve public understanding arises from a desire to help society make wiser decisions or from the desire for better opportunities for scientists, the unsatisfactory — at least to scientists — state of public understanding of science continues. Three decades after these AAAS efforts were made Sigma Xi was preparing for its centennial. In 1986 that scientific honorary society asked a sample of its members about the matters that were of most concern to scientists. "Lack of public understanding" was essentially tied with "interruption of funding" as the most worrisome problem for American scientists in the mid-1980s (44).

## Television

The association's relations with the print media provided a model of the relationships we would have liked to develop with the electronic media. We never succeeded. The association had cooperated effectively with Broadcast Music, Inc., and the Columbia Broadcasting System in producing radio programs and the "Conquest" television programs. There were occasional TV broadcasts from the annual meetings. And the association had earlier been represented on the National Citizens Committee for Educational Television that helped launch educational television in the United States (45). But a permanent and effective relationship with the medium as a whole simply did not develop despite various efforts and the high priority Sherburne gave to television.

For several years after the demise of the "Conquest" program and completion of the "World of the Mind" radio series, there was relatively little active cooperation, although AAAS meetings and publications sometimes made news for the electronic media and *Understanding* went to many people in the radio and television fields. That relatively low level of activity seemed quite insufficient, and a grant of $196,650 from the Sloan Foundation was intended to support studies and demonstrations designed to improve the quality of science programming on television (46).

Richard Heffner, manager of Channel 13, New York City's educational TV station and a member of the association's Committee on Public Understanding of Science, and Sherburne went together to Hollywood to talk with TV executives about how AAAS might help them. They returned reporting greater interest in science than they had expected and saying they had received a number of requests for help (47). Operating at a cross-country distance was difficult, however, and they later recommended that the association seek funds to establish a small office in Los Angeles or Hollywood to provide information and advice to script writers and producers of films and TV programs. The board tentatively approved that proposal, but financial support proved elusive and the office was not established. However, the Sloan Foundation grant did support a series of luncheon seminars for members of the Writers' Guild of America. Warren Weaver initiated that series in Los Angeles in March of 1964 and two months later offered the first of eight seminars for members in and near New York (48).

Working with TV script writers was similar to working with newspaper science writers. Their interest was necessary, but their interest did not guarantee comparable interest on the part of their bosses. The Minnesota meeting of scientists with newspaper editors and publishers provided an attractive model for trying to arrange a similar meeting with top-level executives from the television industry (49). Eventually, and discouragingly, that meeting got postponed out of existence. Frank Stanton and Dan Seymour accepted and would have been top-level representatives from television and advertising. The scientists who were invited were enthusiastic about the prospects, but the other invitations all drew blanks. A long post mortem of regrets and of wondering what other approaches might be tried ended that effort (50).

Earlier, in 1964, the Sloan Foundation had contemplated making a large grant to the association for the general support of efforts to improve the public understanding of science. However, as they considered that possibility, the foundation officers decided to augment their own staff and develop this area as a major new foundation responsibility (51). Of course there were some regrets on receipt of that information, but several members of the AAAS board thought that outcome might be more desirable than a large grant to AAAS. After all, the basic AAAS policy was to help other agencies interested in improving public understanding of science; perhaps the Sloan Foundation decision was an unintended example of the working of that policy.

In any event AAAS soon benefited from that decision, for the Sloan Foundation, along with the Ford Foundation and other supporters, soon provided funds to support extensive television broadcasts from the association's annual meetings (see Chapter 3).

## Seminars for Members of Congress

Senior members of the AAAS staff were constantly on the lookout for scientists able to speak clearly and interestingly about their fields of study. We searched for them throughout the United States to lead seminars for the Writers' Guild members, to speak at the symposia for school administrators, or to deliver the Holiday Science Lectures (see Chapter 8). We searched for them abroad to bring to the United States as lecturers at summer institutes for teachers of science or mathematics (see Chapter 8). Beginning in 1960 our talent search also included finding leaders of seminars for members of the U. S. Congress. Talent scouts we were, and we often used the same scientist for several different assignments.

The idea of conducting seminars for members of Congress originated with the American Institute of Biological Sciences, which in 1958 wrote to AAAS recommending that some scientific organization arrange for seminars at which scientists and members of Congress might engage in discussions of the relations between science and government. The board liked the idea, but it languished for a year until we found the ideal partner in discussions I had with George Graham and James Mitchell of the Brookings Institution. They were enthusiastic about a recent series of evening discussions with groups of 18 relatively new members of the House of Representatives concerning their duties in the House. On the basis of that experience they were eager to join with AAAS in sponsoring a series of seminars on scientific topics (52).

Because of the long-established record of the Brookings Institution in governmental affairs its cooperation seemed highly desirable. The board agreed and we quickly came to a working arrangement that Brookings would invite about 25 members of the House of Representatives to attend a series of dinner meetings. Brookings would provide the audience and the dinners while AAAS provided the topics and the speakers. We considered seeking a foundation grant to support the activity, but decided that we preferred to keep the seminars wholly an activity of the two organizations. It was a relatively inexpensive one with the annual costs to each organization about $3,000 (53).

The seminars began at a time when there was considerable congressional interest in scientific and technological matters. The first Soviet sputniks had flown in 1957. Congress had adopted the National Defense Education Act in 1958. The space race was starting; candidate John F. Kennedy was about to emphasize the missile gap and President Kennedy would soon afterwards announce that the United States intended to send astronauts to the Moon and back within the decade. Those attitudes and actions no doubt helped make the prospective seminars attractive, but they did not set the agenda. As a matter of principle we decided that the seminars would not consider topics of pending legislation, but rather would deal with scientific or technological issues that were developing, that members of Congress should know about, and that might later become matters of congressional legislation.

Members of Congress were interested in that prospect. Brookings invited them individually, deliberately selecting relatively younger members who came from different parts of the country and who were serving on different congressional committees. Among them were Al Ullman, who later became chairman of the Committee on Ways and Means; John Brademas, who later became president of New York University; Emilio Q. Daddario, who later became president of AAAS; Albert Quie, who later became governor of Minnesota; and others who continued in the House of Representatives, moved to the Senate, or left Congress for other leadership roles.

The sessions for the first year illustrated the variety of topics and the high caliber of seminar leaders. That year Thomas F. Malone spoke on meteorology, George Beadle on mechanisms of heredity, Roger Revelle on oceanography, Walter Orr Roberts on astronomy and space, B. F. Skinner on learning, William O. Baker on materials, and Frank Horsfall on viruses (54).

When the seminars were started we planned to hold them every other year. But when the second (1962) series ended the participants asked that we give them every year (55). We agreed and as we planned a new schedule each year, Brookings invited some previous participants and some new ones. Whatever the topic, the half hour or so of introductory talk by the visiting scientist was generally followed by lively discussion. In 1962 Martin Schwarzschild held a rapt audience as he discussed the better view of the history of the universe that would become available when we could put a telescope above the earth's blanketing atmosphere. In 1967, six years before the OPEC oil crisis, Ali Cambel described long-term trends in energy resources and in that same year Milton Harris talked about American versus European technology trends and the future implications for the balance of payments (56).

The speaker was never the only contributor to the evening's discussion. The members of Congress added different points of view, keen questions, and sometimes supplementary information. One evening Richard Bolling of Missouri challenged Stanley Cain to distinguish his all-encompassing definition of ecology from a universal concept of Plato. On another evening attendance was small because the House was in session to discuss a pending motion to expel Representative Adam Clayton Powell. The scientific topic was almost completely abandoned that evening while the speaker and the representatives of AAAS and Brookings sat fascinated as Thomas Curtis of Missouri led a discussion of the constitutional, legal, moral, and political aspects of whether they should vote "Yea" or "Nay" the next morning. The members already named, Catherine May of Washington, Charles Mathias of Maryland, Charles A. Mosher of Ohio, Robert Stafford of Vermont, Jessica Weis of New York, and many others helped make the seminar discussions lively and profitable for all.

After those seminars had been going for a few years Sherburne was pleased to report that similar activities were occurring elsewhere. The National Academy of Sciences had given a dinner for chairmen of congressional committees with a scientist as speaker. There were tentative moves to develop similar seminars for members of the state legislatures in Arizona and California. The Department of State had instituted a series of luncheon meetings for 15 to 20 top-level officials with a scientist as speaker at each meeting (57).

AAAS also expanded the idea. As staff members of individual congressmen or congressional committees learned of the seminars their bosses were attending they began asking for a similar privilege. Accordingly, in several years we invited staff members instead of members of Congress or gave a series for each group (58). And in 1967, with René Dubos as the first speaker, we began getting double duty from speakers by having them repeat their talks for an audience of science attachés from other countries who gathered in the AAAS board room on the morning after each seminar for members of Congress (59; also see Chapter 6).

## Emphasis on Youth

In 1963 the Southwestern and Rocky Mountain Division (SWARM) of AAAS met in Albuquerque, New Mexico, shortly after that city had been host to the National Science Fair for high school students and also to the First National Science Seminars conducted by the association's Academy Conference. Those events were one of the topics of discussion at the SWARM meeting and much of the discussion was critical. The critics regretted that the science fairs tended to neglect fields of science that could not easily be portrayed graphically. They thought there was too much enlistment of adult help in preparing exhibits. They disapproved of the monetary awards that were prominent in some of the fairs and the emphasis on winning those awards. And they disliked the loss of control to commercial sponsors of some of the fairs. All of that discussion led to a resolution asking AAAS to investigate the science fairs and to make a public report of the findings of that investigation (60).

The science fairs that were held annually in many locations had a variety of sponsors, but the prime sponsor of the culminating national fair was Science Service, one of the association's friendly allies in attempting to improve public understanding of science. A few months earlier Watson Davis, director of Science Service, had proposed to AAAS that the two organizations jointly plan a "National Youth Science Congress" and a "National Amateur Science Program" (61).

Programs for young people interested in science were by no means foreign to AAAS. The association had student members. The annual Junior Science Assembly had begun meeting with AAAS in 1946. And even before that the association had been giving encouragement to junior science programs conducted by its affiliated state academies of science. Thus in response to Watson Davis' proposals the board had asked the staff to survey what was being done in the United States and abroad in support of amateur interests in science and to consider what role the association might assume in that effort (62).

The resolution from SWARM arrived before that survey was completed and the resolution was soon followed by other signs of conflict. Some of the staff members at Science Service complained that holding the National Science Seminars of the AAAS Academy Conference at the same time and place as their National Science Fair detracted from the latter and caused confusion. In contrast, officers of the Academy Conference thought the late spring a better time for those seminars than Christmas week when AAAS met and so asked AAAS to appoint a committee consisting of representatives of the several organizations involved to develop plans for holding the annual National Science Fairs, the National Science Seminars, and the Junior Scientists Assembly all at the same time and place (63). That request was shortly followed by resolutions from the Academy Conference and the directors of the junior academies asking AAAS to endorse and sponsor the National Science Seminars (64).

Faced with that variety of requests and complaints the board of directors did what boards of directors are often prone to do: They appointed a new committee, a Youth Science Activities Committee that was asked to investigate science fairs and other science activities for students. Wallace Brode, a former board member and president of AAAS and a trustee of Science Service, agreed to serve as chairman. Fortunately, he was willing to devote a good bit of time to the committee's study and for that purpose was appointed as a temporary consultant to the staff. Working with a committee that included members from several colleges, a high school, a county board of education, a science museum, and an educational research organization, he embarked on a study that went quite beyond the science fairs and their apparent conflict with the new National Science Seminars (65).

By October of 1965 the final report of the committee was available. Brode and the committee had completed an extensive study of a wide variety of science activities for youth: science fairs, the National Science Seminars, science clubs, junior academies, visiting lecturers, summer institutes, summer camps, museum tours, weekend and summer research employment, individual student research projects, prize competitions for scholarships, presentation and publication of papers, and tours of industrial and other research laboratories.

The board accepted the spirit of the committee's recommendations that AAAS establish a more active program for youth and agreed that it would be desirable to add to the staff one or more persons to help local groups improve science activities for young people. However, before proceeding, it seemed desirable to discuss plans and possible cooperation with Science Service and specifically with its new director, for Watson Davis was retiring from that position and the Science Service trustees were then searching for his successor (66).

We soon learned that Science Service wanted Ted Sherburne as its new director. That was a fine opportunity for him and with reluctant but pleased congratulations we saw him move two blocks away to his new office (67). Within a few days Kneeland Godfrey, his major assistant at AAAS, resigned to become assistant editor of *The Civil Engineer*, the journal of the American Society of Civil Engineers, with the expectation that he would probably succeed to the editorship in about three years. With regrets and congratulations we saw him off to his new responsibilities.

Despite the loss of those two men most of the then-existing activities directed toward improving the public understanding of science continued. Thelma Heatwole managed the annual press room. William Kabisch assumed responsibility for the Holiday Science Lectures for high school students. He and I arranged the seminars for members of Congress. The Council for the Advancement of Science Writing took full responsibility for *Understanding*. And the seminars for the Writers Guild East were about to come to an end (68).

Thus the established activities continued, but the prospective new activities for young people and the direct replacements for Sherburne and Godfrey fell victims to what at the time appeared to be matters of greater urgency: efforts to improve the annual meeting (see Chapter 3) and several new programs:—the Committee on Science in the Promotion of Human Welfare, the Committee on Environmental Alteration, and the study of the use of chemical herbicides in Vietnam (see Chapter 11). Although without replacements for Sherburne and Godfrey, the Committee on Public Understanding of Science remained active. With Walter Orr Roberts as its new chairman the committee reviewed past activities and considered new ones, encouraged the National Science Foundation to enlarge and strengthen its own activities in the field of public understanding of science, supported Walter Berl's plans for extensive television coverage of AAAS annual meetings, and for a time continued the frustratingly unsuccessful effort to arrange a meeting with senior representatives of television broadcasting (69).

In 1968 the National Broadcasting Company offered AAAS a contract to provide services and advice in planning a series of TV programs on topics of scientific interest — a relationship somewhat similar to that with CBS for the earlier "Conquest" series. However, NBC wanted the association to promise that it would not give advice or help to any other producer of television programs on science. That restriction led the board to decide that the contract had to be declined (70).

For another medium, however, and without a restrictive contract, the association could be helpful. AAAS, together with the American Chemical Society, the American Institute of Biological Sciences, and the American Institute of Physics, cooperated in preparing a monthly news column on science that was sent without charge to small weekly newspapers (71).

In 1970 it seemed that AAAS and Science Service might move into a closer and more cooperative relationship. AAAS was becoming more interested in science activities for youth and Science Service was already established in that field. Sherburne had moved from AAAS to Science Service and the association always maintained an interest in Science Service through its nominees for the latter's board. A merger with Science Service would have given AAAS a quick entry into the field of science activities for youth. Interest in the possibility of merger peaked in 1970 with an offer to AAAS to assume the assets and liabilities of Science Service. However, Science Service was losing money and absorbing its assets would have required assuming its deficits. Attorneys for the two organizations explored legal possibilities for a union. The Science Service finances were audited and its prospects analyzed. For a time it appeared likely that the merger would be accomplished, but in the end a motion to that effect failed to secure a majority vote from the AAAS board of directors. The reason was primarily financial. AAAS was facing a large deficit at the end of 1970 and anticipated a larger one for 1971; Science Service was losing about $100,000 a year. What would have been an attractive merger in better times seemed too risky at that time. A motion to accept the assets and liabilities of Science Service and to commit up to $500,000 of association reserves to the new responsibilities lost by a vote of 6 to 3 (72). Fortunately for young people interested in science, Science Service recovered from its financial problems and was able to continue all of its activities.

Concurrently, the board authorized the Committee on Public Understanding of Science, which was then under the new chairmanship of Gerard Piel, to spend up to $25,000 to determine present achievements and AAAS opportunities in four areas: (1) cooperation with teaching associations and other organizations available to youth, (2) TV and radio programming, (3) science publications for young people and, (4) kits of materials with which young people could explore scientific principles and relationships (73). By December 1970 the committee could report that it was making progress on those studies. Thus the decade ended with AAAS looking forward to a new start on scientific activities for young people.

## Prizes

On October 14, 1965, Warren Weaver was in Paris to receive what was expected to be the last Kalinga Prize awarded by the United Nations Educational, Scientific, and Cultural Organization (UNESCO). That annual prize of £1,000 was supported by the Kalinga Foundation of India and was awarded each year to recognize someone for outstanding contributions to the public understanding of science. Nine days later Weaver was in Seattle to receive the first Arches of Science Award of the Pacific Science Center, an award intended to recognize sustained and continuing "contributions to furthering the understanding in America of the meaning of science to contemporary man" (74).

It would have been difficult to find anyone more worthy of those two awards. In the 1930s the Sunday afternoon concerts of the New York Philharmonic Orchestra were broadcast by many radio stations throughout the land. Those were live broadcasts, each with a real intermission. Each Sunday afternoon Weaver had a scientist fill that intermission with a talk on some topic of scientific interest. Later, when Gerard Piel, Dennis Flanagan, and Donald H. Miller, Jr., rebuilt the faltering *Scientific American* into the excellent magazine it quickly became, Weaver was one of their advisors and guides. As vice-president of the Rockefeller Foundation he was involved in awarding funds to the National Association of Science Writers for the 1957 landmark survey of public information and attitudes toward science. He was one of the advisors in organizing the Council for the Advancement of Science Writing. And he was the primary architect of the association's Arden House statement and its reorientation of AAAS activities. He had well earned those two prizes.

As things turned out, however, Weaver was not the last recipient of the Kalinga Prize. In 1965 B. Patnaik, the donor of the prize, had informed UNESCO that his Kalinga Foundation could no longer support the prize, but a year later he reinstated support. Annually UNESCO invited nominations from many organizations, of which AAAS was one. In most years the association made a nomination and over the years had a quite peculiar record of success: Every AAAS nominee from outside the United States was awarded the prize, but only one of five winners from the United States had been nominated by AAAS. The association's out-of-the-country nominees were Karl von Frisch from Austria and Ritchie Calder and Fred Hoyle, both from the United Kingdom. Winners from the United States were Waldemar Kaempffert, George Gamow, Gerard Piel, Warren Weaver, and Eugene Rabinowitch, of whom Weaver was the only AAAS nominee. The association's unsuccessful nominees from the United States were George W. Gray, Watson Davis, Margaret Mead, and Harlow Shapley. Isaac Asimov's nomination was never completed for Asimov declined in advance to go to Paris to receive the award if it were offered to him or to go to India to give the addresses expected of all Kalinga Prize winners (75).

After the Pacific Science Center had become well established as the successor to the United States Science Exhibit at the Seattle World's Fair of 1962, it announced that it would make an annual award to recognize outstanding contributions to the betterment of general understanding of the meaning of science to the nation. The Arches of Science Award, a gold medal and $25,000, was intended to focus attention on the importance of communicating and understanding both the benefits and the limitations of science and the ways in which science functions.

Selection was by a committee on which a total of two dozen members served during its six years of life, usually for three-year terms. About a third of the members were presidents or senior officers of commercial or industrial companies, mostly from the Northwest. A third were university or college presidents. Most of the remainder were from foundations, research institutions, or other organizations. There were a couple of officers of federal government agencies and one distinguished Hollywood director, Frank Capra.

AAAS never responded to invitations to submit nominations for this award and for a very good reason: I had the pleasure of serving as chairman of the selection committee for the entire period, which ended when the Northwest Bell Telephone Company, the primary sponsor of the award, and its helpers, the Boeing Company, Textronics Incorporated, and Seattle University, decided they could no longer bear the cost. In the meantime the award had been made to six persons: Warren Weaver (1965), René Dubos (1966), James B. Conant (1967), Glenn T. Seaborg (1968), Gerard Piel (1969), and Margaret Mead (1970). Of those six Conant and Weaver had been presidents of AAAS. Seaborg, Piel, and Mead later became president and Dubos would have been had he outpolled Bentley Glass in the election of the association's 1968 president-elect.

Improving the public understanding of science has long been one of the association's objectives. That such a diverse selection committee chose the list of award recipients they did was a tribute to the extent to which top officers of AAAS had accepted personal responsibility "to increase public understanding and appreciation of the importance and promise of the methods of science in human progress."

Chapter Ten

# International Activities

National boundaries pose problems for scientific societies. Language, custom, and geography all encourage national or regional organizations. Yet science itself is international and the interests of scientists keep spilling over national boundaries. Sometimes that dilemma leads to international unions or committees and sometimes to international exchanges, joint ventures, or other cooperative arrangements. AAAS has used most of those possibilities in furthering its international interests.

Whatever the organizational structure may be, scientists also act as individuals, and it was individual initiative that started AAAS on the road toward its international activities. When early American scientists traveled abroad, the meetings of scientific associations provided good opportunities to meet scientists from other countries. Because of the common heritage and common language, attendance at meetings of the British Association for the Advancement of Science (BA) was particularly attractive and before AAAS was established American scientists were reading papers and making reports at BA meetings (1).

After the American association was established, visitors from other countries had similar opportunities in this country. From 1848 on they were welcomed at AAAS meetings; and in 1876 when the country was celebrating its 100th birthday "numerous foreign savants" who were representing other countries at the Centennial Exhibition in Philadelphia came to Buffalo to attend the AAAS meeting (2).

Meetings themselves could sometimes be moved outside national boundaries and in its first century AAAS met five times in Canada. Canada could hardly be considered a foreign country, for the association has always had Canadian members, and it was at a summer meeting in Ottawa in 1938 that the British and American associations agreed upon a regular exchange of guest lecturers. That pact came into effect a few months later when Sir Richard A. Gregory, distinguished astronomer and retiring editor of *Nature*, "delivered a warmly received lecture ... 'Religion in Science' " at the association's Christmas meeting in Richmond, Virginia. In the following year Isaiah Bowman, American geographer, addressed the British association on the topic "Science and Social Planning" (3). World War II interrupted the exchange, but it was resumed in 1946 when C. B. Fawcett came to Boston to give an address on "The Numbers and Distribution of Mankind" (4).

In 1954 when I became executive officer of AAAS, Sir George Allen, formerly vice-chancellor of the University of Malaya, became secretary of the British association. We had much to learn about our new responsibilities, and we usually got together to discuss association business whenever I visited Europe to attend a conference or committee meeting of UNESCO or the Organization for Economic Cooperation and Development or to recruit talent for the AAAS Visiting Foreign Lecturer Program (see Chapter 8). Those friendly discussions helped continue the exchange relationship. Nearly every year an American scientist delivered one of the general lectures at the BA meeting, a British scientist gave a comparable address at the AAAS meeting, or both associations had a guest lecturer. The exchange continued on a more or less regular basis until 1972 when Richard L. Gregory of the Institute for Brain Research in Bristol gave the last exchange lecture at a AAAS meeting.

The friendly relations between the British and American associations sometimes made it easy to take advantage of special opportunities. For example in 1957 the BA made a last-minute addition to its program by inviting Mary Catherine Bateson to describe some of the anthropological and sociological observations she had recently made in Israel. Her mother, Margaret Mead, had first addressed the British association at an unusually early age; to have her 17-year-old daughter address another BA audience was a source of parental pride for a AAAS director and future president.

For a few years AAAS also had an exchange agreement with Znaniye, the Soviet organization of some two million scientists and teachers of science, engineering, and other technical subjects, whose members were giving several million lectures a year to school groups, young people, and other audiences. In 1969 Bentley Glass, then president of AAAS, spent 10 days visiting officers and members of Znaniye in several Soviet cities. They expressed the wish for continued visits between the two organizations; in 1969 AAAS invited a delegation from Znaniye to attend its annual meeting (5) and at the 1970 meeting the association had two addresses by guests from the Soviet Union. Since then there have sometimes been addresses by foreign visitors, but no continuing exchange agreement comparable to the one that long existed between the AAAS and the BA.

## Publications

Publications could also be used to foster foreign relations, and AAAS early began sending copies of its annual proceedings to academies, museums, and other appropriate organizations in European countries. For example, the 1872 proceedings volume—containing the AAAS constitution, list of officers and committees, the presidential address, summaries of many and titles of all of the papers read, and a report of the business meeting—went to 79 recipient in 56 European cities (6).

Soon after AAAS became the owner and publisher of *Science*, Willard Valentine, the new editor, recommended that the magazine start a new section to be called "International News" and proposed that the association try to find a way to finance 2,000 foreign subscriptions "so that the publication might play a role as a world journal of science" and be of service to the newly formed United Nations Educational, Scientific, and Cultural Organization (UNESCO). Moulton supported those proposals, pointing out that the foreign circulation of *Science* then totaled only 3,303, of which 794 copies went to Canada (7). Valentine's untimely death ended consideration of his recommendations for a while, but as the size and quality of the magazine increased so did its attractiveness to foreign scientists. In 1956 the International Cooperation Administration (later renamed the Agency for International Development) offered to subsidize membership in AAAS for foreign scientists who had come to the United States as students and then had returned home. Membership then cost $10 a year, of which the International Cooperation Administration contracted to pay $8, leaving $2 to be paid by each new foreign member (8).

In 1963 Philip Abelson and I told the board that we wanted to add a European reporter to the staff of *Science*, and in April of 1964 the staff list printed in the magazine began carrying a new name: Victor McElheny was the magazine's European correspondent. When he returned to the United States in 1966 he became a "Contributing Correspondent," and John Walsh, a senior member of the news staff, moved to the Oxford region, where he had earlier been a Rhodes Scholar, and began filing stories about European science. When Walsh came home in 1968 Daniel S. Greenberg from the news staff took his place. After his two years abroad, which ended in August of 1970, no replacement was named. Instead, the editors decided to use free-lance reporters for a trial period and to send members of the staff for short visits, such as the trip Philip Boffey had recently made to Japan (9). A dozen years later, however, the position was reestablished with the appointment of David Dickson as European correspondent.

With increasing interest in European news, a larger foreign circulation seemed achievable. Sarah Dees, AAAS librarian, compared the magazine's mailing list with lists of major research institutions and universities throughout the world, especially in developing countries, to identify those that were not receiving *Science*. The list of nonreceivers turned out to be manageably short, and the board of directors approved offering initially free subscriptions to the nonreceivers so that AAAS could boast that its magazine went to essentially every major university and research institution in the world (10). Most of the invited institutions wanted to receive *Science*, and most of those continued after the free trial period. As the total circulation of *Science* increased year after year the foreign component by 1970 had reached 16,000, a level that pleased the staff, for that was almost as much as the total circulation of *Nature* (11).

## International Conferences

Most international scientific meetings are sponsored by the appropriate international scientific union or some other international body. However, AAAS assumed primary responsibility for arranging international meetings in two fields of study in which no international union existed. First was the Arid Land Conference of 1955; later came the International Oceanographic Congress of 1959 and then the second International Arid Lands Conference in 1969.

### The 1955 Conference on Arid Lands

In the 1950s the UNESCO Arid Zone Programme Committee usually sponsored an arid lands conference in even-numbered years and in odd-numbered years encouraged some other organization to sponsor some type of arid lands meeting. When the Southwestern Division of AAAS met in 1954, Peter Duisberg of the Desert Products Corporation in El Paso, then chairman of the division's committee on arid lands, recommended that the Southwestern Division of AAAS arrange an international meeting for 1955, only one year away. The idea caught fire quickly. Duisberg and other members of his committee were so confident that UNESCO would support them that they started immediately to plan a program and work out detailed arrangements.

While they were doing that, Gilbert White was preparing to go to Paris to attend a meeting of UNESCO's Arid Zone Programme Committee. As part of that preparation the National Academy of Sciences–National Research Council Committee on the UNESCO Arid Zone Programme held a meeting. When the members learned of the Southwestern Division's proposal they were enthusiastic, but recommended that AAAS as a whole be the sponsor. A few days later when White got to Paris, the UNESCO committee approved the plan, agreed to help by convening a meeting of the Arid Zone Programme Committee at the time and place of the proposed conference, and offered an additional $3,000 toward the expenses of the meeting.

All of that was accomplished in just over a month. When the AAAS board of directors met on May 23–24 they were presented with a fait accompli. About all they could do was endorse the plan, which they did unanimously, and ask me, the association's very new executive officer, to seek the necessary funds (12).

Eleven months later the International Arid Lands Meetings opened in Albuquerque, New Mexico. The meetings consisted of three rather different and successive events. First was the symposium, which included three evening lectures for the public and technical daytime sessions for the registrants (13). The symposium followed immediately after the annual meeting of the Southwestern Division in order to make it easy for interested members of the division to attend. Many did; in fact a little over half of the 454 registrants came from Arizona, New Mexico, and Texas. However, interest was much more widespread; 23 other states were represented; 26 registrants came from the District of Columbia; and 62 registrants came from 27 foreign countries.

After the meeting in Albuquerque ended, 71 invited participants from 18 countries were taken on a two-day field trip through southwestern New Mexico to give them an opportunity to get better acquainted with each other and to observe a specific arid region and the measures that were being used to increase its productivity. James Swarbrick, chief of staff of the UNESCO committee, rated it by far the best field trip ever arranged for that committee, not just a sight-seeing tour but an opportunity to study arid lands problems and the remedial and developmental operations being applied.

After the field trip the 71 participants settled down in Socorro, New Mexico, at the New Mexico Institute of Mining and Technology, to formulate the recommendations they wanted to make to scientists working on arid lands problems throughout the world. The 31 resulting recommendations and the major papers from the symposium in Albuquerque were published in *The Future of Arid Lands*, one of the association's symposium volumes. The 3,500 printed copies were quickly sold out, and a larger body of readers found a brief conference report and all of the recommendations in *Science* (14).

The conference in Socorro broke an association tradition by being open only to invited participants: all members of the UNESCO advisory committee and other selected scientists from the United States and abroad. The board was somewhat uncomfortable with that arrangement, but as it turned out, no problems or objections were raised. The Southwestern and Rocky Mountain Division — its name had just been changed — adopted a resolution specifically "endorsing the principle of closed meetings when the expectations are that the purpose of such a meeting will be best served by limited attendance" (15). The board agreed and followed that precedent in planning some later meetings such as the Parliament of Science held in 1958 and the Symposium on Basic Research held in 1959 (see Chapter 2).

Most of the expenses of the arid lands meetings were met by grants of $10,000 each from the National Science Foundation and the Rockefeller Foundation, $3,000 from UNESCO, and $700 from the Department of State. In addition UNESCO paid the traveling costs of members of its own advisory committee; several countries helped with travel grants for their delegates; and the local finance committee raised over $8,000 from local sources with the understanding that any money left over after the meetings would be used for studies of critical arid lands problems in the southwestern region. As things turned out, almost $7,000 of that last fund was left, as intended, for local use.

The meetings attracted substantial press attention. The grant of $700 from the Department of State was specifically for the purpose of employing Carl O. Hodge, a science writer from Tucson, for publicity. Some 9,000 copies of the preliminary program were distributed worldwide "and some 40 news writers and the 150 news and magazine outlets that were sent copies of papers and other news materials blanketed the world with news about the deliberations" (16).

Many of the foreign participants took advantage of the opportunity to visit other parts of the United States, and a good number were invited to visit colleagues or give addresses at universities or other institutions. One calls for special mention: W. F. J. M. Krul, professor at the University of Delft and director of the Netherlands Government Institute of Water Supply. Holland can surely not be thought of as an arid country, but no other country has had as much experience in ridding its soil of excess salt. Krul's experience with that problem was of particular value to countries in which continued irrigation was increasing the salt content of agricultural lands. After the meetings ended Krul gave invited addresses at the University of California in Berkeley, at Northwestern, Chicago, and Harvard universities, and at the Massachusetts Institute of Technology.

All in all, the feeling among members of the Southwestern and Rocky Mountain Division was that they had initiated a most successful meeting. A few years later UNESCO was developing its "Major Project on Arid Lands Research," a project that was in part an outgrowth of the New Mexico meetings of 1955. To help in its planning UNESCO asked Gilbert White, who had served as chairman of the New Mexico meetings, to visit arid lands research institutes in the belt stretching across North Africa and the Middle East as far as Pakistan and India. Everywhere he went, he later reported, he found positive effects of the 1955 meeting: new collaborations, adaptation of ideas or practices discussed at Albuquerque and Socorro, or other outcomes of those discussions (17).

AAAS officers and staff were pleased with the outcome and experience of the Arid Lands Conference (18), and that experience gave AAAS both precedent and encouragement in planning later international meetings.

## First International Congress of Oceanography

Successful experience with the arid lands meeting soon led John Behnke to recommend that AAAS sponsor a similar meeting on oceanography, which, like arid lands, had no international union to sponsor meetings. Oceanographers and other advisors endorsed the idea, and early in 1956 the board told the staff to start planning an international meeting on oceanography. At the same time some of the board members suggested that an international meeting on pollution would also be appropriate (19).

With one international meeting accomplished and two others under consideration, the board decided it was time to consider the circumstances under which it was appropriate for AAAS to sponsor international conferences. Although members and *Science* subscribers were to be found on every continent, AAAS was a national organization, not an international one, and it was the National Academy of Sciences, not AAAS, that usually represented the United States in international scientific affairs. Nevertheless, the board decided that AAAS sponsorship was sometimes appropriate if the topic of the meeting required interdisciplinary consideration (meetings on topics that fell within individual disciplines could more appropriately be sponsored by organizations that specialized in those disciplines); if the topic was of significant interest to the United States; and if it was in a field in which the state of knowledge was such as to make a conference useful. Finally, although cosponsorship by an international organization should not be required, such cosponsorship was desirable (20).

The arid lands meetings met those requirements. Pollution and oceanography both seemed to be appropriate topics, but they had very different fates. A conference on pollution was never held. The one on oceanography quickly outgrew the board's initial expectations.

One of the most critical consultations concerning the oceanographic meeting was with UNESCO to determine the interests of that organization. Instead of raising that question by letter I took advantage of attendance at a conference on science education held by the UNESCO Institute for Education to visit M. Yoshido, the secretary of the UNESCO Committee on Marine Sciences, to tell him of the association's interest and to inquire about UNESCO cooperation and support. He was enthusiastic and took me to meet Viktor Kovda, the Russian soil scientist who was director of the Department of Natural Sciences of UNESCO. (Thirteen years later Kovda was one of the major contributors to the success of the association's second international conference on arid lands.) Kovda told me he was just learning about oceanography and then, grinning at Yoshido, added, "and he is my tutor." With the tutor's enthusiastic support UNESCO agreed to cooperate in making arrangements for the meeting and to provide some financial help (21).

With UNESCO cooperation assured, it was time to appoint a committee to plan the meeting and begin corresponding with oceanographers in other countries. William Rubey was of special help in selecting members of the steering committee. He was the member of the board whose research interests were closest to oceanography, and as the recent chairman of the National Research Council he was well informed about the interests and special abilities of many scientists. It was his good advice that led us to Mary Sears, senior scientist at the Woods Hole Oceanographic Institute, who agreed to serve as chairman of the committee, and with her help we selected the other members (22). With the committee in place I sent the members a letter describing the arid lands meeting of 1955, suggesting that the board had something similar in mind for oceanography, and asking them to start making plans. I did not attend the first meeting of that committee, but the minutes of that meeting and Mary Sears' accompanying letter made it plain that the session had been a lively and productive one. Instead of the modest meeting the board had expected and I had outlined in my letter, the committee proposed a full-scale international congress to last two weeks. They wanted it to be held at the United Nations headquarters in New York City because of its excellent facilities for simultaneous translation. In addition to the scientific sessions the committee also made plans for field trips, receptions, and public lectures or other evening programs of interest to a wide popular audience. They selected the specific themes to be used in organizing the technical sessions and began corresponding with oceanographers in other countries to get their reactions to the plan and to help fill in program details and select major speakers and other program participants. When they added up the prospective costs, they told us the association should raise $105,000 to pay the bills (23). Because that first committee meeting had been attended by and the report approved by four members of the UNESCO Advisory Committee on Marine Sciences who were in Woods Hole at the time, we knew that the committee's plans already had de facto approval of the UNESCO committee.

The board was surprised by the enlarged scale of the committee's plans, and the members were not as enthusiastic as were the committee members. But they deferred to the committee's judgment, for it was clear that we had a strong and excellent committee which was confident that the time had come for a large general meeting at which oceanographers with many interests could come together to assess progress and plan future work. A few weeks later Roger Revelle returned from Bangkok and a meeting of the UNESCO Advisory Committee to report that that international body had endorsed the committee's proposals. Thus the board told the committee to proceed as planned and asked the committee and me to try to find the necessary $105,000 (24).

We did find the money and without much difficulty, for it was the right time for a large international meeting on oceanography. IGY — the International Geophysical Year — was then encouraging scientists from many countries to work together on global geophysical problems. The International Advisory Committee on Marine Sciences of UNESCO and the Special Committee on Oceanic Research (SCOR) of the International Council of Scientific Unions had held their first meetings in 1955 and 1957 respectively. There was increasing governmental and political interest in ocean studies and engineering in the United States (25). The formal concepts of plate tectonics had not yet been expressed, but new discoveries concerning the mid-ocean ridges were stimulating new thinking about the history of the world's oceans and land masses. It was, in short, a time when the world's oceanographers wanted to get together to review progress and to think about what should be done next (26).

The planned congress was endorsed by the International Union of Geodesy and Geophysics as well as the the Special Committee on Oceanography of the International Council of Scientific Unions and UNESCO, both of which also agreed to be listed as co-sponsors. The co-sponsorship relationship had an interesting definition for some members of the UNESCO staff, as I learned during a visit to Paris when I was told "UNESCO is delighted to co-sponsor the congress as long as you do all the work."

UNESCO also joined other organizations in providing more tangible support. For three years its Advisory Committee on Marine Sciences sponsored no other meetings, so that it accumulated $15,000 with which it could cover the travel costs to the congress of its own members and members of SCOR. The U.S. Office of Naval Research gave the association a grant of $50,000, by far the largest single grant received. Other grants came from the Atomic Energy Commission, the National Science Foundation, the International Oceanographic Foundation, and the EDO, Arthur D. Little, Rockefeller, and Sloan foundations. Fourteen industrial and oil companies provided $10,000. A modest registration fee brought in $11,750. All told, we obtained $130,500, well above the budget of $105,000. In addition, the Military Air Transport Service brought a few of the foreign participants without charge to them or us; UNESCO paid the United Nations for the use of their facilities; the United Nations made no charge for the interpreters who worked throughout the whole two-week period; and the work on congress affairs of various members of the AAAS staff was not charged to the congress budget.

With the United Nations building as the preferred meeting site we asked our co-sponsor UNESCO to request permission to meet there. The UN granted that request, but warned that we could not be absolutely assured that all of the facilities would be available for the whole two weeks, for world conditions might require calling an extraordinary session of the General Assembly or for convening the regular session earlier than normal. However, the UN official added, if an emergency did develop, he thought it could be dealt with in the afternoons so we could plan on using the General Assembly room for the morning plenary sessions of the congress. (As international affairs developed, no crisis inconvenienced the Oceanographic Congress, but it was only a few days later that Premier Khrushchev made his famous visit to Washington, New York, San Francisco, an Iowa cornfield, Camp David, and a few other places.)

The AAAS and the National Academy of Sciences informed their sister organizations in other countries of the meeting and invited interested oceanographers to attend, and AAAS also informed foreign embassies in Washington, D.C., of the meeting. The generously cooperative Department of State sent announcements of the meeting to diplomatic posts in 76 other countries; asked that assistance be given to anyone requesting a visa to attend the congress; and added that if anyone who for political reasons might not normally be considered admissible wanted to come to the meeting the visa should not be denied but the case referred promptly to Washington, D.C., for decision.

When the congress opened on August 31, 1959 — under the presidency of Roger Revelle — registration totaled 1,175, well over the committee's early expectation of 500 to 800. There were 840 from the United States and 335 from 53 other countries. The Soviet Union sent the largest foreign delegation: 63 who came on board the *Lomonosov,* the Soviet Union's new oceanographic research vessel, which, along with Jacques Cousteau's *Calypso* and several other research ships, was moored in New York harbor for the duration of the congress.

One aspect of the registration list that pleased the steering committee was the number of young oceanographers present. That was no accident; the committee had used about $25,000 to assist young oceanographers to come to New York for the meeting.

The registrants came for wholly scientific reasons. Unlike the arid lands conference, no resolutions or official recommendations were planned or adopted. Instead, the meeting was intended solely " ... to provide a common meeting ground for all sciences concerned with the oceans and their contained organisms" (27). Two days were devoted to each of five themes: history of the oceans, populations of the sea, the deep sea, boundaries of the sea, and cycles of organic and inorganic substances in the sea. Each morning was devoted to a plenary session with three addresses on the day's theme and each afternoon offered several concurrent sessions for submitted papers and further discussion of particular aspects of the day's theme. At the time of registration participants were given a 1,000-page book containing abstracts of the papers to be presented in the afternoon sessions, with summaries in French, German, or Russian (28). After the congress adjourned Mary Sears edited an AAAS symposium volume containing the addresses given at the morning plenary sessions (29).

There were several evening sessions to which the public was invited including a film session by Jacques Cousteau. During the intervening weekend several field trips were available to the registrants. Remembering the pleasant amenities of international congresses abroad the steering committee made a special effort to be equally cordial hosts. The chief social events were a reception on the evening before the congress opened, parties at the American Museum of Natural History and the Columbia University Faculty Club, a concluding banquet, and a memorable party aboard the *Lomonosov* with the Soviet delegation as host.

The difficulty of bringing some technical sessions to a close, the frequent continuation of discussions and debates in the corridors or over meals, the extensive newspaper coverage (30), and the many compliments to members of the steering committee and AAAS officers all indicated that the congress had been a great success. In his opening address the congress president Roger Revelle had challenged young oceanographers "to ask questions of the ocean; to think no small thoughts about their work; to fan the flame of controversy" that would bring the knowledge and points of view of the many kinds of scientists interested in the oceans to a more complete understanding of their common subject matter. Later, in his preface to the proceedings volume, Revelle concluded: "The International Oceanographic Congress was a great day for oceanographers. ... One of the lasting values ... was that a number of bonfires of controversy were lighted and some of them were fanned to bright flames" (31). The board of directors discharged the steering committee with an enthusiastic vote of commendation for having arranged a meeting of which the association could be proud (32).

When all the bills were paid we informed the sources of financial support that we had several thousand dollars left over. We offered to return any donor's proportionate share of the surplus, but added that any of the fund left with the association would be used for purposes wholly consistent with the original grant, perhaps to help finance the next international oceanographic congress. Most chose to leave their portions with AAAS. We invested the total of $14,722, and when SCOR was planning the international congress on oceanography held in Moscow in 1966 we had nearly $20,000 to contribute. That money was used to help bring 29 oceanographers from 17 countries to attend the Moscow congress (33).

So ended the largest and most elaborate international meeting the association had sponsored. There was no need for AAAS to sponsor further oceanographic congresses, for SCOR assumed that responsibility. In contrast, arid lands became a more enduring interest of AAAS, for there was no comparable international body in that field.

## Return to Arid Lands

The U.S. portion of the earth's arid and semiarid lands lies within the region of the Southwestern and Rocky Mountain Division (SWARM) of AAAS. It was that Division's Committee on Desert and Arid Zone Research that initiated plans for the 1955 arid lands meetings in New Mexico; arranged annual symposia on arid lands problems; published the *Arid Lands Research Newsletter* and mailed copies to recipients in over 40 countries; and helped form citizens groups in Texas and New Mexico to take local action on arid lands problems.

Yet the committee felt limited by its divisional status and in 1960 began to press AAAS to establish a national committee on arid lands issues. Gilbert White, who had been president of the 1955 meetings, had recently returned from his UNESCO study tour of arid lands institutions in North Africa, the Middle East, and South Asia. UNESCO had reconstituted its arid zone program, and by then the report of the association's 1955 arid lands meeting was in its third printing. With all of that evidence of widespread interest, the committee asked AAAS to establish a national committee that could work with UNESCO and with committees or organizations of other countries having arid lands interests (34).

That recommendation came to the board of directors at about the same time that the National Academy of Sciences–National Research Council established an arid lands committee consisting of representatives of interested government agencies and professional societies. The board was reluctant to have AAAS appear to compete with the new NAS–NRC committee, but SWARM's committee insisted that a national committee under AAAS auspices was desirable. An ad hoc conference of persons from NAS, the UNESCO advisory committee, SWARM, and other organizations evidenced considerable support for a national committee under AAAS auspices and specifically recommended that proposed new committee as the appropriate body to prepare the U.S. contribution to a large Latin American congress on arid lands that UNESCO planned to hold in Buenos Aires in 1963. The outcome of all of this was agreement that AAAS would raise the division's Committee on Desert and Arid Zone Research to a national committee, increase its membership, and have it prepare the U.S. contribution to the Argentine meeting (35).

The National Science Foundation provided $40,000. The committee went to work with Carl O. Hodge as the temporary full-time editor coordinating the work of a large number of chapter authors and contributors to produce *Aridity and Man*, a comprehensive account of American experience in research and management of arid lands problems. The Organization of American States translated the volume into Spanish, and copies were ready for participants as they registered at the meeting in Buenos Aires in September 1963.

With the Latin American meeting over, the AAAS committee was to revert to divisional status. But was a national committee also needed? The divisional committee thought so and described a variety of tasks that it could begin working on immediately. The board of directors was not so sure of the need, and it took more than a year for the board, SWARM, and an interim committee appointed to consider the matter to come to agreement that there should be both regional and national committees. The board stipulated that the new AAAS Committee on Arid Lands that was to be established would be examined after three years to decide whether it was working well enough to be continued. As a guarantee of close cooperation between the two committees the board appointed Terah L. Smiley of the University of Arizona, a staunch member of the division's committee, as chairman of the new national committee (36).

The new AAAS committee went promptly to work on several fronts and soon recommended that AAAS plan another international conference on arid lands to be held somewhere in the Southwest in 1969 (37). Staff consultations with UNESCO, the World Meteorological Association, the Department of the Interior, and other interested bodies brought approval of that proposal, and the board authorized the committee to begin planning the proposed meeting and the AAAS staff to start collecting the necessary funds (38).

With its own members as a nucleus the Committee on Arid Lands expanded into an organizing committee, appointed an international program committee, and proceeded to plan the "International Conference on Arid Lands in a Changing World" that was held at the University of Arizona in Tucson on June 3–13, 1969.

Reductions in the budgets of the National Science Foundation and the Agency for International Development made it impossible for those two agencies to provide financial help. But the Ford and Rockefeller foundations, the Departments of Commerce and the Interior, and the Atomic Energy Commission gave major grants. Standard Oil of New Jersey, E. I. du Pont de Nemours, Deere and Company, El Paso Natural Gas, and the Four Corners Regional Commission provided smaller amounts. The University of Arizona promised money and services worth $11,000 to $12,000 (39). With those monies and promises in hand, plans went forward.

With participants from 32 nations in the audience the conference presented five plenary sessions with invited papers and some two dozen smaller sessions with both invited and contributed papers. UNESCO and the academies of science of Arizona and New Mexico helped with planning and arrangements. The American Society of Range Management and the Commission on Geography of Arid Lands of the International Geographical Union held special meetings in conjunction with the conference. Two special lectures were presented, one honoring the memory of John Wesley Powell, 19th-century scientist-explorer of the Southwest, and the other sponsored by the American Geographical Society in honor of Isaiah Bowman, American geographer and former president of the Johns Hopkins University. At the end of the conference summary papers were presented by Gilbert White, then at the University of Chicago, and Viktor Kovda of the Soviet Academy of Sciences. Both stressed the need for continuing international collaboration in working on arid land problems.

Some of the papers and discussion dealt with technological issues, but the chronicler of the meeting concluded that "perhaps of more basic importance were those papers dealing with the economic, the sociological, the administrative, and the cultural problems of areas whose people for many centuries have had few resources at their disposal" (40).

That theme was emphasized in the 17 resolutions adopted by the conference participants following recommendations submitted by a resolutions committee chaired by J. S. Kanwar, deputy director-general of research of the Indian Council of Agricultural Research. The resolutions were largely addressed to issues and methods of aiding the world's developing countries, for one of the differences between those countries and the technologically developed countries was that they were more largely located in arid or semiarid parts of the world and had not had the benefit of the kinds of research on agriculture, animal breeding, and related issues that had proven so useful to developed countries in the world's humid regions (41).

With the 1969 conference concluded and a proceedings volume published (42), the Committee on Arid Lands was continued as a sponsor of symposia, as an advisory body to the University of Arizona Press on publications concerning arid lands issues, and as organizer of the association's third international conference on arid lands, held at the University of Arizona, October 21–25, 1985.

Thus concern with arid lands problems became a permanent interest of the association. In oceanography the Special Committee on Oceanic Research of the International Council of Scientific Unions assumed responsibility after AAAS had organized the world's first large congress of oceanographers in 1959. In contrast, scientists engaged in work on problems of arid lands had no comparable international organization. In 1955 and 1969 AAAS had served that area of scientific, practical, and humane concern well and chose to make the arid regions of the world one of its continuing interests.

## An International Federation?

After World War II, as a number of new associations for the advancement of science were organized in other countries, the AAAS board considered ways in which AAAS might become more helpful to some of those new associations, especially in Latin America. To further that inquiry they appointed a subcommittee on international relations consisting of Walter Miles, Harlow Shapley, and Elvis Stakman (43). Simultaneously, UNESCO was developing its role in scientific affairs and was considering the possibility of bringing the national associations together into an international federation of associations for the advancement of science.

Reactions to that proposal were mixed. The French and several other associations were favorably inclined, but the British and American associations were unenthusiastic. Nevertheless, the AAAS board and council agreed that AAAS should keep informed about the plan, so for several years it was a frequent item of board meeting discussions. In 1950 Kirtley Mather and Karl Lark-Horovitz discussed the plan with David Lowe, then secretary of the British association, and with Børge Michelsen and Pierre Auger of the UNESCO staff. In that same year Wallace Brode attended a meeting in Paris called by Gerald Wendt, formerly a science consultant to *Time, Life,* and *Fortune,* and then the UNESCO staff member who was leading the effort to establish the international federation. That meeting led to no decision other than to meet again in 1953, when AAAS was represented by Detlev Bronk (44).

The 1953 meeting developed alternative "strong" and "weak" proposals. The strong proposal was for an organization with a secretariat in Paris and annual regional meetings on a continental basis. The weak proposal was to appoint an advisory committee on international scientific matters (45).

After consultation with officers of the British association, the AAAS board recommended that any available funds be used "to facilitate international travel of working scientists instead of on the administration of an international federation" (46). The British association had already established a good precedent for that proposal, for with support from the Commonwealth Fund their meetings were providing annual opportunities for the exchange of ideas and information among representatives of associations in the Commonwealth countries, and usually also from France and the United States, and less frequently from some other associations.

A year later the world federation idea came to an end, at least temporarily. The UNESCO proposal had been approved by the national associations of Brazil, Ceylon (now Sri Lanka), France, Rhodesia (now Zimbabwe), South Africa, and Venezuela. It was opposed by the British and American associations. The other 12 countries that then had national associations for the advancement of science had not responded to UNESCO's inquiry. So the idea was dropped and as a kind of substitute UNESCO decided to form a committee to advise the director-general on "matters that concern the teaching and dissemination of science to the public" (47). Pierre Auger, director of UNESCO's Department of Science, already had an advisory committee on scientific matters, one appointed by the National Academy of Sciences with Bart Bok of Harvard as its chairman. The association's temporary Committee on World Federation of Science reviewed the terms of reference of the proposed International Advisory Committee on the Public Understanding of Science, agreed that it would be a good addition to the UNESCO structure, and asked the board to inform UNESCO of the association's general approval (48).

Yet the idea of an international federation was not quite dead. Chauncey Leake once again raised that issue at a meeting of the board of directors. It was briefly considered at a meeting of representatives of several associations who came together at the 1961 meeting of the British association (49). And in 1969 at the joint conference of the AAAS and the BA on the future of associations for the advancement of science the possibility of an international federation was brought up for discussion. It met with so little favor the topic was quickly dropped.

## An Office of International Science for AAAS?

In the period when the council was establishing study committees, one was a Study Committee on Cooperation with Developing Countries. Its 1966 report called upon AAAS to establish an Office of International Science. As activities that the proposed office might conduct or sponsor, the committee suggested symposia on the organization of research or the improvement of education in developing countries; presentation of awards and medals for contributions to techniques for science development in other countries; increasing the coverage of international news in *Science*; and the development of regional science centers and other aids to science in the newer countries (50).

As the board considered that sweeping proposal in light of the association's responsibilities and resources, they consulted with officers of the National Academy of Sciences and the National Science Foundation. It appeared that those organizations would welcome a AAAS Office of International Science and that there would be many opportunities for that office to be useful (51). However, there were already enough other activities, including a number with international aspects, to keep the board and staff busy, and the formal decision concerning a new office was postponed from meeting to meeting until the de facto decision was negative (52).

Several years later, however, as the association's international activities became more extensive, an Office of International Science was established (53), thus institutionalizing a range of activities that had led a national association into a variety of international activities.

## Chapter Eleven

# Science in Society

World War II and the following years brought widespread recognition that science had become inescapably linked with technology, with health, with national prestige and economic well-being, and with warfare. Those linkages also brought increasing signs of trouble, as science began to suffer from some of its successes. The military contributions of engineers and of scientists turned engineers "for the duration" were widely acclaimed, but the amount of academic research being supported by military funds was a source of considerable worry (1). As nuclear engineers turned from building bombs to building power plants, radiation hazards became worrisome. The DDT that had worked wonders as a disinfectant was threatening some bird populations; it was also losing its potency as DDT-resistant strains of disease vectors developed. As such unhappy consequences became evident, relations between society and its scientists and engineers became more strained.

What should AAAS do about this increasingly troublesome situation? That question came to a head at the 1955 annual meeting when Ward Pigman, a biochemist at the University of Alabama Medical Center, submitted a resolution to the council. He was concerned about the harm to society that would or might result from some technological developments and also about the harm to science that could result from mistrust, restrictive requirements, or rejection. He asked AAAS to create an "Interim Committee on the Social Aspects of Science" to examine the sociological problems being created by science and to present to the council such recommendations for action as it found desirable.

The Council Committee on Resolutions was impressed but not convinced. It's chairman noted that the committee "recognized the importance of the matter discussed but did not agree that the appointment of a committee would be able to render assistance in solving it." Pigman was not put off; he asked for permission to submit his resolution directly to the council and when he did, the council overrode its Committee on Resolutions and adopted the resolution. Not surprisingly the council president asked Pigman to serve as chairman of the interim committee he had proposed (2).

That was seven years before publication of Rachel Carson's *Silent Spring* and 15 years before Earth Day. Yet the evidence that later led to those and other manifestations of increasing concern was already becoming apparent. National expenditures for research and development were increasing rapidly. Major industries based directly on research were burgeoning, especially the chemical, electrical, nuclear, and pharmaceutical industries. As the interim committee considered the probable consequences of these and other changes, it seemed to the members that both the problems for science and those for society would surely become more pressing in the years ahead. Thus they recommended that the association accept responsibility for analyzing and presenting the views of scientists on matters of interaction between science and society. As the next step toward that end they asked that the interim committee be enlarged and continued for another year so that it might define the issues more fully, assemble the facts, and suggest practical programs to be submitted to the board of directors for implementation (3). The board agreed and in expanding the committee asked Chauncey Leake to assume the chairmanship (4).

Pigman's resolution and the resulting Interim Committee on the Social Aspects of Science were not the first examples of association interest in the social and environmental consequences of scientific and technological developments. In the 1930s AAAS had had an exploratory committee on science and society and in the 1940s one on conservation and land utilization. But the 1955 decisions constituted the starting point of a long-continuing AAAS involvement with some of the social aspects of science, both the social consequences of scientific and technological developments and the influence of society on the practice of science. Moreover, the Interim Committee quickly got support from an ad hoc meeting Warren Weaver had arranged between representatives of AAAS and representatives of the National Association of Science Writers. That ad hoc group "agreed that the AAAS ought to be a vigorous spokesman for positions which meet with widespread support among scientists" and that in situations in which there was substantial disagreement among scientists "the AAAS might do a valuable service by providing opportunities for open discussion of the questions" and by "presenting the relevant scientific facts" (5).

## The Committee on Science in the Promotion of Human Welfare

Two themes marked the history of the Interim Committee on the Social Aspects of Science and its permanent successor, the Committee on Science in the Promotion of Human Welfare that was established in 1958. One was substantive: the actual work of the committee, which was substantial. The other was procedural. What were the committee's rights and its relations to the board of directors and council and how were its reports to be handled? Basic to both themes was a major policy question: What are appropriate circumstances and ways for an organized body of scientists to participate in the determination of social policy? Scientists are citizens and each should feel free to participate in matters of social policy. But AAAS is an organization of members whose political and social interests and values range widely on almost every issue. With that diversity of membership how far should AAAS go — just studying and stating the relevant scientific facts about a social issue or actively participating in societal decisions concerning that issue? That basic question was posed in an editorial introducing the first report of the Interim Committee (6).

Both the substantive and the procedural themes of the new committee quickly became evident. On the substantive side the committee arranged a symposium on radiation problems for the 1957 annual meeting and planned a symposium on air and water pollution for the 1958 meeting in Washington, D.C. The latter topic was pre-empted by the U.S. Public Health Service, which arranged a large conference on air pollution in Washington a month before the AAAS meeting. The committee interest may have helped stimulate that conference and the committee helped with its conduct. But the committee had to switch topics for its 1958 symposium, and it chose a national program for health eduction as its new topic. The committee also helped plan the association's Parliament of Science of 1958, and it helped stimulate and organize some local groups of scientists and citizens to consider matters of mutual interest (7).

The Interim Committee's plans for a public symposium on radiation problems raised no procedural issues; everyone understood that each participant in an association symposium spoke as an individual and not as a spokesman for AAAS. But when the committee prepared a paper stating its own members' views on the handling of radiation problems the procedural issue had to be faced. To complicate matters, members of the committee were not of one mind on the content. In addition, the association had not developed customs for reporting majority and minority views and had not had much occasion to deal with statements prepared by committees (8). Resolutions adopted by the council had a clear status, for they had gone through the formal procedure of discussion and voting by the association's legislative body. But what was the status of a report prepared by a committee? Need the board or council approve publication? Need they agree with the content? It took quite a while to answer those questions.

On the immediate question of publication of its statement on radiation hazards, the Interim Committee itself decided that publication might be unwise,

> since it might suggest an authoritarianism which is repugnant to the ideals of scientific effort. The committee holds that scientists are within their competence in furnishing scientific data on social problems and in suggesting the consequences of the facts as far as they are verifiable. However, scientists are probably no wiser than other intelligent citizens in reaching policy decisions on major social issues (9).

That decision did more than dispose of the statement on radiation hazards; it helped the committee decide upon the kind of statements it should publish. In 1958 the committee spent three days debating its own method of operation and the nature of its relationship to the board of directors and other parts of the association. The first outcome was the recommendation that AAAS establish a standing committee on social aspects of science, with a rotating membership as was customary for other standing committees. However, the social aspects of science was a domain so broad that no committee of moderate size could be expert on all topics. Thus the proposed standing committee should be expected to select and define questions needing analysis and then to recommend that other agencies or ad hoc bodies of experts study particular issues. The proposed standing committee should also serve as an advisory body or recommend programs of action to the board of directors. Procedures and outcomes might vary, but in general the end result of a study was expected to be an analysis or report in a form intelligible to the general public or to the appropriate legislative body (10).

The recommendation that AAAS establish a permanent committee on the social aspects of science met a somewhat wary reception by the board of directors. The problems involved were surely important, but any proposed solutions would probably be controversial. The substance of science had been the association's primary interest; should it also become involved in issues of social action? If so, would creating the new committee give too much attention to one area of scientists' interests? Paul Klopsteg solved this dilemma by reminding the board that the association's constitution stated three objectives:

1. to further the work of scientists [and] to facilitate cooperation among them;

2. to improve the effectiveness of science in the promotion of human welfare; and

3. to increase public understanding and appreciation of the importance and promise of the methods of science in human progress.

The Interim Committee was asking for a permanent committee to give attention to the second of those three objectives. Shouldn't the association give equal emphasis to all three? Thus Klopsteg proposed a new "Committee on Cooperation Among Scientists" to focus on the first objective, a new standing "Committee on Science in the Promotion of Human Welfare" to continue the work started by the Interim Committee, and continuation of the existing Committee on Public Understanding of Science to continue activities aimed at the third objective.

The board liked that structural symmetry. So did the council. The three committees were established on a parallel basis as regular standing committees of the association and each was provided with a budget to carry out its work. Constitutional equivalence had led to organizational symmetry (11).

With its new name and Barry Commoner as its chairman the Committee on Science in the Promotion of Human Welfare developed further the rationale for its existence and outlined its mode of action. It would seek to stimulate discussion within the scientific community on selected issues; assemble the facts, as far as they were known; prepare a report for the general public; and seek to develop liaisons between scientists and citizens, often on a local basis where the issues could be openly debated and recommendations concerning public action formulated. A committee report in *Science* described those plans so they might be given widespread consideration by scientists (12). With good reason the short-hand name for this committee soon became "the Commoner committee" for Barry Commoner had served on the Interim Committee that preceded it. He was the committee's chairman longer than anyone else — seven years — and was clearly its leader in formulating and stating its operating policies and practices.

The committee's report entitled "Science and Human Welfare" had an unusually wide public reception. The *New York Times* gave it first-page prominence and printed almost all of the text. The Associated Press story was published by many newspapers. And a variety of organizations wrote to the committee to ask for or to offer help. The committee even had an invitation from a film producer.

The board agreed that AAAS as an organization was properly interested in and should conduct studies on carefully selected issues of the type being considered by the Committee on Science in the Promotion of Human Welfare, and as the committee and board considered such studies, discussion gradually led to agreement on rules of procedure. Thus the board stated that

> AAAS should engage in a series of independently selected and motivated studies of specific problems involving science and technology in relation to society. These studies will include, but will not be limited to, those suggested by the Committee on Science in the Promotion of Human Welfare and approved by the Board. Recommendations for studies will be welcomed from all members of and groups within the AAAS. Final decisions as to the studies to be undertaken as well as the composition of the study groups will be made by the Board, in consultation with the Committee on Science in the Promotion of Human Welfare.
>
> In undertaking these studies the AAAS should be willing to deal with subjects broader than those which are normally undertaken as research projects in universities, either because they call for the application of a wide range of scientific disciplines or because they involve problems not presently susceptible to solution by purely scientific methods. The study should seek to report to the interested layman, the political leader, and the government official as well as to the world of science the ways in which science and technology touch on the important problems of society, even when such problems involve aspects which cannot be dealt with by quantitative or experimental methods (13).

Publications resulting from these studies, the board and committee agreed, should appear as statements by the committee, with a clear disclaimer that the views and recommendations expressed did not necessarily represent those of the board or of the association as a whole. Nevertheless, publication required approval by the board, and the disclaimer statement should state that the board had approved publication as a contribution to the discussion of an important topic, but not as an expression of AAAS policy. A formal statement of these points was developed for publication with each report by an association committee (14).

## Committee Studies

The committee was an active one, and without waiting for all of its procedures to be fully agreed upon it began to carry out its substantive responsibilities. Discussions within the committee of a wide range of issues sometimes led to symposia presented at annual meetings, sometimes to a small conference to consider an issue in greater detail, and sometimes to appointment of a study commission to explore the issue in depth and prepare a report for publication.

For the 1965 annual meeting the committee sponsored two symposia, one on new developments in the control of conception and one on civil defense. The former was not published as a whole, but one of its major papers later appeared in *Science* (15). The symposium on civil defense consisted of two panels. One, headed by Barry Commoner, opposed civil defense activities; the other, headed by Eugene Wigner, favored those activities. Henry Eyring edited the resulting report, which was published in the association's symposium series (16).

In 1967, at the request of the board of directors, the committee discussed the problems of secrecy in science and then arranged a symposium on secrecy, privacy, and public information. The symposium itself was not intended for publication, but those discussions, two years of committee debate and review of the literature, and several case histories received in response to an open invitation to AAAS members all culminated in a report published in *Science* (17).

In 1969 the selected topic was, "Is Military Support of Academic Research Justified?" At that time scientific meetings were sometimes being disrupted by protests over U.S. participation in the war in Vietnam or other issues. Special precautions were therefore taken to minimize trouble, and the chairman later learned that the session had in fact been targeted for disturbance (18).

A small conference as well as a public symposium could lead to a committee publication, as illustrated by "Science and Human Survival," a report developed by a conference of half-a-dozen participants. The general tenor of the report was indicated by its subtitle: "Unlimited war is self-defeating and an alternative must be found by a new science of human survival" (19).

One of the most provocative of the committee's reports resulted primarily from internal discussions of the committee. As early as 1960 the members began to express concern over what they called "the integrity of science." That title did not mean conscious fraud in faked data or misused funds, but rather the use of science for improper ends. The report explained:

> As the success of science becomes more evident, its trappings and its personalities begin to make their appearance outside the laboratory; at international conference tables, in legislative chambers, in advertising appeals, and in the hue and cry of partisan politics. Are these appropriate uses of the methods of science? Can political life benefit from the viewpoint of science? Can the integrity of science safely withstand the invocation of science in political issues (20)?

In 1966, as Commoner completed his term of chairmanship of the committee, he published *Science and Survival* (21), a personal statement rather than a committee report, but as he explained,

> Most of the ideas expressed in this book were developed during the course of a number of activities ... in which I have engaged in the last few years. Among them are the Committee on Science in the Promotion of Human Welfare of the American Association for the Advancement of Science, the St. Louis committee for Nuclear Information (CNI) [which Commoner and the AAAS committee had helped to establish] and the Scientists' Institute for Public Information [which he had also helped to establish as an umbrella organization over several city organizations such as the CNI of St. Louis].

In this book, as in the committee's statement "The Integrity of Science," emphasis was on the proposition that scientists themselves are largely responsible for preventing the erosion of scientific integrity and for protecting the world against the misuse of science and technology, for that was one of the basic tenets of the committee.

## Commission on Air Conservation

The best example of the association's use of an ad hoc group of experts to prepare a report on a special issue was the AAAS Air Conservation Commission. One of the topics given early attention by the Committee on Science in the Promotion of Human Welfare was air pollution or, to use the more positive term the committee wanted to emphasize, air conservation. The committee explained its choice:

> Specialists in the field of air pollution have come to the conclusion that the only effective means of dealing with the problem is to develop both a scientific basis and public acceptance for a new concept of air conservation. In effect, the atmosphere can no longer be regarded as an unlimited dumping ground for the rapidly increasing wastes of industrial society, and there is an urgent need for developing principles of management which will conserve this limited and essential natural resource. The efforts of a wide range of scientific disciplines will need to be coordinated for the proper development of a science of air conservation. This need, in turn, requires that scientists generally become aware of the problem and become interested in contributing what they can toward its solution. Public

education in the area is of critical importance, since solutions will require considerable economic and social planning (22).

The board of directors approved that reasoning and, with committee advice, appointed the members of the Air Conservation Commission and asked it to analyze the problems, formulate recommendations, and prepare a public report. James P. Dixon, a specialist in public health who was then president of Antioch College, was chosen as chairman. Other members were selected to give the commission strength in the relevant disciplines of organic chemistry, meteorology and other atmospheric sciences, industrial medicine, ecology, economics, and regional planning (23). The result of the work by those members and 55 "authorities who contributed to this report by writing or reviewing papers, by providing needed information," or in other significant ways was a volume entitled *Air Conservation* that was published in 1965 and distributed as one of the AAAS symposium volumes (24).

Had the recommendations of *Air Conservation* been promptly implemented the nation would have been spared much of its later worry about air pollution and acid rain, but the forces arrayed against one report, no matter how sound, were too strong. Some 20 years later Commoner, who had helped stimulate establishment of the Air Conservation Commission, reported that although much had been learned about air quality and individual air pollutants since the early 1960s, lead was the only air pollutant that had been substantially reduced in that time (25).

## Race

In 1962 the American Anthropological Association (AAA) asked AAAS to consider appointing a commission on race to prepare "an authoritative statement of what is known and not known as a result of research on race as the concept applies to the human species, and a clarification of the issues involved in the scientific study of race." The emphasis of the expected statement was to be on scientific evidence and problems rather than on political questions concerning segregation, discrimination, or public policy. One reason for that request was the contention by a few authors that recent research had brought new evidence of inherent differences in intellectual ability of the different races. The board took the request under advisement and asked that a number of authorities be consulted to determine whether a new statement was needed or would be useful (26).

William Kabisch accepted responsibility for talking with geneticists about the problem and returned to the next board meeting with a mixed report (27). The Committee on Science in the Promotion of Human Welfare then gave their advice that "we believe that no useful purpose would be served by the establishment of a commission to review the scientific evidence on race" because there was not enough new evidence to justify a review and report. That advice and the earlier report from

Kabisch led the board to decide against mounting a new study. That decision and its reasons were reported to AAA. The AAA replied that they deeply appreciated the thoroughness with which AAAS had explored their original request; moreover, the finding that new materials on the question of race differences was not new in principle or qualitatively different from earlier information was of great importance and a highly effective reply to claims that new material demonstrated significant genetic differences in the intellectual abilities of different races (28).

The Committee on Science in the Promotion of Human Welfare went on to prepare a paper discussing some of the scientific evidence on racial differences and the relationships between social and judicial decisions concerning the rights of minority groups. The general conclusion was that "... the available evidence on measurable differences among racial groups cannot properly support a challenge to the principle of human equality" (29). There was some disagreement with that finding (30), but the committee concluded that the letters of opposition did not include any new information (31).

In 1966, with Margaret Mead as its new chairman, the committee returned to the race question and arranged a symposium on "The Utility of the Concept of Race" for that year's annual meeting. Several of the papers were published in abridged form in the *Columbia University Forum,* and in 1968 the entire symposium was published in book form by the Columbia University Press (32).

## Future Tasks

After Walter Modell succeeded Margaret Mead as chairman in 1968, the committee arranged a symposium on "Unanticipated Environmental Hazards Resulting from Technological Intrusions" for the 1968 meeting and one entitled "Is Military Support of Academic Research Justified?" for the 1969 meeting.

The committee also gave much time to two more comprehensive issues. One, a new problem, was what the members called the "drastic reduction" in research funding from the National Institutes of Health. Although neither the committee or most scientists yet realized it, what they were seeing was the end of the period of large annual increases in federal funds for research and development. The committee adopted a resolution asking the board of directors "to take appropriate action to study and deal with the future of federal support for science" (33).

There was little the board could do to "deal with" the end of the period of rapid increases in federal funding for research. Economic conditions were changing. Federal support for research and development, which had grown at an average rate of 12 percent a year since 1953, simply could not continue to increase so fast. Neither committee nor board members knew that the amount would remain essentially flat for several years or that, when measured in terms of constant dollars of equivalent purchasing power, the 1968 and 1969 level would not again be reached until 1983. Nevertheless, it was obvious that a major change was taking place, and about all the board could do on that problem was to share the committee's concern.

The committee had another large topic to consider. John Platt, one of its members, presented a draft version of his paper "What We Must Do" in which he described a number of crises facing the world, among them housing, food, energy, education, population, and perhaps atomic warfare. All involved a scientific or technological component and all involved matters of public policy and management. Scientists alone could neither prevent nor solve any of the crises. But scientists could foresee them, and task forces should therefore be organized in the immediate future to begin to analyze and plan ways to confront each. This, too, was a matter that should be brought to the attention of the board.

Although the committee was not yet ready to make specific recommendations, it informed the board that it would later recommend appointment of task forces on "five large problem areas: food and population; shelter, transportation, and environment; government and peace; health, well-being, and recreation; and education and communication." Each, they suggested, should have a budget of about $20,000 a year, although in committee discussions expenditures of perhaps $85,000 had been considered (34). In the meantime *Science* had published Platt's "What We Must Do" article with the subtitle "A large-scale mobilization of scientists may be the only way to solve our crisis problems" (35) so that it might be widely discussed and perhaps stimulate action.

The proposed task forces were for the future, but as 1970 drew to a close the committee could take some credit for establishment by the board of two new committees working in areas related to its own: the Commission on Population and Reproduction Control and the Committee on Environmental Alteration (36). The members had hoped that the latter committee would serve as a continuing "early warning system to alert all who might be affected by or in a position to control the possibilities of adverse effects resulting from the introduction of new technologies." However, the new Committee on Environmental Alteration wanted to consider other possibilities. In an announcement in *Science* the chairman described two thrusts to be emphasized: one, the watch-dog function over possible adverse consequences of the introduction of new technology; the other, the crisis prevention task forces they hoped would be established to confront issues discussed in "What We Must Do." Having briefly described those plans, the committee's statement, and the committee's year, the announcement ended with an open invitation for reactions and suggestions from AAAS members and other readers of *Science* (37).

## Cooperation Among Scientists

As noted earlier the board created the standing Committee on Cooperation Among Scientists in 1958 to work in parallel with the new Committee on Science in the

Promotion of Human Welfare and the already existing Committee on Public Understanding of Science. The latter two committees contributed usefully, as reviewed in this chapter and Chapter 9, but the Committee on Cooperation Among Scientists demonstrated that logical symmetry may not be the best basis for organizational structure. The committee carried out a few minor assignments for the board of directors, but the members early began to feel frustrated by the fact that cooperation among scientists was so basic to the whole association that most activities directed toward that end were already under the direction of other committees or staff offices. Cooperation among scientists was a function of the annual meetings, of some of the association's other meetings such as the Gordon Research Conferences, of *Science* and other publications, and in general of so much of the whole AAAS program that the committee had little opportunity to be influential or useful.

Nevertheless, at the end of 1962 the board gave the committee its most important assignment. The immediate reason was a request from the American Institute of Physics. The executive committee of that organization had received a proposal to establish an ethical practices committee for physics. Instead of doing that, they recognized that most of the questions to be considered were as relevant to other scientists as to the physicists and so asked AAAS to consider taking up ethical issues for the whole scientific community (38).

The board referred that request to the Committee on Cooperation Among Scientists, which soon received further encouragement to consider ethical issues. The board had invited all past presidents of AAAS to join the board in a review of all of the association's activities and to give advice on how AAAS might do better whatever it should be doing. James Conant (president in 1946), Harlow Shapley (1947), Detlev W. Bronk (1952), Edward U. Condon (1953), Paul Sears (1956), Laurence H. Snyder (1957), Wallace R. Brode (1958), Paul E. Klopsteg (1959), Chauncey D. Leake (1960), and Thomas Park (1961) accepted that invitation. They and the board members had an extensive discussion centering on four broad areas: science education, international activities, the role of state and regional academies of science and other local groups, and ethical problems of scientists (39).

The Committee on Cooperation Among Scientists reviewed the literature and what the former AAAS presidents had said concerning the ethical problems of scientists, examined ethical codes that had been adopted by several professional groups, invited readers of *Science* to submit concrete example of ethical issues that had been handled well or poorly (40), and ended by writing two papers. One, written primarily by James Mitchell but based on preliminary papers written by other members of the committee, was entitled "Problems of Customs or Manners Arising in the Major Relationships of Scientists." The other, written primarily by Lynn White, was entitled "The Etiquette of Research and Publication" and was planned to be suitable for use as a leaflet to "provoke both graduate students and their mentors to more detailed discussion and reflection" concerning their relations and responsibilities to one another. The general flavor of the committee's approach to getting scientists to think about such matters was well expressed in early paragraphs of the etiquette paper:

> Today unprecedented funds from government, industry, and foundations are going into research; the peril is cupidity. Far larger numbers of men and women than formerly are involved in research organizations; the temptation is bureaucratic empire building. The demand for talented researchers often leads to a high velocity of the individual from job to job; the danger is impersonality toward others and exploitation of colleagues for personal ambition.
>
> The AAAS Committee on Cooperation Among Scientists is in no mood to cry havoc over this developing situation, nor is it about to descend from Sinai bearing Tables of the Law. Scientists and scholars are invariably human beings who therefore incorporate that quantum of inherent cussedness which theologians call Original Sin. The Committee, however, feels that the scale of graduate schools and laboratories has grown so greatly that apprentices no longer have adequate opportunity to learn by repeated observation of mature investigators what the pattern of decent relationships is among scholars in matters of research and publication.

The paper then went on to consider the relations involved between professors and graduate students in the publication of cooperative projects, in matters of security and consultation, and in publicity. On the first of these topics professors were reminded:

> The core of courtesy is generosity. ... An honorable professor will never purloin a student's original findings. A good teacher recognizes his own traits to some extent in the thinking of his students, and normally by the time a student's initial work reaches the point of publication, the professor may properly feel that his guidance and materials account for a large part of the results. Nevertheless, especially in the case of theses, the widely accepted practice of joint publication by master and student raises questions of good taste. Independent publication, perhaps with a note acknowledging guidance, is the student's rite of passage, an occasion for rejoicing by the elders of the tribe (41).

Old fashioned? Conservative? Yes, and perhaps even unrealistic as pressures for successful competition for grants to support one's laboratory have increased the apparent value of a longer and longer list of publications to attach to grant requests. But nevertheless an admonishing reminder that civility, courtesy, and generosity are traits to be encouraged, appreciated, and valued. The board of directors liked the two papers and readily approved their publication (42).

Completion of those papers ended the work and life of the Committee on Cooperation Among Scientists, but the association's involvement in studying and writing about the ethics of science continued, for the AAAS council was at that time establishing a number of council study committees, one of which was on ethics.

## The Council Study Committee on Ethics

In May of 1963 the Committee on Council Affairs wrote to all members of the council asking advice concerning the responsibilities and ethical problems of scientists vis à vis institutions or employers, students or assistants, colleagues, and sources of support. The committee knew the parallel interests and study of the AAAS Committee on Cooperation Among Scientists but thought that the council should also discuss those issues (43). To start that discussion, the council asked the Committee on Meetings to schedule a full-day symposium on ethical issues at the 1964 meeting (44).

Chaired by Robert K. Merton the morning session of the resulting symposium included four papers on the sociology of science. The afternoon session included four papers prepared by members of the Committee on Cooperation Among Scientists and the Committee on Science in the Promotion of Human Welfare. Following that symposium 25 members of the council said they would like to serve on a council study committee on ethics. With that much interest the committee was soon appointed (45). William Wildhack was the initial chairman, but was soon succeeded by Anatol Rapaport, who started the committee's inquiry by sending a questionnaire to a random sample of 1,000 members of AAAS. The major objective was to identify the questions or issues with which scientists were most concerned. No single paragraph can describe the variety of information gained from analysis of the 680 replies, but of the 34 questions posed in the questionnaire the respondents seemed to give greatest importance to the following half-dozen topics: drug safety, the use of human subjects, allocation of research grants, allocation of resources to pure and applied research, neglect of duty, and dishonesty (46).

Study of the respondents' answers and comments led to development of a more detailed questionnaire asking for responses to 22 questions or described situations. For some, respondents were asked whether they agreed or disagreed; for others, which of several indicated actions they would take; and for still others, what action they thought should be taken by others. The questionnaire was sent to a random sample of 5,000 AAAS members, of whom 2,502 responded, usually with comments added to their responses. Replies and comments were analyzed in rich detail in a report to the council (47). Rapaport's report concerning a single item can give the flavor of the study. The first of the 22 items read:

> It has been argued that scientists should assume responsibility in urging conservation on other than utilitarian grounds; for example, saving economically unimportant species or "backward" cultures from extinction.

Eight-five percent of the respondents either agreed or agreed strongly with that statement — a slightly higher level of agreement than was found for any of the other 21 items. Nevertheless, 15 percent disagreed, gave more than one answer, or gave none. And the 26 percent who added personal comments expressed a wide range of attitudes, for example, "there are no economically unimportant species in science" and "backward cultures should not be kept 'backward' just out of curiosity." One of the relatively few respondents who disagreed pointed out that "we [mankind] had historically created our own ecology without disastrous results."

The submission of the report ended the study committee's active life, although it was not formally disbanded until some months later.

## Environmental Alteration: The Herbicide Assessment Commission

As the war in Vietnam dragged through the 1960s AAAS became involved in a study of the environmental impacts of the chemical agents that the United States and South Vietnam were using to defoliate forest cover and to destroy some crops thought to be grown for enemy use. How much the association's activities reduced the military use of chemical defoliants or improved the health of the forests of Southeast Asia may be debated, but in the United States the effort had two clear effects. First, it helped get AAAS more firmly engaged in studies of the environmental effects of technological developments. Second, the effort provided a case study of how politico-technical questions were handled in the United States and contributed to the establishment of the congressional Office of Technology Assessment (48).

AAAS involvement began with a resolution introduced by E. W. Pfeiffer of the University of Montana at a meeting of the Pacific Division of AAAS. He asked that division to establish a committee to investigate the use of "chemical and biological warfare agents ... in operations against enemy forces in Vietnam." When the Pacific Division decided to refer that resolution to the national office of AAAS, Pfeiffer requested the Committee on Council Affairs to submit it to the council for action at the 1966 annual meeting. The Committee on Council Affairs agreed, but thought the resolution should not be limited to Vietnam. With Pfeiffer's approval it was broadened to include civilian as well as military use of biological and chemical agents that substantially altered the environment or the ecological balance of the planet (49). After the council made a few further changes the resolution that was adopted read:

> *Whereas* modern science and technology now give man unprecedented power to alter his environment and affect the ecological balance of this planet, and

*Whereas* the full impact of the uses of biological and chemical agents to modify the environment, whether for peaceful or military purposes, is not fully known;

*Be it resolved* that the American Association for the Advancement of Science:

(1) expresses its concern regarding the long-range consequences of the use of biological and chemical agents which modify the environment; and

(2) establishes a committee to study such use, including the effects of chemical and biological warfare agents and periodically to report its finding through appropriate channels of the Association; and

(3) volunteers its cooperation with public agencies and offices of government for the task of ascertaining scientifically and objectively the full implications of major programs and activities which modify the environment and affect the ecological balance on a large scale (50).

Those changes broadened the scope of the proposed inquiry, but the immediate task and the aspect that got attention from the press was Vietnam. Was AAAS about to send a team of scientists to investigate the ecological consequences of weapons in active and violent use? The United States was using chemical herbicides and defoliants in Vietnam "as a means of saving lives — protecting troops, shipping, and aircraft from ambush from the jungle cover; ... [and] to complicate the adversary's logistics — to deny guerrillas their sources of food from remote garden plots by the spraying of these crops from the air with crop-destroying chemicals" (51).

The first task for the AAAS, the board decided, was to find out what had already been learned and what others were doing to study the effects of biological and chemical agents in altering environments or ecological balances. René Dubos agreed to serve as chairman of an ad hoc committee for that purpose (52).

The committee responded with a statement that life had been interacting with the environment for three billion years, that environmental changes were inevitable, but that human influence was accelerating and that the "effects of human activity upon terrestrial space, soil, air, and water have now become matters of grave import." The committee therefore recommended establishment of a continuing "Commission on Environmental Alteration," a new arm of AAAS to conduct or arrange for studies of environmental issues (53).

As for the war in Vietnam, the committee recommended that AAAS "consider requesting the National Academy of Sciences to arrange a continuing study and scientific record of the effects of chemical and biological warfare agents on soil, biota, and human health." Pfeiffer, a member of the committee, added a minority statement disagreeing with that recommendation and urging AAAS to take direct action as soon as possible, including making field studies in Vietnam.

## Assessing the Effects of Herbicide Use in Vietnam

AAAS could not possibly send off a delegation that would arrive unexpectedly in an active war zone and announce that they were there to assess the impact of the chemical and biological weapons being used. Discussions and preliminary arrangements with the military authorities were essential. Don Price and I therefore talked with Donald MacArthur and Rodney W. Nichols, who were members of the staff of the Secretary of Defense; Donald Hornig, director of the President's Office of Science and Technology; and Frederick Seitz, chairman of the Defense Science Board and president of the National Academy of Sciences. On the basis of those discussions Don Price wrote to Robert McNamara, Secretary of Defense, giving an account of the association's interest in the problem, recommending that "the Department of Defense authorize and support a study by an independent scientific institution or committee of both the short- and long-range effects of the military use of chemical agents which modify the environment;" saying that the National Academy of Sciences had been established to aid the federal government on such technical issues; and promising that AAAS would offer any assistance it could in such an undertaking (54).

A prompt reply came from John S. Foster, Jr., then director of Defense Research and Engineering, stating that the Department of Defense had considered the possibility of long-term ecological impacts but that "qualified scientists, both inside and outside the government, and in the governments of other nations, have judged that seriously adverse consequences will not occur. Unless we had confidence in these judgments, we would not continue to employ these materials" (55). Nevertheless, Foster continued, a more complete understanding of the effects of herbicides and defoliants was desirable. The Department of Defense had therefore contracted with a research institution for a thorough review of all current data on the issue and had arranged with the National Academy of Sciences–National Research Council to review that study and make appropriate recommendations concerning it. Finally, the letter offered AAAS the opportunity to review and comment on the forthcoming report and the NAS–NRC review of it.

The research institution involved was not identified in that letter, but soon became known: the Midwest Research Institute. Its report appeared a few months later under the title "Assessment of Biological Effects of Extensive or Repeated Use of Herbicides" (56). The report was a comprehensive review of the existing literature, but its title promised more than the contents supplied, for no previous use of herbicides had been as extensive, applied in as large amounts, or repeated as often as had some of the applications in Vietnam. Moreover, the authors had not been able to make any observations in Vietnam. Copies of the report and the NAS–NRC comments were sent to the board members, members of the new AAAS Committee on Environmental Alteration, members of the Committee on Science in the Promotion of Human Welfare, and several individuals whose advice the board wanted (57). Replies, sometimes in much detail, were mostly to the effect that the available evidence was quite inadequate to justify any statements about long-range effects, a reaction perhaps most succinctly expressed by one ecologist: "The outstanding fact that emerges is that we don't know" (58). There was much agreement that the confidence expressed in John Foster's letter that "seriously adverse consequences will not occur" was quite unjustified.

That agreement did not answer the question of what should be done. Some members of the board, together with the military authorities and some other observers considered the ecological damage a reasonable price to pay for the lives saved by the spraying program. Others in and out of AAAS disagreed. Spraying had greatly intensified in 1967 and was at a high level in 1968 (59). The well-known Agents Orange and White and the less well-known Agent Blue (which contained arsenical compounds) were not well enough understood to know just how damaging they were. Even so, there was much worry about ecological damage, about harm to the people of the region from the direct impact of the arsenical sprays and perhaps the other agents, and also from indirect damage due to destruction of the lands and forests upon which the residents depended (60).

What was agreed on was the desirability of immediate and continuing impact studies which could be made only in Vietnam. The board therefore published a statement explaining its interest, indicating the great amount of uncertainty concerning the impacts of herbicidal and defoliating agents being used, and making three recommendations:

> 1. That a field study be undertaken under the auspices and direction of the United Nations, with participation by Vietnamese and other scientists, and with support and protection by the military forces;
>
> 2. That the maximum possible amount of relevant data be released from military security to aid the scientists in conducting that study; and
>
> 3. That the use of arsenical compounds be stopped until the ultimate fate of the degraded compounds could be more reliably determined (61).

Copies of that statement were sent to the United Nations and the Departments of State and Defense, but none of those agencies initiated the kind of study that was needed.

At last, after unsuccessfully recommending that a field study be conducted under the auspices of the National Academy of Sciences or the United Nations, in December 1968 the board informed the council that

> It is the sense of the Board that the association, looking not only to the effects of the wartime use of herbicides, but also to the opportunities for the peacetime reconstruction of the agriculture and economy of the affected areas:
>
> (1) determines that it shall be a purpose of the Association to bring into being the most effective possible field study of the poten-

tial long-term and short-term ecological risks and benefits to the areas affected; and

(2) specifically directs the AAAS staff to convene, as soon as possible, an ad hoc group involving representation of interested national and international scientific organizations to prepare specific plans for conduct of such a field study and with the expectation that the AAAS would participate in such a study within the reasonable limits of its resources (62).

A whole year was consumed in communicating with Ellsworth Bunker, United States Ambassador to South Vietnam; planning the organization of the effort; and searching for a leader — a search that ended at the time of the 1969 annual meeting when Matthew Meselson agreed to serve as chairman of the association's Herbicide Assessment Commission.

In preparation for a trip to Vietnam the commission corresponded with several hundred scientists and other interested people whose information and suggestions they thought might be useful. In June of 1970 the commission held a conference at Woods Hole of 22 participants, including several who had direct and recent experience with the defoliation program or with other aspects of Vietnam. The participants identified specific questions that the team would want to investigate and the kinds of observations to be made, and prepared the interview and questionnaire instruments the team would use in collecting information (63).

In August of 1970 the team went to Vietnam for five weeks of intensive investigation. The four participants were Matthew Meselson, a molecular biologist on the faculty of Harvard University and active participant in other activities concerning the use of chemical and biological agents of war; Arthur W. Westing, a forestry specialist who was chairman of the Department of Biology at Windham College, Vermont; John D. Constable, professor of surgery at the Harvard Medical School, who had earlier made health surveys in Vietnam; and Robert E. Cook, Jr., a graduate student in ecology at Yale. They were given substantial information and help by the armed forces and were flown over many areas that had been subject to defoliants and crop-destroying sprays. They had little opportunity to make examinations on the ground, however, for most of the area, other than Saigon and a few other strongholds, was battle ground largely controlled by enemy forces. To add to the difficulties, some of the information about application of the chemical sprays was classified (64).

Nevertheless, the team came home with much information. They arranged a small conference on analytical techniques for use in analysis of specimens that they brought with them, gave a briefing at the State Department, and began to prepare their report. Their first, incomplete report was a letter to Ellsworth Bunker, with copies to Secretary of State William P. Rogers and to the commanding general of U.S. forces in Vietnam, Creighton Abrams. That letter included photographic and other evidence that nominal guidelines or restrictions on spraying were sometimes being violated.

A few months earlier the Department of Defense had told AAAS that herbicides were used "for crop destruction of small, isolated crop patches along infiltration routes," that crop spraying was limited to areas of low population (defined as less than eight persons per square kilometer), and that all usage was carefully restricted to areas approved by U.S. and Vietnamese authorities (65). In contrast, and after thanking the ambassador for the generous assistance the embassy had provided, the letter showed that crops had been sprayed in an area with an estimated population of 180 persons per square kilometer and that nearly all of the food being destroyed would have been used by mountain-dwelling Montagnard civilians instead of by enemy troops (66).

During Christmas week of 1970 the commission gave a public report at an annual meeting symposium that included several other participants who had experience with the defoliation program. The account of that symposium in *Science* summarized: "Their formal reports ... were guardedly conservative in tone, but their findings added up to a charge that the military use of herbicides has been considerably more destructive than anyone had previously imagined" (67). The "anyone" in that statement might have several referents, for there had been several earlier missions to Vietnam in addition all the military observers. Among them were Harold Coolidge from the Office of the Foreign Secretary of the National Academy of Sciences and representing Pacific Science Board, who had visited Saigon briefly in January of 1968. In talking with Vietnamese scientists and with American and Vietnamese officials he found deep concern over the fact that thousands of acres of crop and forest land were being affected and he urged Vietnamese scientists to organize a study of the ecological effects of the spraying program, an effort Coolidge indicated would have the support of scientists in the United States (68).

In March and April of 1968 Fred H. Tschirley, assistant chief of the Crops Protection Research Branch of the U.S. Department of Agriculture, spent a month in Vietnam at the request of the Department of State. He concluded that while the program had environmental impacts, the extent of forest destruction would not significantly change air moisture and rainfall patterns; that the agents used did not kill soil microorganisms and so did not significantly change soil ecology; that perhaps 20 years would be required for regrowth of mangrove forests (an area of great susceptibility, for a single spraying killed the mangrove trees); that the aquatic food chain was apparently not disturbed; that the time for restoration of hardwood forests was quite unknown; and that the agents used were probably not toxic to human beings or animals (69).

In March of 1969 E. W. Pfeiffer and Gordon Orians, professor of zoology at the University of Washington, visited Vietnam on a trip sponsored by the Society for Social Responsibility in Science. They had good cooperation from the military forces, as had other visitors; were taken on spraying missions; made observations by helicopter, airplane, and patrol boat; and conferred with a number of Vietnamese scientists. They reported that they had seen no evidence of regrowth of the dead mangrove forests and that they had found severe damage to hardwood forests and also to rubber plantations and fruit trees from windblown defoliants. They cited some evidence that 2,4,5-T was teratogenic in rats and mice at dosage levels possible of ingestion by human beings in the sprayed areas, and they concluded that there was severe damage not only to targeted areas but also to other areas through windblown or jettisoned spray (70).

Reports by the association's assessment team confirmed and extended much that had been reported before, but because the AAAS team had been there longer, searched more thoroughly, and brought back more information than had previous visitors, their reports received more attention. Briefly summarized, they told the symposium audience at the 1970 AAAS meeting that:

1. From one-fifth to one-half of the mangrove forests were utterly destroyed and that areas that had been killed as much as six years earlier were not yet showing any regrowth of mangrove trees;

2. Perhaps half of the trees in hardwood forests that had been sprayed were dead and the open areas were being replaced by worthless species of grass and bamboo instead of by hardwoods;

3. Contrary to the intent that the crop-destruction program would destroy only crops intended as food for guerrilla forces or troops infiltrating from North Vietnam, nearly all of the destroyed crops would have provided food for Vietnam's minority group, the Montagnard tribes of the highlands; and

4. Although there were some indications of possibly serious health effects there was not yet definite evidence on that issue (71).

Two years later when Pfeiffer and Westing (of the AAAS team) returned to Vietnam to investigate other types of damage, they reported that of a total forested area of about 25 million acres some three million acres had been completely destroyed by military action. A little over a third of that damage resulted from the spraying program. Nearly two-thirds resulted from giant bulldozers that could knock over forest trees and from shells, rockets, and bombs, including the 15,000-pound "Daisy Cutter" that could level a two-acre area to create an instant landing field for helicopter use (72).

Bombs and shells, however, were conventional weapons, and the emotional concern was focused on defoliation, a new tactic using a new kind of weapon. On December 26, while the AAAS meeting was in session, the White House announced that authorities in Saigon were "initiating a program for an orderly yet rapid phase-out of the operation" (73).

When the council met a few days later Meselson gave a brief report and said that the written report of the commission should be completed in about three months. The council adopted a resolution commending the U.S. government for its intention to phase out a practice that had "seriously damaged the ecology of that country and may be a serious threat to the health, livelihood and social structure of Vietnam's hill tribes (Montagnards)" (74).

That ended AAAS participation in the herbicide problem, but the National Academy of Sciences — which had been the board of directors' initial nominee for the purpose — was later supported by the Department of Defense in sending another group of scientists to assess the impact of chemical agents that had been used to clear forest cover and destroy crops (75).

How much the association's activity influenced the decision to terminate the use of chemical defoliants and herbicides in Vietnam or how much influence the findings of the Herbicide Assessment Commission may have on the future use of those agents are matters for speculation. Their use stopped, but their usage still remained a matter of argument. At the 1970 symposium at which the Herbicide Assessment Commission gave its report, Arthur Westing, a member of that commission, represented one point of view in saying that the program should be terminated immediately. Kenneth V. Thimann, then a member of the board of directors, represented a different view in saying that the defoliation program "probably represents a military device for saving lives that has an unprecedented degree of harmlessness to the environment." And one enthusiastic member of the audience exclaimed that the herbicide study was "the greatest service the AAAS has ever performed for the human race" (76).

## New Committees

The special committee that advised the board on how to respond to Pfeiffer's original resolution on the use of herbicides in Vietnam also recommended that AAAS establish a standing Commission on Environmental Alteration to serve as a group of advisors who would keep alert to environmental issues and from time to time recommend studies of particular issues. The board liked the idea; in fact it had earlier asked the Committee on Science in the Promotion of Human Welfare to accept that alerting responsibility. However, that committee preferred to devote its energies to its own studies and reports and after initiating the Air Conservation Commission had made no further recommendations of the kind the board had hoped to receive. Even so, the board again asked the Committee on Science in the Promotion of Human Welfare to function as their advisor on needed environmental studies (77). The committee again declined. They feared that the new function would require too much of their time and recommended that the alerting function be assigned to a new committee, which, they proposed, also be given responsibility for maintaining a continuing review of problems of population and of the relations between population changes and environmental changes (78). The board therefore returned to the original recommendation and established a new Committee on Environmental Alteration (79).

First under the chairmanship of Jack Ruina and later that of Barry Commoner the committee also preferred to select and work on specific topics. An early choice was an inquiry into the effects of dumping large amounts of the newly popular enzyme detergents, such as AXION, into sewers. However, primary attention soon turned to the relationship between power production and environmental pollution. The committee arranged a symposium on that topic for the 1969 annual meeting and then submitted a proposal for a two-year study of that topic. The board approved and gave the chairman authority to seek support for the study (80).

In 1969, as the Committee on Environmental Alteration was getting started, the board also established a Commission on Population and Reproduction Control (81), with Garrett Hardin as its chairman.

Thus in 1970 the association had a Committee on Science in the Promotion of Human Welfare, a Committee on Environmental Alteration, and a Commission on Population and Reproduction Control — three groups with related and sometimes overlapping interests. That situation was nicely illustrated at the 1970 annual meeting when the Committee on Environmental Alteration sponsored a symposium on the question "Is Population Growth Responsible for the Environmental Crisis in the United States?" while the Commission on Population and Reproduction Control sponsored a seven-session symposium under the title "Reducing the Environmental Impact of a Growing Population."

## Toward a More Active Role for AAAS

The initial actions that led to creation of the Committee on Science in the Promotion of Human Welfare and later to the Herbicide Assessment Commission were taken by individual AAAS members, not by the board, council, or staff. To be sure, the board approved, authorized financial support, and encouraged the activities of those bodies, although not as rapidly as the initial advocates would have liked. But no officially constituted body of AAAS could say "we started that."

Through the 1960s, however, that situation gradually changed. Some council members wanted to become more involved with issues of policy and societal relations instead of having the council deal only with such administrative chores as electing officers, committee members, and new affiliates. One indication of that restiveness was appointment of council study committees and another was the decision to hold symposia on substantive topics just for council members — a custom that continued for a few years until the council decided that it should concentrate on legislative responsibilities instead of competing with the Program Committee. The board of directors moved in the same direction through several different kinds of activities.

## Studies of Science Policy

In the early 1960s a number of universities began to offer courses or to develop research and degree programs on science policy, science and society, or science, technology, and society. In different ways these programs were responding to growing concerns of students and faculty about the unintended and sometimes undesirable effects of technological developments and the growing dependence of research on federal funding, raising the possibility of greater governmental control over science.

Faculty members from some of those programs came to Don Price, who became president-elect in 1966 and whose *The Scientific Estate* (1965) was used in some of those programs, or to me to ask if AAAS would help them to exchange information and ideas about their programs. We arranged a meeting in 1966 to which we invited representatives of all of those programs we knew about. The participants wanted a continuing relationship with AAAS and continuing opportunities such as were provided at that initial meeting, but did not want a formal organization. Eugene Skolnikoff of MIT, who was then secretary of the AAAS Section on Social, Economic and Political Sciences, volunteered to serve as the coordinator or secretary of the group and thus was born the informal Science and Public Policy Study Group (SPPSG). The group met together for one three-day session in Washington, D.C., and beginning in 1967 held annual symposia and meetings at the association's annual meetings.

To help pay expenses eleven universities each contributed $1,000 and the Sloan Foundation provided a grant of $10,000. With those funds the group began publishing a useful newsletter. In 1970 Harvey Sapolsky of MIT succeeded Skolnikoff in both AAAS and study group duties and the board of directors agreed to continue AAAS support (82). The Massachusetts Institute of Technology also gave support to the group, and the MIT Press published its newsletter, originally called *SPPSG Newsletter* and later *Public Policy*, for about five years. By then the academic programs on science and technology policy were well established and special support from AAAS was no longer needed.

## Joint Meeting with the British Association

In 1969 the British and American associations for the advancement of science held a joint conference for a few invited participants to consider the state of science in the world, the relations of science to society, the obligations of scientists, and the future role of national associations for the advancement of science. Meeting under the general theme of "Science and the Future," 20 participants from the two associations and nine guests from countries as different as the Soviet Union and Trinidad spent April 13 through 19 at the National Center for Atmospheric Research in Boulder, Colorado.

Two addresses set the tone of the meeting. Robert Morison reviewed the history of public attitudes toward science and summarized recent changes in those attitudes:

> Although the general public is grateful to science for some of its more tangible benefits, it is increasingly skeptical and even frightened about its long-term results. The anxiety centers on the concept of science as the prototype — the most magnificent and most frightening example of the rational system which men make to control their environment and which end by controlling *them*.

He concluded by stating what seemed to him to be the most important lesson for the scientific community:

> Science can no longer be content to present itself as an activity independent of the rest of society, governed by its own rules and directed by the inner dynamics of its own processes. Too many of those processes have effects which, though beneficial in many respects, often strike the average man as a threat to his individual autonomy. Too often science seems to be thrusting society as a whole in directions which it does not fully understand and which it has certainly not chosen (83).

Bentley Glass, in another major address, reversed the conference title "Science and the Future" to consider the problems of "The Future of Science." He concluded that scientists and their organizations had to change, not only because of the lack of understanding of the long-term effects of technological change and the changes in public attitudes — the factors discussed by Robert Morison — but also because the scientific effort had grown so large, so specialized, and was changing so rapidly that the nature of research and teaching could not continue much longer in the modes familiar to the conference participants (84).

Lively discussion of those themes, summaries of what the British and American associations had been doing, and speculation over what national associations for the advancement of science should be doing filled the days. No formal resolutions were proposed or adopted, but that was not the purpose. Yet the rapporteur, Marcel Roche of Venezuela, could write that there was general agreement that the growth of science in the more highly technological countries had reached or was approaching the inflection point of a growth curve; that the allocation of limited funds for research should be substantially based on the nature of human needs; that the future of the world depended largely upon scientific knowledge; that advanced and wealthy countries should increase their assistance to the developing and poorer ones; and that one of the responsibilities of national associations for the advancement of science was to "make it possible for government leaders and others who help to develop public policy to understand the probable long- and short-range consequences of technological development, so that technology may better serve the long-range interests of society" (85).

## Committee of Young Scientists

As the directors were considering that joint meeting and its implication for AAAS, Walter Orr Roberts, then chairman, resurrected an idea that, as I recall, had first been proposed by Margaret Mead. From time to time the board had discussed the desirability of appointing young scientists to AAAS committees. However, when a specific young scientist was suggested, it was likely that only the proposer knew the nominee, and the suggestion was likely to get lost as more familiar names were proposed. To get around that difficulty Mead suggested that each member of the board bring a junior counterpart to one of the board meetings. The members of the board could in that way get acquainted with a few young scientists and appoint them to some of the association's committees.

In June of 1969, two months after the joint BA–AAAS meeting, Chairman Roberts suggested that each member bring a junior counterpart to the October meeting to listen to and join in a discussion of what was planned as the major topic on the agenda: "The Responsibilities of the Scientific Community for the Values and Uses of Science." The junior colleagues could then meet by themselves and later return to give the board such advice as they chose.

The board agreed (86) and at the October meeting seven members were accompanied by young colleagues. By December four more members had selected their young colleagues. Most of the 11 were in their thirties, had earned their doctorates, and were well started on professional careers, but the group also included an undergraduate, a medical school student, and a graduate student. Their fields ranged from mathematics and physics to economics and political science.

Both the board members and their young colleagues agreed that their discussions had been useful and that the dialogue should be continued. As the young colleagues met by themselves they decided that they wanted to establish themselves as a Committee of Young Scientists to serve as a continuing advisory body to the board. The board accepted that proposal and provided the new committee with a budget to enable the members to meet and to prepare recommendations for the board to consider in 1970 (87).

By March the Committee of Young Scientists had prepared a 30-page report describing the problems of the association as they perceived them and recommending 15 changes in AAAS structure, governance, membership, and activities (88). Some of those recommendations were adopted by the board either immediately or later, but others seemed unrealistic or unnecessary (89).

The board's written response explaining why it approved or did not approve each of the recommendations was a disappointment to the Committee of Young Scientists. One called it a "masterpiece of evasion and equivocation," and several seemed to feel that the elders of the tribe had let them down and failed to recognize the value of their advice (90). Nevertheless, the board wanted the relationship to continue and so appointed a continuing Youth Council whose six initial members were drawn from the Committee of Young Scientists. Each member of the Youth

Council was also appointed to one of the association's committees: AAAS Meeting, Environmental Alteration, Governance, Public Understanding of Science, Science Education, and Science in the Promotion of Human Welfare. Thus they had opportunities to make further recommendations concerning those activities of AAAS in which they were particularly interested (91).

## To Improve the Effectiveness of Science in the Promotion of Human Welfare

Through the 1960s much attention was given to one of the association's three long-standing objectives: to improve the effectiveness of science in the promotion of human welfare. Ward Pigman's resolution that led to the Committee on Science in the Promotion of Human Welfare; C. W. Pfeiffer and the resolution that led to the Herbicide Assessment Commission; the studies and reports of those and other new committees and commissions; discussions at the joint conference with the British association; and the discussions of and with the Committee of Young Scientists all focused on that objective. By 1969 it seemed time for a systematic review, and the board asked Gerald Holton, one of its members, to prepare a document that could be used as a basis for a comprehensive discussion of AAAS activities and responsibilities (92).

Holton responded with a systematic matrix of areas of responsibility and methods which might be used in carrying out each. His matrix listed 10 topical areas in which the association was or might be engaged: (1) war, militarism, violence; (2) population pressures; (3) hunger, malnutrition; (4) quality of life (environment, the "Gaps," urban problems); (5) promoting the understanding and contributions by scientists on interdisciplinary subjects (within the sciences, and also outside science itself, for example, the humanistic side of science); (6) promoting understanding of the scientific approach (including science education, attraction of science for the young); (7) justification of support for scientific activity; (8) quality of scientific life (including criteria or priority in scientific choice, integrity of science); (9) technology assessments (including closing of feedback loops); and (10) review of science policy proposals in a "lower House of Science" (including responding to proposals by the President's Science Advisory Committee, the National Institutes of Health, and AAAS itself).

Across the other dimension of the matrix Holton listed six methods that were used for some or all of the 10 topical areas of responsibility: (1) educational and other AAAS commissions and committees; (2) sessions at meetings; (3) journals and books; (4) board or council statements; (5) other presentations (for example, to Congress); and (6) other methods and work arrangements.

The staff prepared a brief description of existing AAAS programs and activities that enabled board members to check appropriate boxes in the matrix, see which boxes were left blank, and start deciding what kind of activities directed toward which areas of potential responsibility seemed desirable to expand or to alter. As one outcome of that exercise the board adopted this statement:

> It is the sense of the Board that for the coming decade the main thrust of AAAS attention and resources shall be dedicated to a major increase in the scale and effectiveness of its work on the chief contemporary problems concerning the mutual relations of science, technology, and social change, including the uses of science and technology in the promotion of human welfare (93).

That statement gave evidence that a major change in attitude had developed since the somewhat hesitant — some would say timorous — acceptance of Pigman's resolution of 1955 that led to establishment of the Interim Committee on the Social Aspects of Science and its successor, the Committee on Science in the Promotion of Human Welfare. The attitudes of scientists had changed and so had public attitudes toward science. The power of science and technology had increased, and there was wider recognition of the sometimes damaging impact of technological developments. The board of directors had changed and so had the world in which the association operated. An objective of the association that had always been stated in second position had been raised to top priority.

*Chapter 12*

# 1970: Looking Back and Looking Forward

Looking back over the years from 1945 to 1970 shows how much AAAS activities were influenced by several major features of the relations between the scientific community and the surrounding social, economic, and political environment. One of those features — the one that determined the end of the period being reviewed — was the changing national support for research in science and engineering.

World War II and its aftermath brought national expenditures for research and development to a much higher level than before the war, and from early in the 1950s to late in the 1960s those funds increased at an average rate of almost 10 percent a year as measured in constant dollars. That steady increase enabled industry, universities, and other organizations — AAAS included — to expand scientific activities and to start new programs without having to curtail ones.

A growth rate of 10 percent a year was a heady experience for scientists who could remember the Great Depression of the 1930s, and for the generation that came to maturity in the 1950s and 1960s that rate seemed to be the happy norm. Academic scientists of both groups wanted it to continue. Although funds for development of weapons, space hardware, and energy might fluctuate as needs changed, a regular and predictable growth rate for university research seemed desirable. Several proposals to that effect were put forward (1), and in 1965 President Johnson sent Congress a budget for fiscal year 1966 that was actually based on that principle. As Elmer Staats, then deputy director of what is now the Office of Management and Budget, explained, "we agreed to aim for a 15 percent growth in basic research funds going to the universities. We came out of Congress with some 17 percent" (2).

That plan lasted exactly one year. Increases for research at a rate greater than that of the total national budget simply could not be permanently maintained, as William Carey had told research administrators several years earlier (3). In the late 1960s the growth rate slowed down; 1969 brought an actual decline in research and development funds in constant dollars, and 1970 brought a larger decline (4). Research programs had to be reduced and the job market collapsed. At the 1970 meeting of the American Physics Society and the American Association of Physics Teachers, only 121 positions were listed in the placement register; three years earlier 617 had been listed (5). The number of federal government fellowships and traineeships for graduate students which had peaked at over 57,000 in 1968 fell to about 41,000 in 1970 and was headed toward a much lower level. James R. Killian, Jr., who had been President Eisenhower's science advisor and president of the Massachusetts Institute of Technology, told Congress: "I have been in college administration for 30 years, and I recall no time when the financial outlook was so bleak as it is today." *Science News* summarized: "Viewed from almost any traditional angle, 1970 has been a disastrous year for science" (6).

For AAAS, grant funds in 1970 were only half what they had been five years earlier. Advertising in *Science* fell off when national research funds stopped growing and then declined a bit, and the association ended 1970 with a deficit. Research was not the major business of AAAS, but clearly the decrease in research support affected AAAS and its activities without curtailing older ones.

## Trust and Cooperation

Partly because funds were increasing rapidly and partly as a carryover from World War II, the early years of the 1945–1970 period were marked by a high level of trust and close cooperation between government agencies and the grantees that were carrying out scientific programs. Many individual scientists remember the early days of the Office of Naval Research and its cordial and generous relations with grantees as the golden age of federal research support. For AAAS, relationships were closest and most continuing with the National Science Foundation, which, like the Office of Naval Research, carried on some of the traditions of the wartime Office of Scientific Research and Development (OSRD). It had been the practice of this office to discuss a prospective research program with a university, an industrial research laboratory, or some other qualified research institution. If the two agreed on the research to be done, OSRD would say, in essence, "OK; get busy. We'll prepare a contract and will pay all of the expenses, but no more, so you can plan to do the work at no financial loss, but with no financial gain" (7). In the 1950s, initiation of the AAAS Traveling Libraries of Science exemplified that same spirit of mutual trust and cooperation, as did the later advancement of funds by AAAS to expand the program in a year in which the congressional appropriation for NSF was delayed (see Chapter 2).

As time went on, however, the total amounts of money involved became much larger, and formal controls had to be developed to supplant informal ones. Grantees had to negotiate with auditors as well as with program managers — auditors who sometimes seemed to know or care little about the purpose or nature of the research or other activity involved. Excellent work continued to be accomplished, but some of the pleasure had gone out of the process. AAAS was less affected by this change than were some other institutions, but the change could not help but affect the attitudes of scientists who served on association boards and committees. The association also had to modify its initial policy of contributing financially to every activity for which it requested a supporting grant (see Chapter 2).

## Fear and Hysteria

Another important element in the relations of scientists and engineers to society and government resulted from the long-continuing Cold War that started soon after the hot war of 1939-1945 came to an end. Senator Joseph McCarthy of Wisconsin, several other members of Congress, and a number of other witch hunters thought they saw Communists or Communist sympathizers lurking in nearly every government agency, university faculty, or research laboratory. Consistent with those fears was the belief that protection of our national security required great secrecy and therefore required the insulation of anyone who might possibly give away some of the nation's secrets. An opposing position, and one much more popular among scientists, was that national security could best be maintained by keeping ahead of potential enemies in scientific and technological knowledge and competence and that keeping ahead called for educating and utilizing men and women of the greatest talent and keeping them well informed of advances and new ideas from all sources. One area in which the disagreement between these two positions caused lots of trouble was in determining who was and who was not eligible to receive research or development grants and who was eligible for appointment to fellowships, traineeships, or membership on scientific advisory committees to government agencies. The effects of that disagreement colored the relationships of the scientific community with government agencies and agents through much of the period. Although tensions lessened during the following 25 years, they did not disappear, as evidenced by the 1969 charges that the U.S. Public Health Service was blacklisting some scientists that the National Institutes of Health wanted to appoint to the study committees that evaluated basic research proposals in the biomedical sciences (see Chapter 2).

## Increasing Power and Decreasing Prestige

In the 25 years from 1945 to 1970 science and engineering increased greatly in knowledge and in power. New instruments, new ideas, and new molecules tailored for specific purposes provided new opportunities in agriculture, design, medicine, and manufacturing. But some of the new materials also brought some disquieting consequences or side effects. Pollution of air, land, and water all increased. Materials developed for one good purpose sometimes also brought unexpected damage; DDT, for example. Rachel Carson's *Silent Spring* (8), Earth Day, and the environmental movement were strong and obvious reactions to the damage that seemed at least partially due to the work of scientists and engineers. Concurrently, as economic growth proceeded in the United States, national priorities gradually shifted to give greater emphasis to social goals, goals to which science was not seen as likely to make substantial contributions. Consequences included some calm and judicious changes, but also included disturbances and sometimes violent protests over racial, ethnic, or sex discrimination.

## From Arrogance to Ethical Dilemmas

At a deeper and more general level underlying the mounting worries over pollution and ecological damage was a change in the prevailing concept of the relationship of humankind to the rest of nature. The prevailing attitude had been that other species and the rest of nature existed for the use and benefit of the world's human population. That attitude had theological, traditional, and economic support as we sought to "control" or "tame" nature, exploit natural resources, reroute rivers, get rid of forests, and "improve" upon such wastelands as estuarine marshes. But as the damage and pollution became more and more evident, there also came to be greater recognition that the rest of the world could no longer be treated so wantonly; that the laws of nature had to be accepted; and that even in the long-run interests of the human population the old practices of unchecked exploitation had to be abandoned. This recognition was one of the central themes of Rachel Carson's 1962 book, in which she wrote, "The 'control of nature' is a phrase conceived in arrogance, born of the Neanderthal age of biology and philosophy, when it was supposed that nature exists for the convenience of man" (9).

AAAS members could take some pride in the fact that seven years before *Silent Spring* appeared and 15 years before the first Earth Day, the association had established its Committee on Science in the Promotion of Human Welfare and started the sequence of events that led to several symposia on problems of science and society, the publication of *Air Conservation* (10), the work of the Herbicide Assessment Commission in Vietnam, and other efforts "to improve the effectiveness of science in the promotion of human welfare" (see Chapter 11).

The old arrogant, exploitative attitude and practices have not disappeared, however; witness as a particularly disturbing example the destruction of tropical forests in the Amazon Basin and elsewhere. But those practices are no longer as accepted as they once were. Treatment of the natural environment, of other species, and of the poorer and less-fortunate members of the human species have all become matters of much more widespread ethical, moral, and even political and practical concern. This change in attitude has affected the whole climate of scientific research and thinking, sometimes by changing the interests and practices of scientists and sometimes by changing the public's concern with and regulation or control of scientific work. Some of the specific issues — for example, use of animals in research — are quite controversial and sometimes they pit medical and biological researchers against animal rights' activists in ideological and even physical conflict.

These changes and differences in the relations of the scientific community and the world at large were not the subject of most of the association's activities, but explicitly or implicitly they constantly influenced those activities. It is a truism to say that the activities of an individual or of an association are influenced by the conditions under which they live and work. But the specifics of those relationships between science and society are the stuff of science policy, of the nation's attitudes toward and its support for and control of scientific activities. Thus sometimes explicitly and sometimes implicitly the whole history of AAAS activities and growth was constantly influenced by those relationships between the scientific community and the larger world. It could not have been otherwise.

With the scientific and engineering contributions to World War II getting dimmer in memory, with annual expenditures for research and development much larger, with growing worries over pollution and environmental damage, and with increasing attention to social goals, the prestige of science and engineering declined. Presidential attention was less than it had been in the Truman and Eisenhower years, as was congressional favor. In public opinion polls science and engineering received slightly lower ratings on prestige and general standing in 1972 than they had in 1947, as did most professionals (11). For these several reasons, Philip Abelson, editor of *Science*, reported that

> Conversation with academic scientists these days is often a depressing experience. Almost all of them have been affected by one or more of four major adverse developments — student riots, financial problems, job insecurity, and loss of prestige. Student disorders and faculty dissension have shaken the Ivory Tower .... The financial problems of universities have adversely affected job opportunities and job security. ... To many engaged in research the worst blow has been a decline in the prestige of science. For nearly two decades after World War II scientists enjoyed especially high public esteem. In part this was related to the Cold War and to competition with the U.S.S.R. In part it arose from the belief that science and technology were bringing an increasingly affluent society. Now the public has turned its attention away from the Russians and it is bored with, even critical of, affluence (12).

## Dissent and Disruption

To a casual observer perhaps the most obvious effect of those changing attitudes and beliefs was the fact that late in the 1960s protesters found a target in AAAS meetings. The 1969 and 1970 meetings were sometimes disrupted by shouted protests or guerrilla theater. Protesters dressed as witches disrupted the annual Phi Beta Kappa – Sigma Xi address by casting a hex on both of those sponsoring societies and hit AAAS with a "Witches Hex" that ended with an obscene damnation of the whole "male-dominated" association.

During the 1969 meeting Walter Orr Roberts, then chairman of the board, spent considerable time with some of the protesters trying to understand what they wanted. Their actions were sometimes a nuisance, but on the final evening we were glad to assign them one of the meeting rooms for the party they wanted to hold and then a few hours later were saddened to learn that their party had been "busted" by the hotel security force because they brought in some of their refreshments instead of purchasing them from hotel sources.

## Changing the Watch

June 30, 1970, was my last day as the association's executive officer. Fourteen months earlier I had reminded the board of directors that I had been in office for more than 15 years. I said that I thought I had accomplished about all I could for AAAS and that it was time to start looking for a new helmsman. As for myself, I wanted to return to teaching for the few years that remained before retirement and wanted then to retire in a university community.

In 1970 the board persuaded William Bevan, vice-president and provost of the Johns Hopkins University, to become the association's new executive officer. A few weeks later William Kabisch, my right-hand aide, resigned to become associate dean for research in the Medical School of Southern Illinois and Bevan selected William Trumbull as his assistant.

I had not planned it so, but in some ways the summer of 1970 was a better time for leaving than for arriving. As Bevan started to pick up the responsibilities of the AAAS office he faced the problems described earlier in this chapter and at the end of the year had to tell the council that some of those problems — including a sudden reduction in advertising revenue in *Science* during the latter part of the year — had meant that AAAS had incurred a deficit, only its second in over 30 years. The deficit was temporary, however, and the problems of science were challenges as well as troubles. Thus William Bevan arrived at a time when there were troubles for the association but also opportunities for new thinking and new activity.

## Toward a Positive Future

Shouting down scheduled speakers, hexes by witches, and controversy over a nominee for the association's highest office were all troublesome, but those troubles were not the dominant motif of AAAS in 1970. Much more time and effort went into positive preparations for the future. In some ways the mood resembled that of 1951 when the board and some consultants led by Warren Weaver wrote the Arden House statement. The board of directors was again going through an introspective examination of the association's activities, obligations, and opportunities, but without a 1970 Warren Weaver to focus the inquiry and dramatize its conclusions in a new "Arden House" statement.

Nevertheless, there was a substantial amount of forward planning and some of the preceding chapters have concluded with brief accounts of what lay ahead. As 1970 ended the temporary Committee on Governance — a joint committee of the board of directors and the Committee on Council Affairs — was reaching agreement that the association needed a leaner and more effective council and that the president-elect, members of the board, and members of the Committee on Nominations and Elections should all be elected by AAAS members instead of by the council. These two major changes were soon adopted.

In 1970 the Commission on Science Education distributed the association's second set of guidelines for the preparation in science of future elementary school teachers and that year started work on revised guidelines for preparing high school teachers of science and mathematics. More and more schools were adopting "Science — A Process Approach," the association's innovative program for teaching science in kindergarten and the elementary school grades. In addition, the Commission on Science Education was planning several new programs that flourished after 1970, such as the "Chautauqua" type of short courses for college teachers and the widely distributed bibliographies on science and society.

The Committee on Desert and Arid Zone Research of the Southwestern and Rocky Mountain Division — and while it existed the parallel national committee of AAAS — had held the association's second international conference on arid lands and had led the way to acceptance by AAAS of a permanent commitment to improvements in the arid lands of the world. The Committee on Science in the Promotion of Human Welfare, the Committee on Environmental Alteration, the Commission on Population and Reproduction Control, and the Herbicide Assessment Commission all gave form and substance to the board's resolution to devote more the association's time and energy to societal problems of science and technology. As a further effort in that direction AAAS joined with the American Council of Learned Societies and the Social Science Research Council in discussions of the humanistic implications of science and technology (13).

With all that going on *Science* was flourishing; AAAS was publishing several useful newsletters for special audiences; and AAAS meetings were setting good models for the future. Five of the annual meetings during the 1960s had attracted more than 7,000 registrants. By 1970 AAAS had outgrown the new home it built in 1956 and was beginning to face a space problem, but several frustratingly unsuccessful efforts to solve that problem had left the real solution for the future. The amount of actual change and consideration of future change was the central issue on the board's agenda and led the news staff of *Science* to publish a three-part review of the association's activities and its search for its own future (14).

At the end of 1970 when Bentley Glass, chairman of the board of directors, gave his report to the council he summarized the board's view:

> The mood of the board during 1970 has been one of continued self-assessment and increasing concern for involvement in the larger community, both national and international. Three thematic dimensions are visible throughout the board's deliberations and actions:
> (1) commitment to an engagement with the great problems that plague modern society, (2) concern for a growing disaffection with the scientific enterprise on the part of the lay public, especially the nation's youth, and (3) involvement of a larger number of both scientists and laymen in the association and its affairs (15).

Those were the board's major objectives as 1970 drew to a close. In what manner and how effectively AAAS succeeded in achieving those objectives is for some later chronicler to describe and analyze.

# Appendix 1: Origins and Chronology, AAAS from 1848 to 1970

The founding of the American Association for the Advancement of Science in Philadelphia in September of 1848 was the first permanently successful effort to establish in the United States a truly national scientific society embracing all the sciences, but not the first effort to bring such a society into being.

In 1816, John Quincy Adams, John C. Calhoun, Daniel Webster, Edward Everett, Henry Clay, and other national leaders launched the Columbian Institute for the Promotion of Arts and Sciences. Its diverse and ambitious objectives included not only the advancement of science and the establishment of several national institutions, but also the promotion of agriculture and foreign commerce. The Columbian Institute was started off with high hopes, but the founders were too preoccupied with other activities to keep it going. By 1825 it became inactive, and in 1840 it "passed into" the National Institution for the Promotion of Science.

The National Institution was incorporated by act of Congress in 1840. Monthly meetings were held for a year or two; three bulletins were published; and in April of 1844, the National Institution sponsored an elaborate Congress of Scientists in Washington, to which it invited members of the American Philosophical Society and other learned societies. After this congress, the National Institution never held another meeting, although technically it continued to exist until its charter expired in 1861.

In 1831, the British Association for the Advancement of Science was founded. In 1837, John Collins Warren, a prominent medical scientist of Boston, read a paper at one of its sessions and was greatly impressed with the value of one large meeting devoted to all the sciences. Upon his return to the United States, he endeavored to found a parallel association in America. He invoked the cooperation of the American Philosophical Society, but that organization decided his idea was "inexpedient."

Despite these failures, the first half of the nineteenth century was a time of scientific advance. Geology was particularly active. The mineral resources of a vast continent were largely unexplored. Between 1823 and 1839, no less than 17 states provided for geological surveys. In 1819, geologists incorporated the American Geological Society. One of its officers was Benjamin Silliman of Yale, who had started just the year before the long-lived and important *American Journal of Science and Arts*. Although the American Geological Society lapsed in 1826, its spirit reappeared a few years later in the Association of American Geologists, which, again a few years later and after a change of name, was transformed into the American Association for the Advancement of Science.

In 1837, and probably earlier, Edward Hitchcock, professor of chemistry and natural history at Amherst College, who had been a member of the old American Geological Society, advocated a national association of geologists. Several geologists, among them Lardner Vanuxem and James Hall, had thought of a conference to organize such an association. Crediting Hitchcock with the idea, W. W. Mather arranged a meeting on November 20, 1839, at the Albany home of Ebenezer Emmons, Williams College professor of natural history, who was doing a portion of New York's geological survey. At that meeting, Vanuxem, Hall, Mather, Timothy A. Conrad, and Emmons drew up plans to form a national association, and Vanuxem conducted the subsequent correspondence. A second conference of the same group was held in 1839 at the Emmons house. As a result, the Association of American Geologists was organized on April 2, 1840, at the Franklin Institute, Philadelphia, by 18 prominent geologists of seven eastern states. Edward Hitchcock was chairman.

Two years later, at Boston, this society became "The Association of American Geologists and Naturalists." The new name was logical, for most of the geologists of the period were paleontologists and were familiar with the plants and animals they encountered in the field. Moreover, zoological and botanical surveys were being authorized by the states. But as the number of scientists increased and scientific disciplines became more specialized, the need for a national organization that would include all the sciences became more and more apparent. When the Association of American Geologists and Naturalists met in Boston on September 24, 1847, under the chairmanship of William Barton Rogers, the society passed a resolution to reorganize and to change its name to the "American Association for the Promotion of Science ... designed to embrace all labourers in Physical Science and Natural History." Officers for the new American Association were elected, and a committee of three (H. D. Rogers, Benjamin Peirce, and Louis Agassiz) was appointed to draw up a new "Constitution and Rules of Order" and to report at the meeting set for Philadelphia the next year.

On September 20, 1848, the American Association for the Promotion of Science met in the library room of the Academy of Natural Sciences in Philadelphia. William B. Rogers of the University of Virginia, the last president of the Association of American Geologists and Naturalists, called the meeting together at noon and read the report of the committee on the new constitution. The committee proposed a slightly different name for the new organization and recommended that:

> The Society shall be called THE AMERICAN ASSOCIATION
> FOR THE ADVANCEMENT OF SCIENCE. The objects of the
> Association are, by periodical and migratory meetings, to promote
> intercourse between those who are cultivating science in different
> parts of the United States; to give a stronger and more general im-
> pulse, and a more systematic direction to scientific research in our

country; and to procure for the labours of scientific men, increased facilities and a wider usefulness.

The organizational meeting then adjourned to reconvene in the "Hall of the University of Pennsylvania," at 4 p.m. for the first scientific session of the AAAS. Professor Rogers called the meeting to order and introduced the association's first elected president, William C. Redfield of New York. Among the eminent scientists present were Louis Agassiz, Stephen Alexander, Alexander D. Bache, Asa Gray, Joseph Henry, Benjamin Silliman, and John Torrey.

In 1848, the association had two sections, one on natural history and one on general physics. By 1945, there were 17 sections, and by 1970, 21. The charter membership had passed 27,000 in 1945 and 133,000 in 1970. The objectives of the association, although rephrased and somewhat expanded, have remained fundamentally as stated by the founders in 1848.

Some of the events of the early years of the AAAS through 1970 are summarized here:

1840  The Association of American Geologists, lineal ancestor of the AAAS, was organized on April 2 at the Franklin Institute, Philadelphia, by 18 prominent geologists. Amherst professor of chemistry and natural history Edward Hitchcock, who for some years had advocated a general association of scientists, was the first chairman.

1842  It was decided to change the name of the Association of American Geologists to the Association of American Geologists and Naturalists, in recognition of the breadth of interests of the membership.

1847  At its meeting in Boston on September 24, the AAGN passed a resolution to reorganize as the American Association for the Promotion of Science; a committee was appointed to draft a constitution.

1848  On September 20, at the Academy of Natural Sciences of Philadelphia, the AAGN formally became the AAAS, with 461 charter members. Reconvened in the "Hall of the University of Pennsylvania," the first meeting lasted five days; some 60 papers were read on Natural History and General Physics before the two sections, chaired, respectively, by Louis Agassiz and Joseph Henry.

1850  The office of "Permanent Secretary" was established. Spencer F. Baird was elected to begin service the following year at an annual salary of $300.

1851  The "Objects and Rules of the Association," with two amendments, were accepted and thereafter known as the first AAAS constitution. Dues were increased from $1 to $2; purchase of the *Proceedings* for an additional dollar was optional.

1852    No meeting of the AAAS was held because of "the prevalence of cholera along the approaches to Cleveland from the south."

1856    A new (second) constitution was adopted. Since there were no bylaws, the *Proceedings*, for a period, cumulatively listed the resolutions of the legislative body, which was called the Standing Committee.

1861– 1865    Because of the Civil War, there were no meetings for five years, and no presidents were elected.

1874    In March, the association was incorporated under the laws of Massachusetts. A new (third) constitution of 39 articles was adopted; it provided that each section (then two) should have a vice president, one of whose duties would be to give an annual address. Provision was made for the nomination of fellows from among members who were professional scientists or significantly advancing science. Membership dues were increased from $2 to $3, with the *Proceedings* included; new fellows paid $2 upon their election; and life memberships were offered at $50. The "admission fee" of $5, begun in 1867, was continued.

1887    By a constitutional amendment, the Standing Committee was renamed the AAAS Council.

1895    The council appointed a committee to study AAAS policies and make recommendations. The Policy Committee continued as a standing committee for some years, gradually acquiring specific duties, such as responsibility for meetings and publications (1904).

1899    A constitutional amendment provided that members of affiliated societies could attend AAAS meetings if they registered. This was the first mention of "affiliated societies" in the constitution, but the council had used the term in 1895, and a few societies had been meeting with the AAAS at least as early as 1891.

1900    *Science*, owned and edited by James McKeen Cattell, was made the official publication of the association and thereafter was sent to all members.

1901    A constitutional amendment provided that affiliated societies would be represented on the council by one or two persons, according to the number of AAAS fellows in their memberships.

1902    Partly because of the increase of academic summer schools, and partly to accommodate more affiliated societies, the time of the annual meetings was changed from August to late December.

1907    The *Proceedings* ceased to carry the details of papers and addresses, since many of them were being published in *Science*. The headquarters of the association, which had been in the office or home of the permanent secretary, was located in the Smithsonian Institution in rooms generously provided by the Institution.

1908 This was the last year the *Proceedings* were published as an annual volume. *Summarized Proceedings,* covering more than one year's meetings, appeared from 1910 to 1948.

1915 The Pacific Division of the AAAS was established; its first meeting was held in 1916 in San Diego. *The Scientific Monthly* became an official journal of the Association.

1920 The Southwestern and Rocky Mountain Division of the AAAS was established and held its first meeting in El Paso. The Policy Committee was renamed the Executive Committee.

1923 The first AAAS Thousand Dollar Prize (now called the Newcomb Cleveland Prize) was awarded to Leonard Eugene Dickson for his paper on "The Theory of Numbers."

1924 The Annual Exposition of Science (now called the AAAS Exhibition) was established as an organized and integral feature of the annual meetings of the association. There had been a few commercial exhibits earlier.

1934 The first of a series of special monographs, *The Protection by Patents of Scientific Discoveries*, was published in January as a supplement to *Science*. (Since 1938, books in this series have been designated as symposium volumes.)

1938 The Gibson Island Research Conferences, which had been started in 1931 by Neil Gordon, were established on a permanent basis under AAAS auspices. (In 1948 they were renamed the Gordon Research Conferences in honor of their founder.)

1939 The association assumed responsibility for editing and publishing *The Scientific Monthly* (Actual purchase of the magazine from James McKeen Cattell was completed in 1943.)

1942– No meetings were held because of World War II. Beginning in March
1943 1942, a new publication, the *AAAS Bulletin*, was issued monthly to keep members informed of association activities and plans during the period of interrupted meetings. Publication was discontinued in December 1946.

1944– The association purchased *Science* in 1944 from the estate of James
1945 McKeen Cattell. The 1944 meeting was held in September in Cleveland under wartime conditions. Because of travel restrictions, the 1945 meeting was delayed until March 1946.

1946 — A new (fifth) constitution — with no bylaws — redefined the objects of the association, defined the powers of the council, and provided for a new officer, the president-elect. On September 9, the AAAS moved from the Smithsonian Institution into the largest of five old residences on a block off Scott Circle, Washington, DC.

1948 — From September 13 to 17 in Washington, DC, the association celebrated its centenary with a special meeting, the theme of which was "One World of Science."

1951 — The Alaska Division of the AAAS was established and held its first meeting at Mount McKinley National Park. The Executive Committee issued the "Arden House Statement," reassessing the Association's program and policies.

1952 — A new (sixth) constitution of 12 articles and 11 bylaws changed the Executive Committee to a board of directors.

1955 — Ground was broken in April for a new headquarters building on the Scott Circle site; temporary headquarters were rented at 1025 Connecticut Avenue. The association sponsored an International Conference on Arid Lands in Albuquerque and Socorro, New Mexico, April 26–May 4. The AAAS Socio-Psychological Prize, endowed by Arthur F. Bentley, was established on an annual basis. (In 1985, the award was renamed the AAAS Prize for Behavioral Science Research.) The AAAS Science Teaching Improvement Program was initiated.

1956 — The new three-story headquarters building at 1515 Massachusetts Avenue was occupied on May 25.

1958 — *Science* and *The Scientific Monthly* were merged into an enlarged weekly, *Science*. A "Parliament of Science" to consider national science policy was held in Washington, DC.

1959 — The first International Oceanographic Congress was held under AAAS auspices at the United Nations, New York, August 30–September 11. The Symposium on Basic Research was held in New York City. The AAAS-Westinghouse Science Writing Awards were reestablished. (An earlier series had been administered from 1946 through 1953.) The Committee on Public Understanding of Science was established.

1960 — Amendments to the constitution and bylaws defined more clearly the responsibilities and duties of the council and board of directors. A Committee on Council Affairs was established.

1961 — Publication of the *AAAS Bulletin* was resumed on a quarterly basis.

1962   With support from the National Science Foundation, a Commission on Science Education was established to direct programs and develop materials designed to improve science instruction and the education of science teachers.

1965   *Science Books: A Quarterly Review* was established "to improve ... science education ... and the public understanding of science" by critically reviewing books for all educational levels. (The journal is now *Science Books & Films*.)

1966   Membership passed the 100,000 level. A third annual AAAS-Westinghouse Science Writing Award was started, to recognize excellent science writing in small circulation newspapers.

1967   The Commission on Science Education completed developmental work on *Science — A Process Approach* materials for teaching science in kindergarten and the first and second grades.

1969   The annual meeting in Boston had the largest registration of any in AAAS history (7,900). International Conference on Arid Lands in a Changing World was held in Tucson, Arizona. The board of directors appointed the Herbicide Assessment Commission.

1970   Mina Rees took office as president-elect, the first woman to hold that position.

# Appendix 2: Meetings and Presidents

The AAAS president serves three years in the presidential sequence: as president-elect the first year, as president the second year, and as retiring president and chairman of the board the third year.

A list of the eminent people who have served as presidents of the AAAS since its founding in 1848 through 1970, together with their disciplines and the meetings at which they officiated, follows.

| | | |
|---|---|---|
| 1848 Sept. | Philadelphia | William B. Rogers (*geology*) [acting until the installation of first elected President Redfield] |
| | | William C. Redfield (*geology*) |
| 1849 Aug. | Cambridge, MA | Joseph Henry (*physics*) |
| 1850 Mar. | Charleston, SC | A. D. Bache (*oceanography*) |
| 1850 Aug. | New Haven | A. D. Bache (*oceanography*) |
| 1851 May | Cincinnati | Louis Agassiz (*glaciology, zoology*) |
| 1851 Aug. | Albany, NY | Louis Agassiz (*glaciology, zoology*) |
| 1852 | no meeting | Benjamin Peirce (*physics*) |
| 1853 July | Cleveland | Benjamin Peirce (*physics*) |
| 1854 April | Washington, DC | James D. Dana (*geology*) |
| 1855 Aug. | Providence | John Torrey (*botany*) |
| 1856 Aug. | Albany, NY | James Hall (*geology*) |
| 1857 Aug. | Montreal | J. W. Bailey (*chemistry*) |
| | | Alexis Caswell (*astronomy*) [successor to J. W. Bailey, deceased] |
| 1858 April | Baltimore | Jeffries Wyman (*medicine*) |
| 1859 Aug. | Springfield, MA | Stephen Alexander (*astronomy*) |
| 1860 Aug. | Newport, RI | Isaac Lea (*geology*) |
| | (no meetings 1861–1865 and no presidents) | |
| 1866 Aug. | Buffalo | F. A. P. Barnard (*astronomy*) |
| 1867 Aug. | Burlington, VT | J. S. Newberry (*geology*) |
| 1868 Aug. | Chicago | Benjamin A. Gould (*astronomy*) |
| 1869 Aug. | Salem, MA | J. W. Foster (*geography*) |
| 1870 Aug. | Troy, NY | William Chauvenet (*astronomy*) |
| | | T. Sterry Hunt (*geology*) [successor to W. Chauvenet, deceased] |
| 1871 Aug. | Indianapolis | Asa Gray (*botany*) |
| 1872 Aug. | Dubuque | J. Lawrence Smith (*chemistry*) |
| 1873 Aug. | Portland, ME | Joseph Lovering (*physics*) |

| | | |
|---|---|---|
| 1874 Aug. | Hartford | John L. LeConte (*entomology*) |
| 1875 Aug. | Detroit | Julius L. Hilgard (*geography*) |
| 1876 Aug. | Buffalo | William B. Rogers (*geology*) |
| 1877 Aug. | Nashville | Simon Newcomb (*astronomy*) |
| 1878 Aug. | St. Louis | O. C. Marsh (*paleontology*) |
| 1879 Aug. | Saratoga Springs | George F. Barker (*chemistry*) |
| 1880 Aug. | Boston | Lewis H. Morgan (*anthropology*) |
| 1881 Aug. | Cincinnati | George J. Brush (*geology*) |
| 1882 Aug. | Montreal | J. W. Dawson (*geology*) |
| 1883 Aug. | Minneapolis | Charles A. Young (*astronomy*) |
| 1884 Sept. | Philadelphia | J. P. Lesley (*geology*) |
| 1885 Aug. | Ann Arbor | H. A. Newton (*mathematics*) |
| 1886 Aug. | Buffalo | Edward S. Morse (*zoology*) |
| 1887 Aug. | New York | S. P. Langley (*physics*) |
| 1888 Aug. | Cleveland | J. W. Powell (*geology*) |
| 1889 Aug. | Toronto | T. C. Mendenhall (*physics*) |
| 1890 Aug. | Indianapolis | George L. Goodale (*botany*) |
| 1891 Aug. | Washington, DC | Albert B. Prescott (*chemistry*) |
| 1892 Aug. | Rochester, NY | Joseph LeConte (*geology*) |
| 1893 Aug. | Madison, WI | William Harkness (*astronomy*) |
| 1894 Aug. | Brooklyn | Daniel G. Brinton (*anthropology*) |
| 1895 Aug. | Springfield, MA | Edward W. Morley (*chemistry*) |
| 1896 Aug. | Buffalo | Edward D. Cope (*paleontology*) |
| | | Theodore Gill (*zoology*) [successor to Edward D. Cope, deceased] |
| 1897 Aug. | Detroit | Wolcott Gibbs (*chemistry*) [W. J. McGee presided in Gibbs's absence] |
| 1898 Aug. | Boston | F. W. Putnam (*anthropology*) |
| 1899 Aug. | Columbus, OH | Edward Orton (*geology*) |
| | | Marcus Benjamin (*social sciences*) and Grove Karl Gilbert (*geology*) [successors to Edward Orton, deceased] |
| 1900 June | New York | R. S. Woodward (*mathematics*) |
| 1901 Aug. | Denver | Charles S. Minot (*medicine*) |
| 1902 June | Pittsburgh, PA | Asaph Hall (*astronomy*) |
| 1902 Dec. | Washington, DC | Ira Remsen (*chemistry*) |
| 1903 Dec. | St. Louis | Carroll D. Wright (*economics*) |
| 1904 Dec. | Philadelphia | W. G. Farlow (*botany*) |
| 1905 Dec. | New Orleans | C. M. Woodward (*mathematics*) |
| 1906 June | Ithaca | William H. Welch (*medicine*) |
| 1906 Dec. | New York | William H. Welch (*medicine*) |
| 1907 Dec. | Chicago | E. L. Nichols (*physics*) |

| | | |
|---|---|---|
| 1908 June | Hanover, NH | Thomas C. Chamberlin (*geology*) |
| 1908 Dec. | Baltimore | Thomas C. Chamberlin (*geology*) |
| 1909 Dec. | Boston | David Starr Jordan (*biology*) |
| 1910 Dec. | Minneapolis | A. A. Michelson (*physics*) |
| 1911 Dec. | Washington, DC | Charles E. Bessey (*botany*) |
| 1912 Dec. | Cleveland | E. C. Pickering (*astronomy*) |
| 1913 Dec. | Atlanta | Edmund B. Wilson (*zoology*) |
| 1914 Dec. | Philadelphia | Charles W. Eliot (*education*) |
| 1915 Aug. | San Francisco | W. W. Campbell (*astronomy*) |
| 1915 Dec. | Columbus, OH | W. W. Campbell (*astronomy*) |
| 1916 Dec. | New York | Charles R. Van Hise (*geology*) |
| 1917 Dec. | Pittsburgh, PA | Theodore W. Richards (*chemistry*) |
| 1918 Dec. | Baltimore | John Merle Coulter (*botany*) |
| 1919 Dec. | St. Louis | Simon Flexner (*medicine*) |
| 1920 Dec. | Chicago | Leland O. Howard (*entomology*) |
| 1921 Dec. | Toronto | Eliakim H. Moore (*mathematics*) |
| 1922 June | Salt Lake City | J. Playfair McMurrich (*anatomy*) |
| 1922 Dec. | Boston | J. Playfair McMurrich (*anatomy*) |
| 1923 Sept. | Los Angeles | Charles D. Walcott (*paleontology*) |
| 1923 Dec. | Cincinnati | Charles D. Walcott (*paleontology*) |
| 1924 Dec. | Washington, DC | J. McKeen Cattell (*psychology*) |
| 1925 June | Boulder, CO | Michael I. Pupin (*engineering*) |
| 1925 June | Portland, OR | Michael I. Pupin (*engineering*) |
| 1925 Dec. | Kansas City, MO | Michael I. Pupin (*engineering*) |
| 1926 Dec. | Philadelphia | Liberty Hyde Bailey (*horticulture*) |
| 1927 Dec. | Nashville | Arthur A. Noyes (*chemistry*) |
| 1928 Dec. | New York | Henry F. Osborn (*paleontology*) |
| 1929 Dec. | Des Moines | Robert A. Millikan (*physics*) |
| 1930 Dec. | Cleveland | Thomas H. Morgan (*genetics*) |
| 1931 June | Pasadena | Franz Boas (*anthropology*) |
| 1931 Dec. | New Orleans | Franz Boas (*anthropology*) |
| 1932 June | Syracuse | John Jacob Abel (*pharmacology*) |
| 1932 Dec. | Atlantic City | John Jacob Abel (*pharmacology*) |
| 1933 June | Chicago | Henry N. Russell (*astronomy*) |
| 1933 Dec. | Boston | Henry N. Russell (*astronomy*) |
| 1934 June | Berkeley | Edward L. Thorndike (*psychology*) |
| 1934 Dec. | Pittsburgh, PA | Edward L. Thorndike (*psychology*) |
| 1935 June | Minneapolis | Karl T. Compton (*physics*) |
| 1935 Dec. | St. Louis | Karl T. Compton (*physics*) |
| 1936 June | Rochester, NY | Edwin G. Conklin (*biology*) |
| 1936 Dec. | Atlantic City | Edwin G. Conklin (*biology*) |
| 1937 June | Denver | George D. Birkhoff (*mathematics*) |

| | | |
|---|---|---|
| 1937 Dec. | Indianapolis | George D. Birkhoff (*mathematics*) |
| 1938 June | Ottawa | Wesley C. Mitchell (*medicine*) |
| 1938 Dec. | Richmond, VA | Wesley C. Mitchell (*medicine*) |
| 1939 June | Milwaukee | Walter B. Cannon (*physiology*) |
| 1939 Dec. | Columbus, OH | Walter B. Cannon (*physiology*) |
| 1940 June | Seattle | Albert F. Blakeslee (*genetics*) |
| 1940 Dec. | Philadelphia | Albert F. Blakeslee (*genetics*) |
| 1941 June | Durham, NH | Irving Langmuir (*chemistry*) |
| 1941 Sept. | Chicago | Irving Langmuir (*chemistry*) |
| 1941 Dec. | Dallas | Irving Langmuir (*chemistry*) |
| 1942 | no meeting | Arthur H. Compton (*physics*) |
| 1943 | no meeting | Isaiah Bowman (*geography*) |
| 1944 Sept. | Cleveland | Anton J. Carlson (*physiology*) |
| 1946 Mar.* | St. Louis | C. F. Kettering (*engineering*) |
| 1946 Dec. | Boston | James B. Conant (*chemistry*) |
| 1947 Dec. | Chicago | Harlow Shapley (*astronomy*) |
| 1948 Sept. | Washington, DC | Edmund W. Sinnott (*botany*) |
| 1949 Dec. | New York | E. C. Stakman (*plant pathology*) |
| 1950 Dec. | Cleveland | Roger Adams (*chemistry*) |
| 1951 Dec. | Philadelphia | Kirtley F. Mather (*geology*) |
| 1952 Dec. | St. Louis | Detlev W. Bronk (*physiology*) |
| 1953 Dec. | Boston | Edward U. Condon (*physics*) |
| 1954 Dec. | Berkeley | Warren Weaver (*mathematics*) |
| 1955 Dec. | Atlanta | George W. Beadle (*genetics*) |
| 1956 Dec. | New York | Paul B. Sears (*plant ecology*) |
| 1957 Dec. | Indianapolis | Laurence H. Snyder (*genetics*) |
| 1958 Dec. | Washington, DC | Wallace R. Brode (*chemistry*) |
| 1959 Dec. | Chicago | Paul E. Klopsteg (*physics*) |
| 1960 Dec. | New York | Chauncey D. Leake (*pharmacology, history of science*) |
| 1961 Dec. | Denver | Thomas Park (*animal ecology*) |
| 1962 Dec. | Philadelphia | Paul M. Gross (*chemistry*) |
| 1963 Dec. | Cleveland | Alan T. Waterman (*physics*) |
| 1964 Dec. | Montreal | Laurence M. Gould (*geology*) |
| 1965 Dec. | Berkeley | Henry Eyring (*chemistry*) |
| 1966 Dec. | Washington, DC | Alfred S. Romer (*paleontology*) |
| 1967 Dec. | New York | Don K. Price (*political science*) |
| 1968 Dec. | Dallas | Walter Orr Roberts (*astronomy, meteorology*) |
| 1969 Dec. | Boston | H. Bentley Glass (*genetics*) |
| 1970 Dec. | Chicago | Athelstan Spilhaus (*meteorology, oceanography*) |

* 1945 meeting postponed.

## ADMINISTRATIVE OFFICERS

### Principal Administrative Officers

*Permanent Secretaries*

| | |
|---|---|
| Spencer F. Baird | 1851–1854 |
| Joseph Lovering | 1854–1868 |
| F. W. Putnam | 1869 |
| Joseph Lovering | 1870–1873 |
| F. W. Putnam | 1873–1898 |
| L. O. Howard | 1898–1920 |
| Burton E. Livingston | 1920–1930 |
| Charles F. Roos | 1931–1932 |
| Henry B. Ward | 1933–1937 |
| F. R. Moulton | 1937–1946 |

*Administrative Secretaries*

| | |
|---|---|
| F. R. Moulton | 1946–1948 |
| Howard A. Meyerhoff | 1949–1953 |
| Dael Wolfle | 1954–1955 |

### Executive Officers

| | |
|---|---|
| Dael Wolfle | 1956–1970 |
| William Bevan | 1970–1974 |

### Associate Administrative Officers

| | |
|---|---|
| F. S. Hazard, *Assistant Secretary* | 1912–1920 |
| Sam Woodley, *Assistant Secretary* | 1920–1945 |
| Howard A. Meyerhoff, *Executive Secretary* | 1945–1946 |
| J. M. Hutzel, *Assistant Administrative Secretary* | 1946–1948 |
| Raymond L. Taylor, *Assistant Administrative Secretary* | 1949–1953 |
|     *Associate Administrative Secretary* | 1953–1967 |
| John A. Behnke, *Assistant Administrative Secretary* | 1952–1953 |
|     *Associate Administrative Secretary* | 1953–1956 |
| William T. Kabisch, *Assistant Executive Officer* | 1967–1970 |
| Richard Trumbull, *Deputy Executive Officer* | 1970–1974 |

### Treasurers

| | |
|---|---|
| Jeffries Wyman | 1848 |
| A. L. Elwyn | 1849–1870 |
| William S. Vaux | 1871–1881 |
| William Lilly | 1882–1893 |
| R. S. Woodward | 1894–1924 |
| John L. Wirt | 1925–1940 |
| Carroll W. Morgan | 1941–1944 |

William E. Wrather ................................................................... 1945–1953
Paul A. Scherer ....................................................................... 1954–1962
Paul E. Klopsteg ..................................................................... 1963–1969
William T. Golden .................................................................. 1969—

**Editors of *Science***

*Science* was founded in 1880 by Thomas A. Edison. John Michels was the first editor. In 1883, Alexander Graham Bell and Gardiner Greene Hubbard purchased the magazine and established the Science Company, which published *Science* from 1883 to 1894. The magazine was then sold to James McKeen Cattell. In 1900, the AAAS entered into an agreement with Dr. Cattell to make *Science* the official journal of the Association, and, in 1945, became its owner and publisher.

J. McKeen Cattell ................................................................... 1895–1944
Josephine Owen Cattell and Jacques Cattell ..................... 1944–1945
Willard L. Valentine .............................................................. 1946–1947
Mildred Atwood (*acting*) ................................................... 1947–1948
Editorial Board, George Baitsell, *chairman* ....................... 1948–1949
Howard A. Meyerhoff ........................................................... 1949–1953
H. Bentley Glass (*acting*) .................................................... 1953
Duane Roller ........................................................................... 1954
Dael Wolfle (*acting*) ............................................................ 1955
Graham DuShane ................................................................... 1956–1962
Philip H. Abelson .................................................................. 1962–1984

# Notes and References

The board of directors was earlier named the executive committee of the council. In these notes, that body is always given its later, and current, title: board of directors. Whenever the term "executive committee" is used, the reference is to the executive committee of the board of directors, and not to the board of directors under its earlier name.

Wherever in a single note there is a succession of meeting dates and the identification of relevant agenda items or minutes, the source is the last-named committee or other unit of AAAS. For example, in a note reading "Board of Directors, June 2–3, 1956, minute 4; March 5–6, 1960, agenda item 21; and June 4–5, 1960, minute 37," all three citations are to records of the board of directors.

## Preface

1. Mary Sears and Daniel Merriman, eds., *Oceanography: The Past* (NY: Springer-Verlag, 1980), 42–48.
2. Sally Gregory Kohlstedt, *The Formation of the American Scientific Community: The American Association for the Advancement of Science, 1848–1860* (Urbana, IL: University of Illinois Press, 1976).

## Chapter 1

1. Derek J. deSolla Price, *Little Science: Big Science* (NY: Columbia University Press, 1963).
2. Letter from Franklin D. Roosevelt to Vannevar Bush, November 17, 1944, requesting advice on the postwar organization and support of research. Quoted in Vannevar Bush, *Science: The Endless Frontier* (Washington, DC: U.S. Government Printing Office, 1945), vii.
3. Harry S Truman, "Special Message to the Congress Presenting a 21-Point Program for the Reconversion Period" (September 6, 1945), *Public Papers of Presidents of the United States: Harry S Truman, 1945* (Washington, DC: U.S. Government Printing Office, 1961), 292–293.
4. Sally Gregory Kohlstedt, *The Formation of the American Scientific Community: The American Association for the Advancement of Science, 1848–1860* (Urbana, IL: University of Illinois Press, 1976), 79.
5. Quoted by Kohlstedt, 154.
6. Robert C. Miller, "The AAAS on the Pacific Slope," *Science* 108 (September 3, 1948), 220–223.
7. Frank E. E. Germann, "The Southwestern Division of the AAAS," *Science* 108 (September 3, 1948), 224–226.

8. John C. Reed and Harold Coolidge, "The Alaskan Science Conference," *Science* 113 (March 2, 1951), 223–227; and "Newly Organized Alaska Division of AAAS," *Science* 113 (June 15, 1951), 701.

9. "Local Branches of the American Association for the Advancement of Science," *Science* 39 (February 13, 1914), 246–247.

10. "The Lancaster Branch of the American Association for the Advancement of Science," *Science* 81 (March 22, 1935), 286–287; Board of Directors, June 2–3, 1956, minute 4; March 5–6, 1960, agenda item 21; and June 4–5, 1960, minute 37; John M. Hutzel, "The Membership Campaign," *Science* 108 (November 26, 1948), 578; and Board of Directors, July 6, 1957, minute 19.

11. "The Lancaster Branch," *AAAS Bulletin* 2 (March 1943), 21.

12. "The Lancaster Branch of the Association," *AAAS Bulletin* 4 (March 1945), 21–22.

13. Otis W. Caldwell, "Meeting of the Council of the AAAS," *Science* 103 (April 26, 1946), 506–507.

14. 1920 Bylaw III, paragraph 4; see *Science* 50 (November 21, 1919), 474–477.

15. 1920 Bylaw III, paragraph 5, *loc. cit.*

16. Anton J. Carlson, "Forest Ray Moulton: 1872–1952," *Science* 117 (May 22, 1953), 545–546.

17. F. R. Moulton, "Science," *Science* 85 (June 18, 1937), 571–575.

18. F. R. Moulton, "The AAAS and Organized American Science," *Science* 108 (November 25, 1948), 573–576.

19. *Proceedings of the American Association for the Advancement of Science*, 1 (1848), 8.

20. Sources for the figures given in Tables 1 and 2: For AAAS membership and expenditures, see AAAS annual reports for 1945 and 1970. For total college and university faculty for 1945, see *Statistical Abstracts of the United States;* for 1970, see Martin M. Frankel and J. Fred Beamer, eds., *Projections of Educational Statistics to 1982–1983,* U.S. Office of Education, National Center for Education Statistics (Washington, DC: U.S. Government Printing Office, 1974). For degrees awarded, see Douglas L. Adkins, *The Great American Degree Machine* (NY: McGraw-Hill, 1975). The figures reported by Adkins do not always agree exactly with those reported by the U.S. Office of Education or the National Research Council, but Adkins' series through 1971 is the most self-consistent of any available. For full-time equivalent years of time devoted to research and development by scientists and engineers, see Table 3.1 (p. 55) in Nestor E. Terleckyj, *The State of Science and Research: Some New Indicators,* National Planning Association (Boulder, CO: Westview Press, 1977). For federal and national expenditures for research and development for the Department of Defense for 1945, see *Statistical Abstracts of the United States* (Washington, DC: U.S. Government Printing Office, 1960), 583; for 1970, see *National Patterns of Science and Technology Resources* (Washington, DC: National Science Foundation, 1971).

## Chapter 2

1. A. Hunter Dupree, *Science in the Federal Government* (Cambridge, MA: Belknap Press of Harvard University, 1957), 115–116.

Chapter 2                                                                                                    279

2. Vannevar Bush, *Science — The Endless Frontier: A Report to the President* (Washington, DC: U.S. Government Printing Office, 1945).

3. Howard A. Meyerhoff to the members of the council, September 12, 1945.

4. For a history of the hearings and the numerous bills that were considered before the National Science Foundation was finally authorized in 1950, see J. Merton England, *A Patron for Pure Science: The National Science Foundation's Formative Years* (Washington, DC: National Science Foundation, 1982), chapter 1–5. For an analysis of the position of Bush and the role he took in shaping the National Science Foundation, see Nathan Reingold, "Vannevar Bush's New Deal for Research; or, The Triumph of the Old Order," *Historical Studies in the Physical and Biological Sciences* 17 (1987), 299–344.

5. Report of the Subcommittee on Science Legislation of the Executive Committee of the American Association for the Advancement of Science, December 16, 1946.

6. Council, December 27 and 30, 1946, minute 11. When Warren Magee, the association's always cautious legal counsel, learned of the AAAS initiative in creating the Inter-Society Committee, he complained that he had not been consulted, and he warned that "this action may seriously jeopardize the tax exemption of the Association, as that exemption was based upon the premise that the Association was organized entirely for educational purposes." See Warren Magee to F. R. Moulton, January 21, 1947.

7. *Science* 105 (March 7, 1947), 249.

8. Dael Wolfle, "The Inter-Society Committee for a National Science Foundation: Report for 1947," *Science* 106 (December 5, 1947), 529–533.

9. Howard A. Meyerhoff, "Obituary: National Science Foundation," *Science* 104 (August 2, 1946), 97–98.

10. J. Merton England, *A Patron for Pure Science*, 81.

11. Howard A. Meyerhoff, "The Truman Veto," *Science* 106 (September 12, 1947), 236–237.

12. Milton Lomask, *A Minor Miracle: An Informal History of the National Science Foundation* (Washington, DC: National Science Foundation, 1976), 54–56. This documentation is somewhat circular, for Lomask cites a letter from me of July 25, 1973, as his evidence for this account. Elmer Staats, in a personal communication, informed me that there are diary notes filed in the Bureau of the Budget (now Office of Management and Budget) reporting the Carey–Staats–Day–Shapley–Wolfle meeting and its consequences.

13. J. Merton England, *A Patron for Pure Science*, 102–105.

14. Ibid., 96.

15. The officers of the Inter-Society Committee described this pamphlet as a "masterpiece of misinterpretation and intemperate vilification," published a number of quotations from it, and listed the 100 and more people who were described as "members or affiliates of subversive organizations." That list started with Harry S Truman, Vannevar Bush, Harold L. Ickes, Karl T. Compton, Robert L. Patterson, James V. Forrestal, C. F. Kettering, and went on and on. See "The Case Against the National Science Foundation," *Science* 111 (February 24, 1950), 208–210.

16. The American Association of University Professors had been a valiant defender of academic freedom, investigating cases of apparent violation and condemning college and university administrations found guilty of such violations. In the early 1950s, however, its performance had been poor, not primarily because of the timidity of its members in the face of cold war fears and McCarthy excesses, but because of the shortcomings of the

AAUP general secretary. See Walter P. Metzger, "Ralph F. Fuchs and Ralph E. Himstead: A Note on the AAUP in the McCarthy Period," *Academe* 72 (November–December, 1986), 29–35.

17. Board of Directors, May 23–24, 1954, agenda item 14 and minute 35; October 17–18, 1954, minute 55.

18. Board of Directors of the American Association for the Advancement of Science, "Strengthening the Basis of National Security," *Science* 120 (December 10, 1954), 957–959. By mail ballot the council voted 191 to 18 to endorse this statement.

19. National Science Foundation, *Fifth Annual Report, Fiscal Year 1955*, 19–20.

20. Quoted by J. Merton England, *A Patron for Pure Science*, 330.

21. National Science Foundation, *Fifth Annual Report, Fiscal Year 1955*, 20.

22. Council, December 27 and 30, 1954, minute 8.

23. Board of Directors, December 27 and 29, 1954, agenda item 41.

24. Dael Wolfle, "Commission on Security," *Science* 122 (December 2, 1955), 651.

25. "Loyalty-Security Programs," *Science* 124 (August 3, 1956), 210–212, presented excerpts from *The Federal Loyalty-Security Program*, the report of a Special Committee of the Association of the Bar of the City of New York (NY: Dodd Mead, 1956), and Dael Wolfle, "Defense of the Nation," same issue, 201.

26. "Loyalty and Research: Report of the Committee on Loyalty in Relation to Government Support of Unclassified Research," *Science* 123 (April 20, 1956), 660–662; and Dael Wolfle, "Science and Loyalty," same issue, 651.

27. J. Merton England, *A Patron for Pure Science*, 335.

28. Board of Directors, December 27–28, 1956, minute 31.

29. Bryce Nelson, "Scientists Increasingly Protest HEW Investigation of Advisors," *Science* 164 (June 27, 1969), 1499–1504; and Bryce Nelson, "HEW Security Checks Said to Bar Qualified Applicants to PHS," *Science* 165 (July 18, 1969), 269–271.

30. Bryce Nelson, *Science* 164 (June 27, 1969), 1501.

31. Board of Directors, June 27–29, 1969, agenda item 31 and minute 31; October 18–19, 1969, agenda item 28; and December 28–29, 1969, minute 10.

32. The HEW "blacklist" of scientists for study panels and similar appointments was first brought to light in *Science*, June 27 and July 18, 1969. See note 29 and also Philip M. Boffey, "HEW Blacklists: New Security Procedures Adopted," *Science* 170 (October 9, 1970), 142–144, which gives an account of the history of the issue, some specific cases, and the improvements of the new system over previous arrangements.

33. Philip M. Boffey, "Science and Politics: Free Speech Controversy at Lawrence Laboratory," *Science* 169 (August 21, 1970), 743–745; and Philip M. Boffey, "Gofman and Tamplin: Harassment Charges Against AEC, Livermore," *Science* 169 (August 28, 1970), 838–843.

34. Board of Directors, October 17–18, 1970, minute 17.

35. Board of Directors, December 12–13, 1970, minutes 21 and 22; December 28, 1970, minute 7; and *AAAS Bulletin* 16 (September 1971).

36. *AAAS Handbook 1988–1989* (Washington, DC: American Association for the Advancement of Science, 1988), 124–126.

37. J. Merton England, *A Patron for Pure Science*, 165.

38. Dael Wolfle to members of the Executive Committee, September 26, 1958; Board of Directors, October 18–19, 1958, minute 38.

Chapter 2 281

39. Board of Directors, March 17–18, 1962, minute 5; June 22–23, 1963, minute 7.

40. Board of Directors, October 18–19, 1958, agenda items 2 and 24 and minutes 2 and 24.

41. Board of Directors, December 27–28, 1959, agenda item 19 lists the board's recommendations made in 1953, 1955, and 1957, and indicates the three to five names from each of those lists who were appointed. Board minutes for later odd-numbered years show later recommendations.

42. Board of Directors, October 14–15, 1950, agenda item 24 and minute 36.

43. Board of Directors, September 15–16, 1951, agenda item 13 and minute 25.

44. Board of Director, October 17–18, 1952, agenda item 31 and accompanying memorandum entitled "The Proposed Scientific Manpower Commission" and minute 31; March 16, 1953, minute 14.

45. As the Scientific Manpower Commission was getting started, and on behalf of the other founders of that organization, I asked Meyerhoff if he could assume responsibility for the new commission in addition to his AAAS responsibilities, and if the AAAS Board would permit that arrangement. I thought that possible, for Meyerhoff was already doing some of the work that would fall to the commission when it became established. He told me then that he would welcome the Scientific Manpower Commission post because his disagreements with Condon and Weaver had become so acute that he was about to tell the board he was leaving (see chapters 3 and 4).

46. The information concerning housing and the final move into the AAAS structure is from Betty Vetter, personal communication, September 5, 1984.

47. Board of Directors, March 3–5, 1961, agenda item 25, tab D, "Political Responsibilities of the AAAS," and minute 27.

48. Dael Wolfle, "Government Organization of Science," *Science* 131 (May 13, 1960), 1407–1417.

49. A. Hunter Dupree, *Science in the Federal Government* (Cambridge, MA: The Belknap Press of Harvard University, 1957), 215–231.

50. Colin Norman, "A New Push for a Federal Science Department," *Science* 226 (December 21, 1984), 1398–1399; and Colin Norman, "Commission Proposes Science Department," *Science* 227 (March 1, 1985), 1017.

51. "1958 Parliament of Science," *Science* 127 (April 18, 1958), 852–858.

52. Dael Wolfle, ed., *Symposium on Basic Research* (Washington, DC: American Association for the Advancement of Science, 1959). That my name appeared as editor of the volume was due to an unfortunate illness. Dean Mark Ingraham of the University of Wisconsin had agreed to edit the proceedings, but an outbreak of boils kept him from attending the symposium.

53. Board of Directors, June 20–21, 1959, minute 9.

54. Dael Wolfle, *Symposium on Basic Research*, 259.

55. Howard A. Meyerhoff to the members of the council, August 24, 1951.

56. J. Merton England, *A Patron for Pure Science*, 163.

57. Board of Directors, June 21–22, 1958, tab B to agenda item 1, and minute 3.

58. Board of Directors, June 19–20, 1965, agenda item 20 and minute 20.

59. Board of Directors, October 23–24, 1965, agenda item 6 and minute 6; December 28, 1965, agenda item 25 and minute 25; and March 26–27, 1966, agenda item 7 and minute 7.

60. Paul E. Klopsteg, "University Responsibility and Government Money," *Science*, 124 (November 9, 1956), 919–922.

61. Paul E. Klopsteg, "How Shall We Pay for Research and Education?" *Science* 124 (November 16, 1956), 965–968.

62. Robert W. King, "A Tax Credit Plan for Re-establishing Private Support of Pure Science," *Science* 105 (June 6, 1947), 593–594.

63. National Science Foundation, *Basic Research: A National Resource* (Washington, DC: National Science Foundation, 1957).

64. Paul E. Klopsteg, "How Shall We Pay for Research and Education?" *Science* 124 (November 16, 1956), 965–968.

65. Board of Directors, December 27–28, 1956, minute 4.

66. Board of Directors, March 16–17, 1957, minute 35.

67. I had not known Rice and do not recall who proposed his firm be asked to conduct the study. However, his name was familiar. In the year of my birth, he was a high school student taking classes from my father, a teacher of science and mathematics. The student-teacher relationship was evidently a strong and fond one, for Rice was one of several students whose careers my father followed with particular interest. The last time I saw Rice was at his request; he was writing his memoirs and wanted to check some dates and details about my father's teaching years.

68. Board of Directors, December 27–28, 1957, minute 8.

69. "Stimulating Philanthropic Giving by Equalizing Relative Costs to Taxpayers," prepared for the American Association for the Advancement of Science by Stuart Rice Associates, Washington, DC, 1957 (litho).

70. Board of Directors, March 18, 1958, minute 5, and June 21–22, 1958, minute 43.

71. Board of Directors, June 21–22, 1958, minute 43.

72. Board of Directors, October 3–4, 1959, agenda item 26.

73. Board of Directors, October 3–4, 1959, minute 26.

74. Board of Directors, December 27, 28, and 30, 1959, minute 20.

75. Don K. Price to Paul E. Klopsteg, June 24, 1959.

76. Paul E. Klopsteg to Don K. Price, July 1, 1959.

77. *Congressional Record*, February 11, 1959, A1001–A1009.

78. Burton A. Weisbrod, "Alternatives Must Be Found to the Growing Commercialization of Universities," *The Chronicle of Higher Education* 34 (June 29, 1988), A32.

79. Board of Directors, October 8–9, 1949, agenda item 3 and minute 10; March 25–26, 1950, minute 22; June 24–25, 1950, minute 43; and December 26–27, 1951, minute 29.

80. Board of Directors, June 28–29, 1952, agenda item 40.

81. Board of Directors, October 13–14, 1956, minute 34; December 27–28, 1956, minute 34; and October 18–19, 1958, agenda item 6.

## Chapter 3

1. *Proceedings of the American Association for the Advancement of Science, Fifty-eighth Meeting* (Washington, DC: Gibson Brothers, 1908), 341.

Chapter 3

2. AAAS Section FG, for biology, and Section N, for medicine, have each included at least twice as many affiliated societies as have the sections for chemistry, physics, social and economic sciences or any other broad area of scientific interest and AAAS sectional structure.

3. Letter from Paul Weiss to Board of Directors, April 9, 1946.

4. Harlow Shapley, "Some Plans for the American Association for the Advancement of Science" (July 21, 1947, memorandum to the Board of Directors).

5. Members of the committee were Edward U. Condon, National Bureau of Standards; James Gilluly, UCLA; John M. Hutzel, assistant administrative secretary of AAAS; F. R. Moulton, administrative secretary of AAAS; Harlow Shapley, Harvard University and AAAS president; Edmond W. Sinnott, Yale University and AAAS president-elect; Wendell M. Stanley, Rockefeller Foundation; and Roger Williams, University of Texas.

6. Decisions concerning the AAAS centennial meeting no doubt hastened organization of AIBS, but according to Detlev Bronk, "the Institute had its origins on the waters of the Great Harbor of Woods Hole during a pleasant afternoon in the summer of 1944." Two years later, Bronk arranged another meeting that furthered the idea, and at the 1946 meeting of AAAS the possibility was thoroughly discussed by biologists. The first meeting of the new institute's governing board was held in May 1948. See *The AIBS Story* (Washington, DC: American Institute of Biological Sciences, undated).

7. *Science* 108 (September 3, 1948).

8. Harlow Shapley, "The One World of Stars," *Science* 108 (September 24, 1948), 315–321.

9. "Address of the President of the United States," *Science* 108 (September 24, 1948), 313–314.

10. Letter from George M. Elsey to David W. Stowe, Bureau of the Budget, September 11, 1948, copy on file in the Harry S Truman Library, Independence, Missouri; letter from George M. Elsey to Dael Wolfle, September 25, 1985; letters from Benedict K. Zobrist, Harry S Truman Library, to Dael Wolfle, July 5, 1985, and October 4, 1985.

11. It would be interesting to compare the address as given with the original Condon draft. Unfortunately, that draft has not been located. Nevertheless, I am indebted to Benedict K. Zobrist, Librarian of the Truman Library, and to Beth Carroll-Horrocks of the Library of the American Philosophical Society, where Condon's papers are filed, for their diligent searches. The fact that Condon did submit a draft has been independently confirmed by Charles V. Kidd, who was then a member of John Steelman's White House staff; personal communication, July 1985.

12. Personal communication, July 10, 1985, from Charles E. Odegaard, president emeritus of the University of Washington, who was in 1948 executive director of the American Council of Learned Societies, Washington, D.C.

13. *New York Times*, September 14, 1948, 1.

14. See, for example, Congressman Richard B. Vail, Government Loyalty and Security Program, *Congressional Record*, 98, Part 1, 406, January 22, 1952. Illustrative passages are his comment on the AAAS Special Committee on Civil Liberties ("the whining gibberish of the committee, obviously aware that they were precisely the type necessitating strict security measures") and on two AAAS presidents ("... the Red professor of Harvard, Kirtley F. Mather, whose Communist front affiliations detailed by me ... consumed three solid pages of the *Congressional Record*," and Mather's successor

"... the ubiquitous Edward U. Condon described by the Committee on Un-American Activities as 'the weakest link in our atomic security chain' "). Such attacks had been started much earlier. On March 2, 1948, a subcommittee of the House Committee on Un-American Activities denounced Condon as "one of the weakest ...." Simultaneously, however, the Loyalty Board of the Department of Commerce unanimously cleared him. Condon requested a public hearing by the House Committee; that request was never granted although their charges continued for years, as indicated by the quotations above. For analysts of the handling of the Condon case by the New York newspapers, see Joseph T. Klapper and Charles Y. Glock, "Trial by Newspaper," *Scientific American* 180 (no. 2, 1949), 16–21.

15. Letter from George M. Elsey to Dael Wolfle, September 25, 1985.

16. *Science* 108 (November 26, 1948, and several later issues).

17. *Centennial: Collected Papers Presented at the Centennial Celebration, Washington, D.C., September 13–17, 1948* (Washington, DC: American Association for the Advancement of Science, 1950).

18. J. M. Hutzel, "AAAS Centenary—A Preliminary Report," *Science* 108 (October 22, 1948), 428–430.

19. Ibid., 429.

20. Board of Directors, December 27, 1948, agenda section VII.

21. In 1949 AIBS did not hold a meeting of its own and encouraged its member societies to meet with AAAS. See minutes of the May 4, 1949, meeting of the board of governors of the American Institute of Biological Sciences, 5.

22. Board of Directors, October 14–15, 1950, minute 17.

23. Letter to Howard A. Meyerhoff from Anton J. Carlson, A.C. Ivy, and Ralph A. Rohweder, June 1951, reproduced in Board of Directors, June 30–July 1, 1951, agenda item 28.

24. The Arden House statement and the names of its signers were reported in a note entitled "AAAS Policy," *Science* 114 (November 2, 1951), 471–472.

25. Council, December 27 and 29, 1951, minute 4.

26. Memorandum from Kirtley F. Mather to Board of Directors, January 31, 1952.

27. Report of the assistant administrative secretary, March 12, 1952.

28. Earl Ubell, "Scientists Out to Give Ideas Popular Appeal; American Association to Reorganize in Face of 'Intellectual Bankruptcy,' " *New York Herald Tribune*, December 29, 1952, 15.

29. *Science* 117 (February 20, 1953), 3a; *The Scientific Monthly* 76 (March 1953), 195.

30. *Science* 117 (May 29, 1953), 3a; *The Scientific Monthly* 76 (May 1953), 318.

31. *Science* 117 (March 27, 1953), 319.

32. Board of Directors, March 16, 1953, agenda item 21.

33. Council for the Advancement of Negroes in Science, "Memorandum on the Situation with Regard to Negroes in Science in the U.S.A.," sent to Raymond L. Taylor, associate administrative secretary for the American Association for the Advancement of Science, February 17, 1954.

34. Raymond L. Taylor, memorandum to the Board of Directors, undated but presented as agenda item 28, Board of Director, March 6–7, 1954.

Chapter 3

35. Board of Directors, June 11–12, 1954, agenda item 24. The Board's reaffirmation of the decision to meet in Atlanta and the statement explaining that decision are in minute 25, Board of Directors, June 11–12, 1955.

36. Margaret Mead to Gabriel Lasker, March 9, 1955 (file copy in Margaret Mead's papers, AAAS box, Library of Congress, Washington, D.C.).

37. In most years, 50 to 80 percent of registrants came from the state in which the meeting was held and its neighboring states; at Atlanta, only 34 percent came from Georgia and its neighbors.

38. Montague Cobb to AAAS presidents, Board of Directors, and Council, December 22, 1955.

39. Council, December 27 and 30, 1955, minute 12.

40. The board did plan to hold the 1972 meeting in Atlanta, and the necessary arrangements with the hotels and convention bureau were made but later canceled on the advice of the meeting editor, who wanted to hold the meeting in a city where a larger attendance could be expected. See Board of Directors, April 20–21, 1969, agenda item 29. An account of the board's vacillation on the question of whether to meet in Atlanta is given in a memorandum from Dael Wolfle to the board of directors, July 10, 1969. Two decades later the board decided to meet in New Orleans in 1990.

41. There was a pleasant bit of symmetry in information about Meyerhoff's departure and my appointment. Early in the year, he had told me of his intention of leaving AAAS before he gave that information to the board. Late in the year, on the day that Detlev Bronk, the chairman of the board, invited me to become Meyerhoff's successor, he had breakfasted with Meyerhoff and told him the board was offering the appointment to me.

42. Board of Directors, March 6–7, 1954, minute 17.

43. Dael Wolfle to the board of directors, May 13, 1954.

44. Board of Directors, May 23–24, 1954, minute 7.

45. Memorandum from Clarence E. Davies, chairman of Section M, to Warren Weaver, January 28, 1953; submitted to the board of directors, October 17–18, 1954.

46. Board of Directors, December 27 and 29, 1954, minute 14.

47. John Behnke, preliminary memorandum on meetings, December 3, 1952.

48. Jacob Bronowski to Warren Weaver, June 28, 1954.

49. Report of the Committee on Membership Development, a supplement to agenda item 38, Board of Directors, October 29–30, 1955, 4 and 7.

50. Ibid., 6.

51. Dael Wolfle to board of directors, tab B for Board of Directors, February 28–March 1, 1959.

52. Dael Wolfle and Raymond L. Taylor to Committee on Meetings, February 2, 1956.

53. Minutes of the Committee on AAAS Meetings, February 6, 1956.

54. Board of Directors, March 3–4, 1956, minute 18. On several later occasions the board encouraged the Committee on AAAS Meetings to take a more active stance in reviewing programs proposed by sections and affiliates and to accept only those of which it approved. The committee, however, decided it did not wish to have a veto role. See, for example, minute 4 of the Committee on AAAS Meetings, February 19–20, 1965, and minute 27, Board of Directors, March 27–28, 1965.

55. In its original composition this committee consisted of three members who served ex officio, two selected from other AAAS committees, and four chosen individually. The committee worked effectively in bringing about a number of improvements, but after a few years, the board decided that, with procedural and administrative aspects in good shape, the committee should focus more on program content and for that purpose altered its composition so that all nine members were selected individually and on the basis of what they could contribute to program planning. Again, the analogy of a board of editors was invoked. The members were to be chosen as carefully, and for essentially the same kinds of reasons, as were members of the board of editors for *Science*. See Board of Directors, May 27–28, 1961, agenda item 25 and minute 26.

56. Board of Directors, February 28–March 1, 1959, tab B. The Committee on AAAS Meetings had earlier proposed to reserve more time for Association-wide programs, leaving less for sessions arranged by AAAS sections and affiliates. Protests from officers of those units led to a compromise arrangement giving the Committee on AAAS Meetings complete jurisdiction over evening programs and those presented on December 28.

57. Ruth C. Christman, ed., *Soviet Science* (Washington, DC: American Association for the Advancement of Science, 1952).

58. Dael Wolfle to members of the Executive Committee of the Board, April 19, 1960.

59. "Science in Communist China," *AAAS Bulletin* 6 (March 1961), 1–3.

60. S. H. Gould, ed., *Science in Communist China* (Washington, DC: American Association for the Advancement of Science, 1961).

61. Arthur H. Livermore, ed., *Science in Japan* (Washington, DC: American Association for the Advancement of Science, 1965).

62. Board of Directors, March 18, 1958, minute 29; October 24–25, 1964, minute 29.

63. Board of Directors, June 10–11, 1967, minute 19; October 21–27, 1967, minute 21.

64. Board of Directors, April 20–21, 1969, minute 29; June 27–29, 1969, minute 26.

65. *Proceedings of the American Association for the Advancement of Science, Eighth Meeting, Held at Washington, D.C., May 1854* (NY: G. P. Putnam & Co., 1855), 304.

66. Ibid., 308–309.

67. At the October 1961 meeting of the board, Chairman Leake appointed a subcommittee consisting of William Rubey, chairman, and Harrison Brown and Don K. Price to consider how the association could "maintain the interest and make fuller use of the experience and wisdom of its former presidents" (Board of Directors, October 7–8, 1961, minute 23). At the same meeting, plans were made for the first of what became an annual dinner to which members of the board invited past presidents and board members, representatives of other associations for the advancement of science, selected guests, senior members of the AAAS staff, and their spouses. In December the subcommittee reported that they did not recommend any formal procedures to include past presidents in the association's governing structure, but did recommend (1) that from time to time the past presidents be invited to meet with the board when matters of broad policy or plans were scheduled for discussion, (2) that past presidents be appointed to committees dealing with matters in which they had demonstrated special interests and in which an understanding of association history and earlier actions was desirable, and (3) that the dinner suggested

Chapter 4

by Leake be instituted as a tradition (Board of Directors, December 27, 1961, minute 18).

68. Walter Berl, *AAAS Annual Meetings: A Review of Purpose and Objectives in a Period of Change*, July 1, 1974 (multilithed), 1.

69. J. W. Kiermaier, president, Channel 13, WNDT, New York City, "Final Report to the Ford Foundation," February 13, 1968.

70. Walter Berl, *AAAS Annual Meetings* (see note 68), appendix 12, "Report on Television Coverage, 1967 and 1969."

71. Walter Berl, "Whither AAAS Annual Meetings?" *Science* 167 (September 22, 1967), 1379.

## Chapter 4

1. The history of *Science* is given in its centennial issue (*Science* 209, July 4, 1980) published 100 years and one day after the first issue appeared. It includes four accounts: Sally Gregory Kohlstedt, "*Science*: The Struggle for Survival, 1980 to 1984"; Michael M. Sokol, "*Science* and James McKeen Cattell, 1894 to 1945"; John Walsh, "*Science* in Transition, 1946–1962"; and Dael Wolfle, "*Science*, a Memoir of the 1960s and 1970s." Because that history is easily available there will be no effort to duplicate its content. Yet the history of AAAS cannot be written without considering *Science*, its major publication. Although there is inevitably some overlap between this chapter and the centennial issue, emphasis here will be on the relations between *Science* and the association's policies and problems rather than on the history of the magazine itself.

2. For the wording of the original agreement, see "The Journal 'Science' and the American Association for the Advancement of Science," *Science* 44 (October 3, 1925), 342–345. After some negotiation and compromise, the sum paid for *Science* was $166,430.59 plus the additional amount calculated each year on the basis of the declining value of the dollar. (See Price, Waterhouse & Co. to the American Association for the Advancement of Science, June 5, 1945, and Farmers Bank and Trust Company of Lancaster, Pennsylvania, to F. R. Moulton, June 21, 1945, both attached to Agenda, Board of Directors, June 24, 1945; and Board of Directors, September 2, 1945, minute 10.) Over the next ten years, the value of the dollar declined significantly and the total amount paid to the Cattell estate came to $269,832.82 (see the report of C. P. Graham and Company, auditor, for 1954, p. 17). The amount paid for *The Scientific Monthly* was $9,499.59, all paid in one payment in 1943 (see report of C. P. Graham and Company, auditors, for 1946, p. 16).

3. "Ware Cattell vs. AAAS Settled for $7,500 by Consent Judgment," *AAAS Bulletin* 5 (April 1945), 29.

4. The February 25, 1954, issue of *Science* was a special one of tribute to Cattell for his contributions to AAAS, to *Science*, to American science generally, and to his own discipline of psychology. He had been a member of the association's board of directors for almost a quarter of a century, editor and publisher of *Science* for twice that long, and a prominent and stimulating member of the American scientific community for all of that time. Thus articles in that issue of *Science* consisted of memorial tributes from a dozen scientists who had known and worked closely with him.

5. Stephenson attended part of the board's meeting on November 14, 1944, to discuss his plans further, but was told that the action taken in accepting his "resignation" was final. See Board of Directors, November 14, 1944, minutes 27 and 28; F. R. Moulton to the Board of Directors, June 5, 1945; and F. R. Moulton to the Board of Directors, September 13, 1946.

6. F. R. Moulton, "The Appointment of Willard L. Valentine as Editor of Science," *Science* 102 (October 19, 1945), 387–388.

7. F. R. Moulton to the board of directors, April 29, 1944.

8. Board of Directors, November 4, 1945, minute 4.

9. Walter R. Miles, "Willard Lee Valentine, 1904–1947," *Science* 105 (June 20, 1947), 638–640.

10. Thomas Park, "New Editor for AAAS Journals," *Science* 122 (September 23, 1955), 550.

11. John Walsh, "*Science* in Transition, 1946 to 1962," *Science* 209 (July 4, 1980), 52–57.

12. *The Scientific Monthly* 63 (December 1946), 488.

13. Gustav Albrecht, "The Schuss-Yucca (*Yucca whipplei*, var. Schuss)," *The Scientific Monthly* 75 (October 1952), 250–252.

14. The editors also accepted some other articles for *The Scientific Monthly* that resulted in controversy, but for a different reason. See Harry J. Fuller, "The Emperor's New Clothes or Prius Dementat," *The Scientific Monthly* 72 (January 1951), 32–41, a criticism of colleges of education with a title taken from the Hans Christian Anderson story. The general tone can be illustrated by the author's description of the work of schools of education as "the debasement of liberal education and of sound scholarship by these dreary intellectual sinks and their often dismal practitioners" (p. 32). Of the 248 letters prompted by the article, 226 were "enthusiastically for and 22 bitterly against" Fuller's paper. Two replies were published, a letter from D. A. Worcester, secretary of the AAAS Section on Education, "Fuller's Folly?" *The Scientific Monthly* 72 (March 1951), 204–205; and a longer article by Simon Williams and James D. Laurits, "Scientists and Education," *The Scientific Monthly* 72 (May 1951), 282–288. The strained relations between Meyerhoff and some educators were not made any better when he published another highly critical article, written by one of Fuller's University of Illinois colleagues: Arthur E. Bestor, Jr., "Aimlessness in Education," *The Scientific Monthly* 75 (August 1952), 109–116.

15. As AAAS was assuming responsibility for the magazine, Moulton told the board of directors that *Science* should "become at least comparable to *Nature*." See F. R. Moulton to the board of directors, March 6, 1945. Complaints from readers sometimes made the same comparison.

16. Board of Directors, April 27, 1947, minute 37.

17. Gerard Piel to Howard A. Meyerhoff, May 2, 1949.

18. Publications Committee, July 6, 1949, agenda item 3.

19. Publications Committee, July 6, 1949, minute 1.

20. Board of Directors, July 7, 1949, minute 1. See also Board of Directors, April 24, 1949, minute 25. The folder "*Scientific American*, 1949–1955" in Box 9 of Catherine Borras papers in the AAAS files includes the correspondence, memoranda, and proposed articles of agreement.

21. Board of Directors, October 8–9, 1949, minute 26.

22. Publications Committee, August 20, 1949, minute 1; and October 7, 1949, minute 5.

23. Board of Directors, July 7, 1949, minute 8.

24. Uncertain that he would be able to finance a new edition of *American Men of Science*, in 1950 Jacques Cattell appealed to AAAS for help. The board declined, but Cattell persisted. In 1951 he asked AAAS, the American Council of Learned Societies, and the American Council on Education to provide or to secure grants to support publication of *American Men of Science*, *Directory of American Scholars*, and *Leaders in Education*. Howard Meyerhoff conferred with Arthur F. Adams of the American Council on Education and Charles E. Odegaard of the American Council of Learned Societies (see Board of Directors, December 26–27, 1951, agenda item 20). None of the three liked the prospect and the board decided that no action should be taken (Board of Directors, December 26–27, 1951, minute 20). AAAS consideration of this possibility was complicated by the fact that the association had published some directories of its own members. They gave brief information about positions and addresses but not as much biographical information as *American Men of Science*. However, the association's directories gave information about all members, a considerable number of whom were not listed in *American Men of Science*. The AAAS attitude was further complicated by the fact that the last AAAS directory was a supplement giving brief information about the 13,000 new members who joined the association in the centennial year membership drive of 1948. Financially, that directory was virtually a complete loss, and AAAS never published another.

In 1953 Jacques Cattell offered to sell *American Men of Science* to AAAS for $15,000 (or $25,000; there was some confusion about the price) assuming that he would continue to be the editor (see Jacques Cattell to Howard A. Meyerhoff, March 2, 1953). The Board of Directors "emphatically expressed that $25,000 need not and should not, be paid to Jacques Cattell for the name "American Men of Science." There was general agreement that a directory 'Scientists of America' or 'Who's Who in Science' should be produced by the Association both as a service to science and for its probable aid to increased membership in the Association" (Board of Directors, March 16, 1953, minute 24). Accordingly, the board asked John Behnke to give top priority to exploring possible publishing arrangements. That possibility was terminated when Jacques Cattell was able to publish the ninth edition of *American Men of Science*. However, in planning that edition, which came out in 1955, he made the unfortunate decision to split what had always before been one alphabetical arrangement of all of its biographies into three separate directories, one for the physical sciences, one for the biological sciences, and one for the social and behavioral sciences. The board of directors immediately noted that the National Academy of Sciences and the National Research Council were reluctant to see that split. So were many users. From the 10th edition on, the Cattell Press, and later the Bowker Company, reunited the physical and biological scientists in a single directory, which in the 12th edition was renamed *American Men and Women of Science*. The social and behavioral scientists were not included in that recombination, but separate directories for those fields were published to accompany the 11th, 12th, and 13th editions of the physical and biological science directories. Since the appearance of the 13th edition in 1982, there has been no up-to-date directory of social and behavioral scientists.

25. Board of Directors, December 27–28, 1949, minute 24.

26. Board of Directors, October 14–15, 1954, minute 61.

27. Duane Roller to Paul Sears, chairman of the Publications Committee, December 2, 1954.

28. Board of Directors, December 27 and 29, 1954, minutes 40 and 41.

29. Dael Wolfle to the board of directors, November 16, 1954.

30. Board of Directors, December 27 and 29, 1954, minute 59, Dael Wolfle to the board of directors, January 14, 1955.

31. Dael Wolfle to the board of directors, January 28, 1955.

32. Dael Wolfle, "Editing of *Science* and *The Scientific Monthly* (February 4, 1955, memorandum to the board of directors); Editorial Board, March 13, 1955, minutes; and Dael Wolfle, "Publication and Editorial Problems" (April 18, 1955, memorandum to the board of directors). Dissatisfaction with Moulton and Company (no relation to F. R. Moulton) went back several years. At the board meeting of June 30, 1951, members noted that the company was not aggressive in seeking new accounts or in following up suggestions from the AAAS staff, and that a larger commission for new accounts had not proven effective. However, a new advertising representative would want a 25 percent commission instead of the 7.5 or 10 percent commission charged by Moulton and Company. The board therefore voted to continue the contract, but asked Meyerhoff to study other possibilities. Two years later, the board voted (October 18, 1953, minute 39) to terminate the contract with Moulton and Company and a few days later, Raymond Taylor wrote Moulton to inform him of that action. His letter did not invoke the termination clause of the contract, however, but did provide Moulton an opportunity to offer suggestions for improving advertising sales (Raymond L. Taylor to F. A. Moulton, October 30, 1953). As things turned out, the contract was not actually terminated until 1955.

33. Graham DuShane, "New Garb," *Science* 130 (October 2, 1959), 829.

34. Dael Wolfle to the board of directors, November 16, 1954.

35. Editorial Board, May 31, 1953, minutes, 4.

36. See, for example, Editorial Board, February 13, 1954, minutes, 2.

37. Board of Directors, December 27–28, 1956, minute 28, and Council, December 27 and 30, 1956, minute 2.

38. Board of Directors, March 16–17, 1957, minute 24.

39. AAAS Council Newsletter, April 17, 1959.

40. Publications Committee, June 3, 1960.

41. The most detailed account of the magazine on science policy proposed by Daniel Greenberg is given in appendix A to the agenda for a meeting of the Committee on Publications on October 25, 1963. Also given are cost and income estimates and brief excerpts from many letters of support from scientists, science policy makers, and members of Congress and relevant executive agencies. A sample issue was printed and the matter was given much attention by the publications committee and the board of directors, but in the end, the decision was negative.

42. The magazine attained a circulation of over 700,000, but by 1986 its inability to secure sufficient advertising meant that it was losing so much money the association was forced to sell it (see William D. Carey, "Annual Report of the Executive Officer," *Science* 235 (February 6, 1987), 638–641). Time, Inc., the buyer, merged *Science 86* into its own *Discover*, sent *Discover* to all *Science 86* subscribers, and planned to try to satisfy their expectations of the magazine they had lost. However, *Discover* and other popular

Chapter 4   291

science magazines were also losing money, and a year later Time, Inc. sold *Discover* to another publisher.

43. Charlotte V. Meeting, "Man-Made Diamonds," *Science* 121 (February 18, 1955), 228–229.

44. Jerome B. Weisner, "John F. Kennedy: A Remembrance," and Daniel S. Greenberg, "John F. Kennedy: The Man and His Meaning," *Science* 142 (November 29, 1963), 1147–1152.

45. The *"Science* and *Scientific Monthly* (old,I)" folder in box 10 of the Catherine Borras files in the archives contains a number of letters and memoranda that supplement board minutes on these issues, including the letter (November 20, 1963) from Frederick Seitz to Philip Abelson.

46. Philip M. Boffey, "AAAS Presidency: Controversy Flares Over Seaborg Candidacy." *Science* 170 (December 11, 1970), 1177–1180.

47. Columbia University, Office of Public Information, press release, October 7, 1970.

48. Board of Directors, October 21–22, 1967, agenda item 6.

49. Howard A. Meyerhoff, "Advertising in Association Journals," *AAAS Bulletin* 4 (August 1945), 61–62.

50. Results of the surveys were regularly reported to the Committee on Publications, the editorial board, and the board of directors, and were occasionally summarized for *Science* readers. For example, see Dael Wolfle, "The Voice of the Reader," *Science* 126 (August 16, 1957), 285, and Dael Wolfle, "Readers' Judgment," *Science* 156 (June 2, 1968), 1181.

51. George W. Corner to Duane Roller, July 12, 1954. The remarkable George Washington Corner, with a dozen honorary degrees from three countries and numerous other honors, was then director of the Department of Embryology at the Carnegie Institution of Washington. He retired from that position two years later to become historian of the Rockefeller Institute of New York, and then, at age 75, became executive officer of the American Philosophical Society in Philadelphia, where he continued for over a decade.

52. Duane Roller, "On Symbols for Units of Measurement," *Science* 120 (December 24, 1954), 3a; and "Symbols for Units of Measurement," same issue, 1078–1080.

53. Editorial Board, January 30, 1950, minutes.

54. Editorial Board, May 28, 1950, minute 11.

55. Board of Directors, June 24–25, 1950, minute 25.

56. Board of Directors, June 30–July 1, 1951, agenda item 19.

57. Board of Directors, June 30–July 1, 1951, minute 23.

58. Howard A. Meyerhoff to Edward U. Condon, January 14, 1953.

59. Because of the charges made against him by the House of Representatives Committee on Un-American Activities, Condon was inevitably a somewhat controversial figure. During the 1952 annual meeting in St. Louis, a reporter for the *Globe-Democrat* asked Meyerhoff about AAAS support or opposition to Condon, who was about to move up from his position as president-elect to the presidency. Condon did not like the newspaper account of that interview, and point 11 of his January 1, 1953, letter to Meyerhoff took up that point: "A related understanding that I want formally on the record is that you will under no circumstances make any statement about me and the relation of the AAAS to me to the press without explicit clearance with me on the exact text. The

experience with the *Globe-Democrat* in St. Louis shows how easily one can go wrong on these matters, and I prefer in every case to make my own mistakes." On January 7, Meyerhoff replied to that point:

> Must I repeat what I said in St. Louis regarding the statement attributed to me in the *Globe-Democrat*? I told the reporter, Mr. Dudman, that there was a question in the minds of some of our members as to the advisability of electing a controversial figure as the association's president. To indicate that even most of these people are solidly behind you, I invented the figures of 5 percent for those definitely opposed to your presidency and maybe as much as 20 percent for those who thought your election inadvisable. If I am any judge of the situation, both of those figures are low, but for public consumption, they are close enough to the truth. [The name Dudman is apparently a mistake on Meyerhoff's part since at that time Richard Dudman was a reporter for the St. Louis *Post-Dispatch*, not the *Globe-Democrat*; however, Dudman had also interviewed Meyerhoff, and Condon did not like the resulting newspaper account.] ... I am quite sure that I am the only person with sufficient information to have made any statement of the kind demanded by the press, and insofar as the whole import of the statement was that of solid backing from American scientists, I don't see why you of all people should be so steamed up about it.

Condon replied on January 10 that he objected to having Meyerhoff make up figures about him and concluded: "I hope there will not be any more smear publicity about me, but if there is, I shall expect you to refrain from making any comments to the press about it. In view of this entirely reasonable request, any contrary action in this field I shall regard as a deliberately unfriendly act."

60. This letter and the 31 others in the Condon-Meyerhoff exchange were later sent to all members of the council and are filed with the material sent to council members for the December 1953 meeting.

61. Board of Directors, March 16, 1953, preliminary agenda, part I.

62. Gladys M. Keener and Howard A. Meyerhoff to Detlev W. Bronk, March 23, 1953.

63. Members of the delegation were Richard E. Blackwelder, Lyman J. Briggs, R. E. Marsh, and Waldo L. Schmitt. Minute 1, Board of Directors, May 2, 1953, reports that "No formal request was left by this committee with the Board for action." Blackwelder's account of that meeting contains extended comments on the May 2 meeting and a number of disagreements with the statements and actions of some members of the board. See Richard E. Blackwelder to Detlev W. Bronk, May 4, 1953.

64. R. H. Arnett, Jr., R. E. Blackwelder, W. A. Dayton, J. B. Knight, J. P. Marble, A. R. Merz, W. L. Schmitt, and L. E. Yocum to members of the AAAS Council, June 30, 1953.

65. Council, December 27 and 30, 1953, minute 1.

66. Council, December 27 and 30, 1953, minute 12.

67. An appreciative account of Warren Weaver and his qualifications as a leader of AAAS can be found in Chester I. Barnard, "Warren Weaver, AAAS President-Elect," *Science* 117 (February 20, 1953), 174–176.

68. That decision was made on March 16, 1953, even as the board was deciding not to accept the resignation of Keener and Meyerhoff and hoping that both could be persuaded to remain. Even so, "in view of the burden and importance of both administrative and editorial work of the Association" the board decided to create the position of editor of the journals of the American Association for the Advancement of Science and authorized appointment of a search committee to recommend a new administrative

secretary and a new editor, thus returning to the arrangement that had existed prior to 1949 when Meyerhoff became chairman of the editorial board. Although decided on March 16, the decision to separate the two positions was not recorded until May 2, 1953, when minutes of the March 16 meeting were revised to include reports of the actions taken in response to the termination of services by Keener and Meyerhoff.

69. Dael Wolfle, *America's Resources of Specialized Talent* (NY: Macmillan, 1954).

70. Editorial Board, October 25, 1953, minutes.

71. Editorial Board, May 30, 1954, minutes.

72. Board of Directors, October 17–18, 1954, minute 61. That minute records the board's acceptance of Roller's resignation and the members' appreciation for his services and the improvements he brought about in the magazines. The individual discussions with Roller and me were not recorded.

73. Graham Phillips DuShane, *Science* 141 (July 26, 1963), 341; and Philip H. Abelson, "Seven Years of Progress," *Science* 141 (August 9, 1963), 491.

74. Frank L. Campbell, "Philip Hauge Abelson, New Editor of *Science*" *Science* 137 (July 27, 1962), 267–268).

75. Philip H. Abelson and Robert V. Ormes, "*Science* — Report to the Board of Directors, March 2–3, 1963," Board of Directors, March 2–3, 1963, agenda, tab E.

76. Philip H. Abelson and Robert Ormes, "Report on *Science*," Board of Directors, March 7–8, 1970, agenda item 22, tab D.

77. Board of Directors, June 20–21, 1970, agenda item 8. The AMA editorial appeared in the March 23, 1970, issue of the *Journal of the American Medical Association*.

78. Committee on Publications, May 26, 1961, agenda item 9 and minute 9; and Board of Directors, May 27–28, 1961, minute 24.

79. Editorial Board, May 31, 1953, minutes.

80. Bentley Glass, "Report of the Acting Chairman of the Editorial Board," June 22, 1953. At its meeting on October 25, 1953, the editorial board continued the discussion of this matter and asked Glass to write to the Publications Committee (and the board of directors) restating the position that *Science* should feel free to include editorial discussions of controversial issues.

81. D(ael) W(olfle), "Editorial Responsibility," *Science* 122 (November 25, 1955), 1000.

82. Allen L. Hammond, in Philip H. Abelson, *Enough of Pessimism* (Washington, DC: American Association for the Advancement of Science, 1985), 10.

83. F. R. Moulton to the board of directors, March 6, 1945.

84. Announcement published frequently in *Science* after 1984.

85. The Distinguished Public Service Award of the National Science Foundation, May 9, 1984, signed by Edward Knapp, director.

## Chapter 5

1. F. R. Moulton, memorandum to the board of directors, April 4, 1946.

2. Frank L. Campbell, "From Dungeon to Tower," *The Scientific Monthly* 62 (December 1946), 127–130.

3. Ibid., 130.

4. F. R. Moulton, memoranda to the board of directors, April 2, 1946, and April 4, 1946.

5. F. R. Moulton, memorandum to the board of directors, April 30, 1946. For Moulton's account of the purchase, the officers and other consultants who advised him on the purchase, and his hopes for the future, see "Ad Astra Per Aspera, or Acquiring a Home for the Association," *AAAS Bulletin* 5 (May 1946), 33–36. A brief account of the purchase is given in "AAAS Buys Washington Building Site," *Science* 103 (May 10, 1946), 591–592.

6. The leased top floor and AAAS as a landlord proved very satisfactory to the American Psychological Association (APA), and as AAAS planned to construct a building large enough to provide office quarters for several affiliated societies as well as for itself, the APA decided to "buy into" that building. To cement the relationship, APA contributed $5,000 to the AAAS building fund and began setting aside more in order to have a fund with which to pay its appropriate share of the cost of the new building. In return, AAAS reduced the rent charged to APA by $200 a year, or four percent of the APA's $5,000 contribution. The plans being made at that time were never carried out, and the APA later purchased a building of its own. See Donald G. Marquis, "Proceedings of the 54th Annual Meeting of the American Psychological Association, Inc., Philadelphia, September 3–7, 1946," *The American Psychologist* 1 (November 1946), 493–502, minutes 56 and 57; and Dael Wolfle, "Report of the Treasurer for the Year 1946," *The American Psychologist* 2 (November 1947), 484–485. Before the APA made its contribution to the building fund, the American Association of Economic Entomologists—another of the association's affiliated societies—had contributed $2,000 to the building fund. See *AAAS Bulletin* 5 (June 1946), 41–42.

7. Harlow Shapley, memorandum to the board of directors, July 21, 1947.

8. F. R. Moulton, "Basic Information about the American Association for the Advancement of Science for the Committee on Raising Funds for the A.A.A.S. Permanent Home," May 28, 1948.

9. Waldron Faulkner, "The Architect Reviews His Files," *Science* 124 (October 12, 1956), 659–663.

10. Board of Directors, March 25–26, 1950, minute 21.

11. Board of Directors, December 26–27, 1950, minute 18; and Council, December 27 and 29, 1950, minute 6.

12. Board of Directors, March 22, 1952, minute 1.

13. Many details about the building were not yet decided. It would have provided over twice as much office space as did the building that was later constructed, but how that space would have been arranged was something the architect, and AAAS, left for later decision.

14. AAAS Building Committee, July 9, 1953, minutes.

15. Waldron Faulkner, "The Architect Reviews His Files," *Science* 124 (October 12, 1956), 659–663.

16. Dael Wolfle, memorandum to the Board of Directors, August 5, 1954.

17. Council, December 27 and 30, 1954, minute 16.

18. Warren Weaver, memorandum to the council, May 13, 1955.

19. Board of Directors, March 2–3, 1963, agenda item 7.

20. The objecting member of the National Capital Planning Commission was Paul Thiry, a Seattle architect. Thiry and I were not acquainted, but we had graduated from the University of Washington in the same class.

21. Board of Directors, March 26–27, 1966, minute 32.

22. Board of Directors, October 21–22, 1967, minute 24; and December 29, 1967, minute 25.

23. Committee on Cooperation Among Scientists, February 17, 1965, minute 2.

24. Robert L. Calkins, "The National University," *Science* 152 (May 13, 1966), 884–889.

25. An account of the development of those plans is given in Board of Directors, October 19–20, 1968, agenda item 31 and tab B. See also "The National Memorial to Woodrow Wilson: An International Center for Scholars" (Washington, DC: Woodrow Wilson Memorial Commission, 1968).

26. Board of Directors, October 18–19, 1969, minute 30.

27. Alfred Friendly, *The Washington Post*, March 2, 1980, C2.

## Chapter 6

1. The Gibson Island property was purchased for $16,000. When the conferences left Gibson Island at the end of the 1946 season, sale of the property had to be in accordance with the regulations of the Gibson Island Club. No early purchaser appeared and it was not until late in 1948 that the property was sold for $13,500 (see F. R. Moulton to the board of directors, December 8, 1948). Receipts from the sale, plus other monies from chemical companies and operations, provided a fund which has been managed by AAAS wholly for the Gordon Research Conferences.

2. W. George Parks, "Gordon Research Conferences: A Quarter Century on the Frontiers of Science," *Science* 124 (December 28, 1956), 1279–1281.

3. Board of Directors, December 26, 1946, agenda item 3 and minute 6.

4. W. George Parks, Gordon Research Conferences, AAAS, 1948, *Science* 107 (March 26, 1948), 308–312.

5. Board of Directors, June 2–3, 1956, agenda item 12, tab H, and minute 29; Board of Directors, October 13–14, 1956, minute 39.

6. Maxine Singer and Dieter Soll, "Guidelines for DNA Hybrid Molecules," *Science* 181 (September 21, 1973), 1114. That seminal letter to *Science* did not break the rule against publishing conference reports; however, that rule was occasionally relaxed. For example, in 1946 the board of directors approved publication of a monograph on tumor chemotherapy consisting of papers presented at one of the 1945 conferences (see Board of Directors, March 29, 1946, minute 13).

7. The early history of events flowing from that 1973 Gordon Conference — the self-imposed moratorium on some kinds of research, the Asilomar Conference, the National Institutes of Health Guidelines and related developments — is given in June Goodfield, *Playing God* (NY: Random House, 1977); Nicholas Wade, *The Ultimate Experiment: Man-Made Evolution* (NY: Walker, 1977); and Michael Rogers, *Biohazard* (NY: Knopf, 1977).

8. Glenn T. Seaborg, "The Future Through Science," *Science* 124 (December 28, 1956), 1275–1278; and W. George Parks, "Gordon Research Conferences: A Quarter Century on the Frontiers of Science," *Science* 124 (December 28, 1956), 1279–1281.

9. "The AAAS: The Permanent Secretary's Report of the Cincinnati Meeting," *Science* 59 (January 25, 1924), 71–79.

10. Board of Directors, March 18, 1958, minute 28.

11. See, for example, Board of Directors, June 21–22, 1958, minutes 18, 20, 21, and 22. An account of actual and proposed prizes and some of their problems is given in Board of Directors, June 27–28, 1969, agenda item 25.

12. Board of Directors, June 21–22, 1958, minute 19.

13. "Report of the AAAS Council Study Committee on Natural Areas as Research Facilities," October 1, 1962 (submitted to the council with the agenda for the meeting of December 27 and 30, 1962), and Paul Bruce Dowling and Richard H. Goodwin, "Survey of College Natural Areas," May 31, 1962, Annex 1 to report of the study committee.

14. Council, December 27 and 30, 1963, agenda item 4 and accompanying table of contents and first chapter of "Report of the American Association for the Advancement of Science Council Study Committee on Natural Areas as Research Facilities."

15. Committee on Council Affairs, December 7, 1963, minute 8.

16. Council, December 30, 1966, minute 8; and December 30, 1967, minutes 9 and 10; Board of Directors, March 11–12, 1967, agenda item 12 and minute 12.

17. "A Proposal to the National Science Foundation for a Project to Develop the Use of Broadcast Television for Communication Among Scientists and Engineering," June 19, 1962; "Experiment in Communication," *AAAS Bulletin* 7 (December 1962), 1–2; and E. G. Sherburne, Jr., and John K. Mackenzie, "Science and Engineering Television Journal: Interim Report on a Project to Develop the Use of Broadcast Television for Communication Among Scientists and Engineers," February 1, 1963.

18. Board of Directors, December 17, 1963, agenda item 22; June 20–21, 1964, minute 25; October 24–25, 1964, agenda item 5 and tab B; and December 27, 1964, tab A and minute 10. See also "Science and Engineering Television Journal," *AAAS Bulletin* 10 (March 1965), 1–2.

19. Committee on Public Understanding of Science, March 16, 1963, minute 7.

20. *AAAS Proceedings* 36 (1888), 255.

21. "Grants for Research of the AAAS," *Science* 50 (December 19, 1919), 559–561; and William C. Steere, *Biological Abstracts/Biosis: The First Fifty Years* (NY: Plenum, 1976). A few months after the grant to *Botanical Abstracts* was made, Burton Livingston, the first editor of that journal, became permanent secretary of AAAS.

22. Personal communication from Hans Nussbaum, September 25, 1984.

23. Charles M. Goethe to F. R. Moulton, May 4, 1944.

24. Personal communication from Hans Nussbaum, September 25, 1984.

25. Board of Directors, October 17–18, 1954, minute 41; March 20–21, 1955, agenda item 32 and accompanying memorandum from the Committee on Research Grants, and minute 41.

26. Board of Directors, June 11–12, 1955, agenda item 34.

27. Board of Directors, December 27–28, 1955, minute 27.

28. John H. Behnke to the academies of science affiliated with AAAS, March 15, 1956.

29. Board of Directors, October 13–14, 1956, minute 17.

30. *National Science Foundation, 9th Annual Report*, 1959 (Washington DC: U.S. Government Printing Office, 1960), 78–79.

31. Burton E. Livingston, "Grants for Research by the American Association for the Advancement of Science," *Science* 59 (March 28, 1924), 294–295.

32. Dael Wolfle, "Robert H. Goddard," *Science* 146 (December 25, 1964), 1639.

33. Board of Directors, October 23–24, 1965, agenda item 12.

## Chapter 7

1. Otis W. Caldwell, Burton E. Livingston, and F. R. Moulton, "Revision of the Association's Constitution," *Science* 103 (March 1, 1946), 245–250, Article IV.

2. Howard A. Meyerhoff, "Revision of the AAAS Constitution and Bylaws," *Science* 116 (November 21, 1952), 575–578, Article IV, Section 1, and Article V, Section 1.

3. Dael Wolfle, "AAAS Constitution Amendments," *Science* 132 (November 25, 1960), 1558–1559.

4. Dael Wolfle, "AAAS Council Meeting, 1961," *Science* 135 (February 16, 1962), 528–530.

5. Dael Wolfle, "AAAS Council Meeting, 1959," *Science* 131 (February 19, 1960)., 503–506.

6. Council, December 27 and 30, 1960, minute 8. This description of the Committee on Council Affairs was adopted as a new Section 11 of Article V of the then current bylaws.

7. C. E. Davies, L. F. Kimball, and Kirtley F. Mather to members of the council, December 20, 1950, page 2.

8. Council, December 27 and 30, 1950, minute 4.

9. Council, December 27 and 30, 1950, minute 11; and December 27 and 30, 1956, minute 11.

10. Dael Wolfle to members of the council, August 30, 1961.

11. Council, December 27 and 30, 1961, minute 3.

12. Dael Wolfle to members of the council, December 9, 1969.

13. Council, December 30, 1969, minute 5.

14. Council, December 30, 1969, minute 15.

15. Board of Directors, June 20–21, 1970, minute 37.

16. Board of Directors, December 28, 1970, minute 5.

17. Council, December 30, 1970, minute 4 (b).

18. Bush's decision not to be a candidate was on grounds of high principle. That year he also declined a nomination for membership on the Council of the National Academy of Sciences and explained, "I also think it highly desirable that individuals who have occupied prominent positions during the war in connection with scientific matters should stand aside when it comes to the post-war period. This would apply particularly to such individuals as Conant and myself." (Bush to Isaiah Bowman, February 6, 1945, in the Bowman Papers, Special Collections, Eisenhower Library, Johns Hopkins University. I am indebted to Nathan Reingold, senior historian, Department of the History of Science and Technology of the Smithsonian Institution, for this quotation — DW.)

19. F. R. Moulton to the executive committee, December 17, 1945.

20. Board of Directors, May 23-24, 1954, minute 6; Dael Wolfle to members of the council, June 15, 1954; and Board of Directors, October 17-18, 1954, minute 34.

21. Dael Wolfle to members of the board of directors, May 13, 1954, 3-4; and Dael Wolfle to members of the council, December 13, 1954.

22. Council, December 27 and 30, 1954, minute 9.

23. Board of Directors, June 11-12, 1955, agenda item 17 and minute 18.

24. Board of Directors, October 29-30, 1955, minute 37; Council, December 27 and 30, 1955, agenda item 9 and attached report of the Committee on Constitution, Bylaws, and General Operation.

25. Council, December 27 and 30, 1955, minutes 9 and 10.

26. Dael Wolfle (for Wallace Brode) to AAAS Committee on Nominations, July 2, 1958.

27. Alan T. Waterman, "Letter from the President," *AAAS Bulletin* 8 (August 1963), 1.

28. Arnold Prostak, "Is AAAS Democratic?" *AAAS Bulletin* 9 (March 1964), 4.

29. Alan T. Waterman, "Yes It Is," *AAAS Bulletin* 9 (March 1964), 4.

30. Alan T. Waterman to the board of directors, undated, but in 1964.

31. Board of Directors, October 24-25, 1964, minute 34.

32. Council, December 27 and 30, 1964, agenda item 10 and minute 10.

33. Council, December 30, 1965, minute 10.

34. Lawrence Cranberg, "AAAS Election System," *Science* 152 (April 8, 1966), 157.

35. Ellis Yochelson, "AAAS Election System," *Science* 152 (June 10, 1966), 1456.

36. Alfred S. Romer, "A Letter to Council Members," *AAAS Bulletin* 12 (September 1967), 3-4.

37. Some members of the council could remember that AAAS had earlier had membership as well as the council voting in the election of the association's presidents. Lists of all the nominees were sent to members for an advisory vote, the results of which were given to the council, which then held the ruling vote. As an example, early in 1943 members were sent a list of 20 nominees. The 4,809 ballots that were returned included some 500 to 600 each for Roger Adams, Isaiah Bowman, and Harlow Shapley, and not over half as many for any of the other 17 nominees. That year the council elected Bowman, and a few years later each of the other two. Incidentally, that was the first time a AAAS president was elected by mail ballot. Wartime conditions prevented the physical meeting; the council balloted by mail on the 20 nominees, none received a majority, and a run-off ballot among the leaders was needed. Later in the same year, in electing the 1944 president, the three leaders in the membership ballot were James B. Conant, A. J. Carlson, and Harlow Shapley. All three were close together in vote totals, with the next highest candidate receiving less than a third as many votes. The council chose Carlson for 1944, but elected both the other two within the next three years. See "Vote of the Council for President of the Association," *AAAS Bulletin* 2 (March 1943), 18-19; and "Election of President of the Association," *AAAS Bulletin* 2 (December 1943), 93-94.

38. Committee on Council Affairs, April 11, 1968, agenda item 13 and minute 13.

39. Committee on Council Affairs, April 6, 1969, minute 9.

40. Committee on Council Affairs, November 19, 1969, agenda item 14 and attached tabulation of questionnaire returns; Committee on Council Affairs, November 30, 1970, minutes 3 and 4; and letter from Lorrin Riggs — a member of the Committee

on Council Affairs and of the Committee on Governance — to Leonard Rieser, chairman of the latter committee, December 28, 1970.

41. Board of Directors, December 28, 1970, minute 9.

42. Council, December 30, 1970, minute 4 (b); "The New Constitution and Bylaws," *AAAS Bulletin* 17 (November 1972); and William Bevan, "AAAS Council Meeting, 1972," *Science* 179 (February 23, 1973), 821–824.

43. The presidents discussed here are those who served during the time period being reviewed. Direct election by AAAS members did not start until 1973. The following table shows some comparisons between the first dozen elected under that system with the last dozen elected by the council. However, it should be noted that under both systems the slates presented to voters were determined by the Committee on Nominations and Elections.

|  | 1963–1974 Election by Council | 1975–1986 Election by Members |
|---|---|---|
| Field of Science: |  |  |
| Physical sciences and engineering | 4 | 3 |
| Biological sciences and medicine | 2 | 3 |
| Other sciences | 6 | 6 |
| Institutions: |  |  |
| Universities | 9 | 7 |
| Other | 3 | 5 |
| Women | 1 | 3 |
| Median age at time of presidency | 63–64 | 59–60 |

44. Board of Directors, December 26–28, 1951, agenda item 8 and minute 8; March 21–22, 1952, agenda item 3.

45. Board of Directors, May 23–24, 1954, minute 34; Paul Scherer and Dael Wolfle to the board of directors, enclosure number 3 to Board of Directors, October 17–18, 1954, and minute 48 of that meeting.

46. Board of Directors, October 17–18, 1954, minute 49.

47. As this account of AAAS history was being written, William Golden, who in 1969 succeeded Paul Klopsteg as treasurer, reached 19 years of continuous service as a member of the board of directors and was still in office. Earlier in the century James McKeen Cattell had been on the board for 23 years when he died in 1944.

## Chapter 8

1. *Science Education News* (March 1962), 1.

2. *The Organization and Work of the Association* (Washington, DC: American Association for the Advancement of Science, 1925).

3. Historical accounts of the founding, organization, and work of the Cooperative Committee are given by Karl Lark-Horovitz, "The Cooperative Committee for the Teaching of Science: A Report to the AAAS Council" (December 1949), and by Bernard R. Watson, "A Brief History of the AAAS Cooperative Committee on the Teaching of Science and Mathematics" (October 12, 1965), published in *Science Education News*

(November 1965). Most of Watson's history is included in Vivian A. Johnson, *Karl Lark-Horovitz: Pioneer in Solid State Physics* (NY: Pergamon, 1969), 26–31.

4. Proposal to the American Association for the Advancement of Science from the Cooperative Committee on Science Teaching, February 1945.

5. AAAS Cooperative Committee on the Teaching of Science and Mathematics, "The Present Effectiveness of Our Schools in the Training of Scientists," 47–119 in *Manpower for Research*, Volume 4 of *Science and Public Policy, a Report to the President* [The Steelman Report], John R. Steelman, Chairman, The President's Scientific Research Board, October 11, 1947.

6. Board of Directors, March 21–22, 1952, agenda item 8.

7. Herbert M. James, "Karl Lark-Horovitz, Physicist and Teacher," *Science* 127 (June 27, 1958), 1487–1488. See also Vivian A. Johnson, *Karl Lark-Horovitz: Pioneer in Solid State Physics*, 1969.

8. Board of Directors, October 17–18, 1954, minute 35.

9. National Science Foundation, *Fifth Annual Report, Fiscal Year 1955* (Washington, DC: National Science Foundation, 1956).

10. Dael Wolfle, *America's Resources of Specialized Talent* (NY: Harper and Brothers, 1954). The study reported in this volume was conducted under the auspices of the Commission on Human Resources, consisting of members appointed by the American Council of Learned Societies, the American Council on Education, the National Academy of Sciences – National Research Council, and the Social Science Research Council.

11. "An Action Program to Meet the Shortage of Well-Qualified Science and Mathematics Teachers," proposed by the Cooperative Committee on the Teaching of Science and Mathematics working in cooperation with representatives of the Academy Conference, undated but developed in the fall of 1954.

12. Board of Directors, December 27–28, 1954, minute 20.

13. Mark H. Ingraham, "Director Named for AAAS Science Teaching Improvement Program," *Science* 122 (July 22, 1955), 151.

14. John R. Mayor, "The Science Teaching Improvement Program, 1956–1957," a report to the board of directors, October 3, 1957.

15. Ibid.

16. "Study on the Use of Science Counselors," Final Report, AAAS Science Teaching Improvement Program, 1959.

17. Statement of Laurence H. Snyder, chairman of the Board of Directors, before the Senate Committee on Labor and Public Welfare, February 21, 1958.

18. Statement of Dael Wolfle, executive officer, AAAS, for the House of Representatives Committee on Education and Labor, April 2, 1958. Provisions of the two bills are described in Dael Wolfle, "Science Legislation for 1958," *Science* 127 (February 21, 1958), 389–392.

19. Panel on School Science of the National Research Council's Commission on Human Resources, *The State of School Science, A Review of the Teaching of Mathematics, Science, and Social Studies in American Schools* (Washington, DC: National Research Council, 1979).

20. Cooperative Committee on the Teaching of Science and Mathematics, November 14–15, 1958, minutes.

21. Annual Report to the Carnegie Corporation of New York on the AAAS Science Teaching Improvement Program, January 25, 1960.

22. Report to the Carnegie Corporation of New York on the AAAS Science Teaching Improvement Program, February 1962.

23. AAAS Cooperative Committee on Science Teaching, "The Preparation of High School Science and Mathematics Teachers," *School Science and Mathematics* 46 (February 1946), 107–118.

24. Alfred E. Garrett, "Recommendations for the Preparation of High School Teachers of Science and Mathematics — 1959," *School Science and Mathematics* 59 (April 1959), 281–289. A revised version of this article taking account of comments and suggestions from a number of reviewers was published as "Preparation of High School Science Teachers," *Science* 131 (April 8, 1960), 1024–1029.

25. Dael Wolfle, "Review of 1961," *AAAS Bulletin* 7 (September 1962), 3–6.

26. *Annual Review of the Scientific Personnel and Education Division of the National Science Foundation, FY 1963*, II- 89–90. For the AAAS report referred to, see "Science Teaching in Elementary and Junior High Schools," *Science* 133 (June 23, 1961), 2019–2024.

27. *Science Education News* (December 1962) 1–4, announced the formation, objectives, and initial members of the Commission on Science Education.

28. Cooperative Committee on the Teaching of Science and Mathematics, October 5–6, 1962, minutes.

29. Board of Directors, June 20–21, 1970, minute 15.

30. "AAAS Cooperative Committee on the Teaching of Science and Mathematics Cumulative Record of Representatives to March 1971," unsigned but compiled by Emory Will, who served as the Committee's secretary from 1963 to 1971. During its 30-year lifetime, the following organizations held membership in the committee, for the dates given or from the single date shown until the committee disbanded in 1971. The number of organizations on this list that joined after the early 1950s indicates the committee's continued viability and usefulness.

American Association for the Advancement of Science, 1945–1970
American Association of Physics Teachers, 1941—
American Astronomical Society, 1945—
American Chemical Society, 1941—
American Geological Institute, 1951—
American Institute of Biological Sciences, as the Union of Biological Science, 1941–1945, and as AIBS, 1964—
American Institute of Physics, 1947—
American Medical Association, 1963—
American Nature Study Society, 1951—
American Psychological Association, 1964—
American Society for Engineering Education, 1953–1967
American Society for Microbiology, 1963—
American Society of Zoologists, 1947—
American Statistical Association, 1963—
Association for the Education of Teachers of Science, 1963—
Botanical Society of America, 1946—
Central Association of Science and Mathematics Teachers, 1947—
Conference Board of the Mathematical Societies, 1963—

Engineering Council for Professional Development (first as the Engineers Joint Council), 1962—
Geological Society of America, 1945–1949
Mathematical Association of America, 1941—
National Association for Research in Science Teaching, 1941—
National Association of Biology Teachers, 1946–1970
National Association of Geology Teachers, 1956—
National Association of Teachers of Mathematics, 1946—
National Science Teachers Association, 1946—
United States Office of Education, 1949–1953

31. Board of Directors, June 16–17, 1962, agenda item 4.

32. *The New School Science: A Report to School Administrators on Regional Orientation Conferences in Science* (Washington, DC: American Association for the Advancement of Science, 1963).

33. William P. Viall, "Report on NASDTEC-AAAS Studies," *AAAS Bulletin* 8 (September 1963), 1–3.

34. Commission on Science Education, October 2–3, 1970, agenda item 3.

35. AAAS Council Newsletter, April 17, 1959.

36. The formal status of the Commission on Science Education was that of an internal AAAS committee. However, its establishment was preceded by considerable discussion with representatives of other organizations interested in science education, including staff members of the National Science Foundation. At the request of NSF, AAAS postponed start of development work on "Science — A Process Approach" in order to allow NSF to use its limited elementary science instruction funds to make a grant to Educational Services Inc., which was also planning to hold an extended workshop in the summer of 1962 to start developing its K–6 materials. The board agreed to delay start of work on the proposed AAAS materials, but insisted on an informal understanding with NSF that the Commission on Science Education would receive continuing funding to permit it to serve as an effective, even though informal, coordinating body for work on science education at the pre-college level (see Board of Directors, December 27, 1961, agenda item 38 and minute 38).

37. The Commission on Science Education was terminated in 1974, but Livermore continued as head of the association's Office of Science Education. In 1981 F. James Rutherford became the association's chief education officer and started AAAS on a new and expanding program of work on science education.

38. Personal communications from Bowen C. Dees, former assistant director of the National Science Foundation's Division of Scientific Personnel and Education, August 12, 1985; and J. Merton England, National Science Foundation Historian, September 13, 1985.

39. Edwards Park, *Treasures of the Smithsonian* (Washington, DC: Smithsonian Books, 1983), 62. The most worn volume in the Smithsonian's set, which is the only complete set known, is *Swiss Family Robinson*.

40. Hilary J. Deason, "Improving Science Collections in School Libraries," *AAAS Bulletin* 9 (March 1964), 2–6.

41. Early plans and the first year's list of books are described in Hilary J. Deason, "Traveling High School Science Libraries," *Science* 122 (December 16, 1955), 1173–

1176. See also *Science* 124 (November 23, 1956), 1013–1017, and reports for later years in issues of the *AAAS Bulletin*.

42. Board of Directors, October 15–16, 1960, agenda tab C.

43. Board of Directors, March 18, 1958, agenda tab G.

44. Board of Directors, October 15–16, 1960, agenda tab C.

45. Hilary J. Deason, "Improving Science Collections in School Libraries," *AAAS Bulletin* 9 (March 1964), 2–6.

46. *The AAAS Science Book List*, prepared under the direction of Hilary J. Deason (Washington, DC: American Association for the Advancement of Science, 1959; second edition, 1964; third edition, 1970).

47. *The Science Book List for Children*, compiled by Hilary J. Deason (Washington, DC: American Association for the Advancement of Science, 1960; second edition, 1963).

48. Kathryn Wolff and Jill Storey, eds., *AAAS Science Book Supplement* (Washington, DC: American Association for the Advancement of Science, 1978); Kathryn Wolff, Joellen M. Fritsche, Elina N. Gross, and Gary T. Todd, eds., *The Best Science Books for Children* (Washington, DC: American Association for the Advancement of Science, 1983); Kathryn Wolff, Susan M. O'Connell, and Valerie J. Montenegro, eds., *AAAS Science Book List: 1978–1986* (Washington, DC: American Association for the Advancement of Science, 1986); and Susan M. O'Connell, Kathryn Wolff, and Valerie J. Montenegro, eds., *The Best Science Books and A-V Materials for Children* (Washington, DC: American Association for the Advancement of Science, 1988).

49. *A Guide to Science Reading*, compiled and edited by Hilary J. Deason (NY: New American Library, 1963).

50. *Science* 140 (April 5, 1963), 42.

51. Board of Directors, October 15–16, 1960, agenda tab C.

52. Hilary J. Deason, "Improving Science Collections in School Libraries," *AAAS Bulletin* 9 (March 1964), 2–6. This article gives a brief concluding account of the libraries and their byproducts.

53. Board of Directors, June 16–17, 1962, agenda item 8.

54. National Science Foundation, *Annual Report, FY 1962* (1963), 87.

55. Hilary J. Deason, "Proposed Publication of a Quarterly Science Book Review," memorandum to the Committee on Publications, Mary 19, 1963; and "Proposal for the Publication of a New Quarterly Review of Science Books for School and Public Libraries," memorandum to the Committee on Publications, October 25, 1963.

56. Cooperative Committee on the Teaching of Science and Mathematics, "You Too Can Help the Science Teacher," October 1950.

57. Joint Commission on the Teaching of Science and Mathematics, "Proposed Studies of the Education of Teachers of Science and Mathematics" (October 1956), 4.

58. Cooperative Committee on the Teaching of Science and Mathematics, "Recommendations for the Preparation of High School Teachers of Science and Mathematics–1959" (1959).

59. *New York Times*, February 22, 1959, 63.

60. Board of Director, June 20–21, 1959, tab D to agenda and minute 36; *Science* 130 (November 6, 1959), 1237; and "Better Education for Science Teachers Is Ahead," *AAAS Bulletin* 6 (July 1961), 5–8.

61. *Guidelines for Preparation Programs of Teachers of Secondary School Science and Mathematics* (Washington, DC: American Association for the Advancement of Science, 1961).

62. William P. Viall, "Report on NASDTEC-AAAS Studies," *AAAS Bulletin* 8 (September 1963), 1–3.

63. *Guidelines for Science and Mathematics in the Preparation Programs of Elementary School Teachers* (Washington, DC: American Association for the Advancement of Science, 1963).

64. National Science Foundation, *Secondary School Science and Mathematics Teachers: Characteristics and Service Loads* (Washington, DC: U.S. Government Printing Office, 1963).

65. William P. Viall, "Report on NASDTEC-AAAS Studies," *AAAS Bulletin* 8 (September 1963), 1–3.

66. *Preservice Science Education of Elementary School Teachers* (Washington, DC: American Association for the Advancement of Science, 1970).

67. American Association for the Advancement of Science and National Association of State Directors of Teacher Education and Certification, *Guidelines and Standards for the Education of Secondary School Teachers of Science and Mathematics* (Washington, DC: American Association for the Advancement of Science, 1971). For a preliminary evaluation of these guidelines and their influence on educational practice, see Howard J. Hausman, "Acceptance of the AAAS-NASDTEC Guidelines and Standards for the Education of Secondary School Teachers of Science and Mathematics," undated, but apparently submitted to the AAAS Office of Education in 1974 or 1975.

68. Board of Directors, October 15–16, 1960, agenda item 23 and minute 23. The "adequate" help required was found internally. For the first year John Mayor and I shared responsibility. After William Kabisch joined the staff at the end of 1961, he and Linda McDaniel took over much of the work. In 1970 when Kabisch and I both resigned, responsibility went to Mayor and the science education staff.

69. R. V. Jones (University of Aberdeen, Scotland) to Dael Wolfle, September 12, 1978. The other honorary initials stand for Companion of the Order of the Bath and Fellow of the Royal Society.

70. Arthur H. Livermore, personal communication, September 26, 1984. The visitor was Herbert Frank Halliwell, a British chemist. His U.S. colleague was Laurence Strong of Earlham College.

71. Commission on Science Instruction in Elementary and Junior High Schools, minutes, August 3–4, 1962; and minutes, September 25–26, 1962.

72. AAAS Commission on Science Education, "Statement of Purposes and Objectives of Science Education in School," foreword by Paul B. Sears, statement by William Kessen, *Journal of Research in Science Teaching* 2 (1964), 3–6.

73. Henry H. Walbesser, "Curriculum Evaluation by Means of Behavioral Objectives," *Journal of Research in Science Teaching* 1 (1963), 296–301. In selecting the educationally most effective order of presentation of the tasks or experiments in the program, and in developing the practical tests to determine how competent a pupil had become on each of the several processes, the commission drew heavily upon the ideas and analyses of one of its own members, Robert Gagné, whose work on these issues is reported in *Psychological Monographs* 75 (no. 14, 1961) and 76 (no. 7, 1962), and in *The Conditions of Learning* (NY: Holt, Rinehart, and Winston, 1965).

74. Arthur H. Livermore, "The Process Approach of the AAAS Commission on Science Education," *Journal of Research in Science Teaching* 2 (1964), 271–282. The processes covered in the K–3 material were observing, classifying, measuring, communicating, inferring, predicting, recognizing space/time relations, and recognizing number relations. For grades 4–6 integrating processes were added: formalizing hypotheses, making operational definitions, controlling and manipulating variables, experimenting, interpreting data, and formulating models. See also John R. Mayor, "Science in the Elementary School," *AAAS Bulletin* 9 (June 1964), 2–4.

75. Commission on Science Education, September 26–27, 1966, agenda item 3; Board of Directors, March 11–12, 1967, minute 13; and "Xerox Corporation to Publish 'Science — A Process Approach,'" *AAAS Bulletin* 12 (March 1967), 3–4.

76. 1969 Report of the Commission on Science Education, December 1969; and "Two Million Youngsters in U.S. Learn Science With a New Curriculum," *AAAS Bulletin* 16 (June 1971).

77. Commission on Science Education, October 2–3, 1970, agenda item 4; and 1970 Report of the Commission on Science Education, 3.

78. Commission on Science Education, April 25–26, 1968, minute 15.

79. I. R. Weiss, *Report of the 1977 National Survey of Science, Mathematics, and Social Studies Education*, National Science Foundation Report SE 78-72, Appendix B (Washington, DC: U.S. Government Printing Office, 1978).

80. James A. Shymansky, William C. Kyle, and Jennifer M. Alport, "The Effects of New Science Curricula on Student Performance," *Journal of Research in Science Teaching* 20 (no. 5, 1983), 387–404.

81. Ted Bredderman, "Effects of Activity-based Elementary Science on Student Outcomes: A Qualitative Synthesis," *Review of Educational Research*, 53, no. 4, 1983, 499–518.

82. James A. Shymansky, William C. Kyle, and Jennifer M. Alport, "How Effective Were the Hands-on Science Programs of Yesterday?" *Science and Children* 20 (no. 3, 1982), 14–15.

83. Paul DeHart Hurd, "Problems and Issues in Precollege Science Education in the United States," a paper presented to the National Science Board Commission on Precollege Education in Science, Mathematics, and Technology, July 9, 1982.

84. National Science Foundation, January 29, 1987 (NSF PR 87-5). A press release announcing three grants totaling $6.6 million to the Biological Sciences Curriculum Study, Colorado Springs, Colorado; the Educational Development Center, Newton, Massachusetts; and the Technical Education Research Centers, Cambridge, Massachusetts.

85. Commission on Science Education, September 26–27, 1965, minute 13.

86. An account of plans and negotiations concerning the proposed junior high school curriculum were given in Commission on Science Education, February 28–March 1, 1966, agenda item 6 and minute 6. For board approval, see Board of Directors, October 23–24, 1965, minute 25.

87. Commission on Science Education, October 26–27, 1966, minute 15, and minutes of following meetings through January 16–17, 1969, minute 10.

88. Committee on Public Understanding of Science, January 20, 1962, agenda item 3.

89. Board of Directors, March 26–27, 1966, minute 5.

90. Edward G. Sherburne, Jr., who was primarily responsible for managing the program until he resigned in 1966 to become Director of Science Service, gave a full account of objectives and procedures in "Holiday Science Lectures, A Progress Report to NSF," February 1, 1963. Brief accounts and lists of speakers, their topics, and their cities were published in several issues of the *AAAS Bulletin* 8 (September 1963), 3–4; 9 (August 1964), 1–3; 11 (March 1966), 3; 12 (June 1967), 1–3; and 13 (September 1968), 2–3.

91. A report on *Science Education News* to the AAAS Publications Committee, March 18, 1969.

92. John R. Mayor and Willis G. Swartz, *Accreditation in Teacher Education: Its Influence on Higher Education* (Washington, DC: National Commission on Accrediting, 1965).

93. *Science Policy Reviews* (Columbus, OH: Battelle Memorial Institute, published quarterly from 1967 through 1972).

94. John A. Moore, *Science for Society, A Bibliography* (Washington, DC: 1970; second edition, 1971; ...sixth edition, 1976).

95. In the 1970s half the colleges and universities in the United States developed programs or courses on "science and society," "science, technology and society," "science and values," or with other titles relating science to societal or cultural affairs. See Paul DeHart Hurd, "State of Precollege Education in Mathematics and Science," a paper presented in Washington, D.C., before a meeting arranged by the National Academy of Sciences and National Academy of Engineering, May 12–13, 1982. Hurd cites as his sources, *Evist Resource Directory* (Washington, DC: American Association for the Advancement of Science, 1978) and *Science, Technology and Society*, the Curriculum Newsletter of the Lehigh University STS Program (Bethlehem, PA: published periodically).

96. "Tomorrow's Teachers: A Report of the Holmes Group, East Lansing, Michigan," Holmes Group, Inc., Michigan State University, 1986. See also National Commission for Excellence in Teacher Education, "A Call for Change in Teacher Education," *Chronicle of Higher Education* 30 (March 6, 1985), 13–21.

97. For a brief review of the ills, see F. James Rutherford, "The Crisis in Science Education: An Overview," paper presented at the annual meeting of AAAS, Detroit, Michigan, May 28, 1983. For more analytical reports and sets of recommendations, see *What Are the Needs in Precollege Science, Mathematics, and Social Science Education? Views From the Field* (Washington, DC: National Science Foundation, 1980); National Commission on Excellence in Education, *A Nation at Risk* (Washington, DC: U.S. Government Printing Office, 1983); National Science Board Commission on Precollege Education in Mathematics, Science, and Technology, *Educating Americans for the 21st Century* (Washington, DC: U.S. Government Printing Office, 1983); Task Force on Economic Growth, *Action for Excellence: A Comprehensive Plan to Improve Our Nation's Schools* (Denver, CO: Education Commission of the States, 1983); College Entrance Examination Board, *Academic Preparation for College: What Students Need to Know and Be Able to Do* (NY: College Entrance Examination Board, 1983); Twentieth Century Fund Task Force on Federal Elementary and Secondary School Policy, *Making the Grade* (NY: Twentieth Century Fund, 1983); Ernest L. Boyer, *High School: A Report on Secondary Education in America* (NY: Harper and Row, 1983); Linda Darling-Hammond, *Beyond the Commission Reports: The Coming Crisis in Teaching* (Santa Monica, CA: Rand Corporation, 1984); Carnegie Corporation of New York Task Force on Teaching as a Profession, *A Nation Prepared: Teachers for the 21st Century* (NY: Carnegie Corporation, 1986);

National Governors' Conference, *Time for Results: The Governors' 1991 Report on Education*, parts of which were published in *The Chronicle of Higher Education* 33 (September 3, 1986), 79–90; and *Science for All Americans: A Project 2061 Report on Literacy Goals in Science, Mathematics, and Technology* (Washington, DC: American Association for the Advancement of Science, 1989).

## Chapter 9

1. Board of Directors, April 22, 1945, minute 13; and June 24, 1945, minute 14.

2. Board of Directors, June 24, 1945, minute 14 and accompanying memorandum "A Plan to Further Public Knowledge and Understanding of Science." The AAAS program that did develop over the next 20 years has recently been described by Bruce V. Lewenstein in "Public Understanding of Science in America, 1945–1964," his 1987 doctoral dissertation earned at the University of Pennsylvania for work in the history and sociology of science. Lewenstein's dissertation covers a wider scope than does this chapter. Within it, his chapter 3 is a thoughtful treatment of the AAAS program as seen from the perspective of an analyst of the whole national interest in and effort to improve the public understanding of science.

3. Board of Directors, June 28–29, 1953, agenda item 43.

4. Board of Directors, December 27, 28, and 30, 1958, minute 30; February 28–March 1, 1959, minute 10.

5. Board of Directors, November 4, 1945, minute 15.

6. Board of Directors, March 3–4, 1956, minute 31.

7. A series of books that bordered on popularization was more successful, but the credit should go to one man rather than to AAAS. In 1924 Gregory Walcott, professor of philosophy at Long Island University, initiated a series of volumes on the history of science intended for small college libraries and undergraduate courses on the history of science, and he saw the project through to publication of the final volume in the 1950s. Included were books on Greek and medieval science, on the major fields of science from 1400 to 1900, and on those fields from 1900 to 1950. A supporting grant from the Carnegie Corporation of New York was made but not used, for McGraw-Hill published the volumes as a purely commercial venture without the need for a subsidy, and later, when some of the books had gone out of print, the Harvard University Press took over responsibility, reissued the out-of-print volumes and published the later ones. Producing the whole series was a labor of love by Walcott, but throughout, he was counseled and assisted by a committee appointed by AAAS, so the board of directors regularly received reports of progress. For an account of plans and of books published and in prospect, see John A. Behnke, "Report: Committee on Source Books on the History of Science," tab C, Board of Directors, October 13–14, 1956, agenda item 24. At that meeting the board voted a resolution of thanks and appreciation to Walcott "for the devotion and excellent judgment with which he has stimulated and brought to circulation a valuable series of source books on the history of science" (Board of Directors, October 13–14, 1956, minute 22).

8. Board of Directors, October 18, 1953, minute 31. See also Board of Directors, June 21–22, 1958, minute 14.

9. Report [of Philip H. Abelson and Robert V. Ormes] to the editorial board of *Science*, March 17, 1980.
10. *Life* 28 (no. 2, January 9, 1950), 17–24.
11. *Life* 38 (no. 2, January 10, 1955), 13–19.
12. *AAAS Bulletin* 8 (September 1963), 1.
13. Board of Directors, March 20–21, 1955, agenda item 10.
14. Kneeland Godfrey, director of public information for the Institute of Technology and the departments of science at Northwestern University, was engaged to work with Sherburne and to replace Negus. See Board of Directors, June 22–23, 1963, minute 10.
15. Board of Directors, March 16–17, 1957, agenda item 4; and July 6, 1957, agenda item 10.
16. This was the project to drill through the ocean bottom to the discontinuity between the earth's crust and its mantle. See Willard Bascom, *A Hole in the Bottom of the Sea: The Story of the Mohole Project* (Garden City, NY: Doubleday, 1961).
17. Board of Directors, December 27–28, 1957, minute 15.
18. Board of Directors, December 27–28, 1956, minute 23; March 16–17, 1957, agenda item 4 and minute 4; July 6, 1957, agenda item 10 and minute 10; December 27–28, 1957, minute 15; and December 27–28, 1958, agenda item 10.
19. Board of Directors, December 27–28, 1958, agenda item 10.
20. Board of Directors, June 4–5, 1960, agenda item 31.
21. Board of Directors, November 4, 1945, minute 15.
22. Board of Directors, July 6, 1957, agenda item 31; Dael Wolfle to members of the board of directors, July 24, 1957; and Board of Directors, October 12–13, 1957, minute 32.
23. Dael Wolfle to the executive committee, February 10, 1959; and Board of Directors, February 28–March 1, 1959, minute 6.
24. Board of Directors, June 20–21, 1965, agenda item 31 and accompanying tabs C and D.
25. John A. Behnke, "Science Information Center for the American Association for the Advancement of Science," Board of Directors, June 2–3, 1956, agenda item 2, tab D.
26. Board of Directors, March 3–4, 1956, minute 24 and accompanying draft of AAAS-NASW meeting report.
27. Board of Directors, October 13–14, 1956, minute 25; and October 18–19, 1958, minutes 36 and 40.
28. Board of Directors, July 6, 1957, agenda item 27; October 12–13, 1957, minute 31; and Laurence H. Snyder, president, and Dael Wolfle, executive officer, to Henry Heald, president, Ford Foundation, November 12, 1957.
29. Board of Directors, March 18, 1958, minute 25.
30. Wallace R. Brode, president, and Dael Wolfle, executive officer, to John W. Gardner, president, Carnegie Corporation of New York, July 2, 1958; a similar letter to the Rockefeller Foundation; and Board of Directors, October 18–19, 1958, minute 40.
31. Board of Directors, February 28–March 1, 1959, minute 21.
32. Board of Directors, October 18–19, 1958, minute 36; and February 28–March 1, 1959, minute 21.

33. "Sherburne to Head New AAAS Program to Improve Public Understanding," *Science* 132 (December 16, 1960), 1823–1824, and "Man for the Job," *AAAS Bulletin* 6 (January 1961), 6.

34. Board of Directors, June 21–22, 1958, minute 41.

35. "U.S. Science Exhibit, Seattle World's Fair," *AAAS Bulletin* 7 (June 1962).

36. Committee on Public Understanding of Science, "Status Report to AAAS Council," December 1964; and Board of Directors, October 13–14, 1962, minute 30.

37. *The Public Impact of Science in the Mass Media: A Report on a Nationwide Survey for the National Association of Science Writers* (Ann Arbor, MI: Survey Research Center of the University of Michigan, 1958). This study and the ones by Schramm and Tichenor were helpful in understanding what kind of information (or misinformation) to expect in the general public, but when the studies by Jon Miller and his colleagues began to appear in 1980, everyone interested in the issue had access to much more detailed and analytical information on the kinds of people interested in science and the kinds of information that interested them. See Jon D. Miller, Kenneth Prewitt, and Robert Pearson, *The Attitudes of the U.S. Public Toward Science and Technology* (a final report to the National Science Foundation under contract number C-SRS78-16839), National Opinion Research Center, University of Chicago, 1980, two volumes; Jon D. Miller, Robert W. Suchner, and Alan M. Voelker, *Citizenship in an Age of Science: Changing Attitudes Among Young Adults* (NY: Pergamon Press, 1980); Jon D. Miller, *The American People and Science Policy: The Role of Public Attitudes in the Policy Process* (NY: Pergamon Press, 1983).

38. Margaret Mead and Rhoda Metreaux, "Image of the Scientist Among High School Students," *Science* 126 (August 30, 1957), 384–390.

39. Board of Directors, June 20–21, 1959, agenda item 42 and minute 41.

40. Board of Directors, October 3–4, 1959, minute 30; and December 29, 1966, minute 22.

41. *AAAS Bulletin* 6 (March 1961), 4.

42. Board of Directors, December 27 and 29, 1960, minute 29; Alfred Friendly and Dael Wolfle to Alan T. Waterman, National Science Foundation, January 6, 1961; and Edward G. Sherburne, Jr., to AAAS Committee on Public Understanding of Science, May 12, 1961.

43. Committee on Public Understanding of Science, January 20, 1962, agenda item 5; and Board of Directors, October 13–14, 1962, agenda item 30. When Sherburne left AAAS four years later, the two organizations agreed that the Council for the Advancement of Science Writing would assume full responsibility for *Understanding*. That arrangement did not work well and publication was soon abandoned.

44. *A New Agenda for Science* (New Haven, CT: Sigma Xi, The Scientific Research Society, 1987), 5.

45. In 1969 the board declined to rejoin that committee because the invitation seemed to indicate that the association would have little influence on the committee's activities. The board did, however, offer to consider an invitation to participate in later plans or recommendations if we could be helpful in their formulation. See Board of Directors, April 20–21, 1969, agenda item 26 and minute 26.

46. Board of Directors, June 16, 1962, minute 32; and December 27, 1962, agenda item 5.

47. Committee on Public Understanding of Science, October 5, 1963, minute 6.

48. Board of Directors, October 23–24, 1965, agenda item 32; and March 26–27, 1966, agenda item 15.

49. Committee on Public Understanding of Science, March 16, 1963, minute 10.

50. Committee on Public Understanding of Science, April 23, 1967, minutes, 2–4; and July 9, 1968, minute 3.

51. Board of Directors, December 27, 1964, agenda item 5 and minute 5.

52. Board of Directors, March 5–6, 1960, agenda item 2.

53. Dael Wolfle to members of the executive committee, February 12, 1960.

54. Board of Directors, June 4–5, 1960, agenda item 7.

55. Board of Directors, June 16, 1962, minute 14. Brookings later polled the 1962 participants and received replies from 90 percent saying they wanted the seminars to be continued. See Board of Directors, March 16, 1963, minute 3.

56. Board of Directors, March 11–12, 1967, agenda item 10.

57. Committee on Public Understanding of Science, March 16, 1963, agenda item 3.

58. Board of Directors, June 20–21, 1964, tab A; March 27–28, 1965, agenda item 13; and March 26–27, 1966, agenda item 14.

59. Board of Directors, March 11–12, 1967, agenda item 11.

60. Resolution adopted by the Southwestern and Rocky Mountain Division of AAAS on May 3, 1963, reported in Board of Directors, June 22–23, 1963, agenda item 40.

61. Board of Directors, October 13–14, 1962, agenda item 35 and tabs S and T.

62. Board of Directors, October 13–14, 1962, minute 35.

63. Board of Directors, October 26–27, 1963, agenda item 29.

64. Board of Directors, March 14–15, 1964, agenda items 27 and 28.

65. Board of Directors, June 22–23, 1963, minute 40; October 26–27, 1963, agenda item 29; December 27, 1963, minute 30; March 14–15, 1964, minutes 2, 26, 27, and 28, and tab F to agenda item 28; June 20–21, 1964, agenda item 19 and minute 19; and March 27–28, 1965, agenda item 28.

66. Board of Directors, October 23–24, 1965, minute 31.

67. Wallace R. Brode, "New Director for Science Service, Inc.," *AAAS Bulletin* 11 (no. 2, June 1966), 2.

68. Committee on Public Understanding of Science, April 23, 1967, minutes.

69. Board of Directors, June 25–26, 1966, agenda item 19; October 15–16, 1966, minute 22; and Committee on Public Understanding of Science, "Report to the AAAS Board of Directors and Council," December 1966.

70. Board of Directors, October 19–20, 1968, minute 12.

71. Committee on Public Understanding of Science, October 2–3, 1970, minutes, 3.

72. Board of Directors, June 20–21, 1970, minute 29; and December 28, 1969, supplementary agenda item 31; March 7–8, 1970, minute 25; and October 17–18, 1970, minute 16.

73. Board of Directors, October 17–18, 1970, minute 13; and December 12–13, 1970, minutes 16 and 17.

74. Board of Directors, October 23–24, 1965, minute 2.

75. Board of Directors, June 20–21, 1970, minute 20.

# Chapter 10

1. Jack Morrell and Arnold Thackray, *Gentlemen of Science: Early Years of the British Association for the Advancement of Science* (Oxford: Clarendon Press, 1981).
2. *Proceedings* 25 (1876), 341.
3. *Science* 89 (February 3, 1939), 341; and Isaiah Bowman, "Science and Social Planning," *Science* 90 (October 6, 1939), 309–319.
4. *Science* 104 (December 6, 1946), 523.
5. Board of Directors, October 18–19, 1969, minute 8.
6. *Proceedings* 21 (1872), 276–277.
7. Board of Directors, March 29, 1946, minute 5; and F. R. Moulton, New Haven Conference, June 3, 1947.
8. Board of Directors, December 27–28, 1956, minute 23; and July 6, 1957, agenda item 16.
9. Editorial Board of *Science*, March 6, 1970, minutes page 2.
10. Board of Directors, June 4–5, 1960, minute 17.
11. Editorial Board of *Science*, March 6, 1970, minutes page 1.
12. Board of Directors, May 23–24, 1954, minute 15.
13. Four lines of future development were considered at the symposium: (1) variability and predictability of water supply in arid regions; (2) better use of present resources; (3) prospects for additional water resources; and (4) better adaptation of plants and animals to arid conditions. At each session, the speakers and discussants — from different countries and different fields of science — were given a common set of questions in order that the whole session would be well focused. For example, participants in the session on "Better Use of Present Resources" were asked: What are the possibilities of increasing and maintaining sustained production from grass and forest lands without accelerating erosion? What are the consequences of utilizing arid lands beyond their capabilities? What constitutes wise allocation of available water supplies among the various needs in arid land drainage areas? How can production be increased from existing water supplies? And, can irrigated lands be occupied permanently?
14. "Recommendations for Development of Arid Lands: International Arid Lands Conference," *Science* 122 (July 8, 1955), 61–64.
15. Board of Directors, June 11–12, 1955, agenda item 10.
16. John Behnke, "Report on the International Arid Lands Meetings," supplement to agenda item 10, Board of Directors, June 11–12, 1955.
17. Board of Directors, February 28–March 1, 1959, agenda item 28.
18. John Behnke's report on attendance, finances, and the enthusiasm of participants at the arid lands meetings and a draft of the conference report published in *Science* (note 14) were given to the board in Behnke's "Report on the International Arid Lands Meetings, Albuquerque and Socorro, New Mexico, April 26–May 4, 1955." Board of Directors, June 11–12, 1955, supplement to agenda item 10.
19. Board of Directors, March 3–4, 1956, minute 20; and June 2–3, 1956, minute 6.
20. Board of Directors, June 2–3, 1956, minute 6; and October 13–14, 1956, minute 30.
21. Board of Directors, December 27–28, 1956, agenda item 10.

22. Other members of the committee were Gustaf Arrhenius, John Cushing, Fritz Koczy, Gordon Lill, George Meyers, Roger Revelle, Henry Stommel, Dael Wolfle, and Lional A. Wolford. Invited guests at the first committee meeting were G. Böhnecke, A. Fr. Brunn, G. E. R. Deacon, G. Dietrick, N. W. Rakestraw, and L. Zenkevitch. Deacon was then chairman of the UNESCO Advisory Committee on Marine Sciences, and Böhnecke, Brunn, and Zenkevitch were members of that committee.

23. Mary Sears to Dael Wolfle, September 6, 1957; AAAS Steering Committee for a Conference on the Marine Sciences, August 26–27, 1957, minutes; AAAS Steering Committee for a Conference on the Marine Sciences, January 21–22, 1958, minutes; and Board of Directors, October 12–13, 1957, minute 26.

24. Dael Wolfle to Mary Sears, November 8, 1957; and Board of Directors, March 18, 1958, minute 30.

25. Edward Wenk, Jr., *The Politics of the Ocean* (Seattle, WA: University of Washington Press, 1972). Chapter 2 reports the many governmental actions and initiatives of the years in which the oceanographic congress was being planned and conducted, and the immediately following years.

26. This account of the oceanographic congress is condensed from a paper presented at the Third International Congress on the History of Oceanography. See Dael Wolfle, "The 1959 Oceanographic Congress: An Informal History," in Mary Sears and Daniel Merriman, eds., *Oceanography: The Past* (NY: Springer-Verlag, 1980), 42–48.

27. AAAS Steering Committee for a Conference on the Marine Sciences, January 21–22, 1958, minutes, page 2.

28. Mary Sears, ed., *International Oceanographic Congress, 31 August–12 September, 1959: Preprints of Abstracts to be Presented in Afternoon Sessions* (Washington, DC: American Association for the Advancement of Science, 1959).

29. Mary Sears, ed., *Oceanography: Invited Lectures Presented at the International Oceanographic Congress held in New York, 31 August–12 September, 1959* (Washington, DC: American Association for the Advancement of Science, 1961).

30. The congress was a newsworthy event. No records were made of how many stories were filed or words printed, but the press room, managed by Sidney Negus, was a busy place. From the United States there were five representatives of wire services, 18 from general circulation magazines such as *Time, Ladies Home Journal*, and *Scientific American*, and 22 from more specialized magazines such as *Fortune* or *Business Week*. Fifteen foreign countries were represented by reporters, most of whom were stationed in the United States but several of whom came to New York especially for the Congress. West Germany had the largest number, with four reporters, and there were 21 others from 14 other countries. The radio, television, and press services of the United Nations were represented by nine correspondents; Radio Free Europe by one; and the U.S. Information Agency by six. Press conferences were held twice a day, with some or all of the preceding session's speakers present, and with Roger Revelle as the moderator and skillful interpreter of technicalities for the benefit of the reporters in attendance. See Dael Wolfle to members of the Committee on Arrangements, International Oceanographic Congress and members of the board of directors, November 5, 1959.

31. Mary Sears, *Oceanography* (note 29), iii–vi.

32. Board of Directors, October 3–4, 1959, agenda item 11 and minute 11.

33. Board of Directors, October 3–4, 1959, minute 4; June 4–5, 1960, agenda item 23; and October 23–24, 1965, agenda item 7.

34. "The Need of a United States National Arid Lands Committee," prepared by the Committee on Desert and Arid Zone Research of the Southwestern and Rocky Mountain Division of the American Association for the Advancement of Science, November 12, 1960.

35. Board of Directors, June 4–5, 1960, agenda item 27 and minute 29; December 27, 1960, agenda item 20 and minute 19; May 27–28, 1961, agenda item 31 and minute 31; and December 27, 1961, minute 13. See also June 22–23, 1963, agenda item 33 and minute 33.

36. Board of Directors, June 20–21, 1964, agenda item 26, tab B, and minute 26; October 24–25, 1964, agenda item 31 and minute 31; March 27–28, 1965, minute 2; June 19–20, 1965, agenda item 22 and minute 22; and October 23–24, 1965, agenda item 13.

37. Board of Directors, December 29, 1966, agenda item 25 and tab C, Report of Committee on Arid Lands, 1966; and June 25–26, 1966, agenda item 22.

38. Board of Directors, June 25–26, 1966, minute 22.

39. Board of Directors, March 9–10, 1968, agenda item 28 and minute 28; June 15–16, 1968, minute 19; October 19–20, 1968, minute 5; and April 20–21, 1969, minute 4.

40. Terah L. Smiley, "International Conference on Arid Lands in a Changing World," *AAAS Bulletin* 14 (September 1968), 7–8.

41. J. S. Kanwar, chairman, Committee on Resolutions, "Resolutions and Recommendations passed by International Conference on Arid Lands in a Changing World," June 3–13, 1969, University of Arizona, Tucson.

42. Harold E. Dregne, ed., *Arid Lands in Transition* (Washington, DC: American Association for the Advancement of Science, 1971).

43. Board of Directors, July 26–27, 1947, minutes 5 and 14.

44. Council, September 13, 1948, minute 10; Board of Directors, December 27, 1949, minute 15a; November 14–15, 1950, agenda item 23 and minute 33; June 28–29, 1952, minute 31; October 18, 1952, minute 34; and December 26–30, 1952, agenda item 33.

45. Dael Wolfle to Board of Directors, February 1, 1954.

46. Board of Directors, May 23–24, 1954, minute 19.

47. Dael Wolfle to members of the Board of Directors, August 5, 1954.

48. Committee on World Federation of Associations for the Advancement of Science, report to the board of directors, December 1954; and Board of Directors, December 27 and 29, 1954, minute 37.

49. "Science and the Future," *BA Record*, September 1969, 6. To several of the associations, the exchange of visitors continued to be more attractive than the formation of an international federation. In 9161, during the annual meeting of the British Association, Thomas Park and I represented AAAS in a meeting to discuss ways of facilitating such visits, or other ways in which the several associations might be mutually helpful. That meeting also included representatives of the Australia–New Zealand, British, French Canadian, Indian, and South African associations. See Board of Directors, March 3–5, 1961, minute 26 and tab D, "Foreign and International Activities of the AAAS"; and October 7–8, 1961, agenda item 39 and minute 39.

50. Final report of the Council Study Committee on Cooperation with Developing Countries, November 28, 1966; and Council, December 30, 1966, minute 8a.

51. Board of Directors, March 11–12, 1967, agenda item 27 and minute 29.

52. At the March 1967 meeting (note 51), the issue was discussed and then deferred to June for decision. In June, the item was tabled until October, and in October until some "later" meeting. "Later" turned out to be several years.

53. Earlier, existing, and prospective international activities of the association are described in a paper prepared by the AAAS Office of International Science, "The AAAS Global Outreach: Major Programs and Challenges," January 12, 1987.

## Chapter 11

1. For several years after World War II, the National Science Foundation was a hope rather than a reality, and the National Institutes of Health had not yet become the major source of support for academic research. During those years, university research programs received most of their federal support from the military services, most notably from the Office of Naval Research, which some old-timers still remember with nostalgia as the most enlightened and generous of government sources for research support. But that is only part of the story; the military services surely had their own interests well in mind in developing their relations with academe and their programs to support academic research. A recent analysis of the whole issue of military-university relations in the field of physics, the amounts of money involved, faculty attitudes toward military support, and the effects of military support on the development of physics itself, all leading up to the conclusion that physicists lost control of their discipline and that military interests and agencies were responsible for the selective development of different fields of physics, is given in Paul Forman, "Behind Quantum Electronics: National Security as Basis for Physical Research in the United States, 1940–1960," *Historical Studies in the Physical and Biological Sciences* 18 (Part 1, 1987), 149–229.

2. Council, December 29–30, 1955, minute 17.

3. "Society in the Scientific Revolution," *Science* 124 (December 21, 1956), 1231; and report to the council of the Interim Committee on the Social Aspects of Science, December 1956.

4. Board of Directors, March 16–17, 1957, agenda item 9.

5. Board of Directors, March 3–4, 1956, agenda item 22, report of the January 24, 1956, meeting of representatives of AAAS and National Association of Science Writers.

6. Dael Wolfle, "Social Responsibility of Science," *Science* 125 (January 25, 1957), 141; and "Social Aspects of Science," *Science* 125 (*January 25, 1957*), 143–147.

7. AAAS Committee on the Social Aspects of Science, report to the AAAS Council, December 30, 1958.

8. Board of Directors, October 12–13, 1957, agenda item 35 and tab H: "AAAS Committee on the Social Aspects of Science, Statement on the Radiation Problem"; and minute 29.

9. 1957 Report of the Committee on the Social Aspects of Science, included as minute 15 in Dael Wolfle, "AAAS Council Meeting, 1957," *Science* 127 (February 21, 1958), 397–400.

10. "Program for A Committee on Social Aspects of Science: Purposes, Policies and Procedures," a report to the Board of Directors of the American Association for the Advancement of Science, meeting at Princeton, New Jersey, May 30–June 1, 1958.

11. Board of Directors, October 18–19, 1958, minute 36; Council, December 27 and 30, 1958, agenda item 4 and minute 4. See also Chauncey D. Leake, "Report of the President," *AAAS Bulletin* 6 (no. 1, January 1961), 1–2.

12. "Science and Human Welfare: The AAAS Committee on Science in the Promotion of Human Welfare States the Issues and Calls for Action," *Science* 132 (July 8, 1960), 68–73.

13. Board of Directors, March 3–5, 1961, minute 27; December 27, 1961, minute 20; March 17–18, 1962, minute 40; June 16, 1962, agenda item 30 and tab N: "Recommendations to the Board of Directors Concerning the Implementation of the Recommendations of the Committee on Science in the Promotion of Human Welfare," prepared by Harrison Brown, Henry Eyring, and Don K. Price, and minute 30; and October 13–14, 1962, minute 33. The quotation is from pages 1 and 2 of the memorandum by Brown, Eyring, and Price.

14. Board of Directors, June 22–23, 1963, agenda item 30 and tab N, a memorandum from the Committee on Science in the Promotion of Human Welfare to the Board of Directors, and minute 30; and October 26–27, 1963, minute 28.

15. Gregory Pincus, "Control of Conception by Hormonal Steroids," *Science,* 153 (July 29, 1966), 493–500.

16. Henry Eyring, ed., *Civil Defense* (Washington, DC: American Association for the Advancement of Science, 1966).

17. "Secrecy and Dissemination in Science: A Report of the Committee on Science in the Promotion of Human Welfare," *Science* 163 (February 21, 1969) 787–790. Plans for this study and a request for appropriate case histories were published in *Science* 155 (February 10, 1967), 679, and in *AAAS Bulletin* 12 (June 1967).

18. AAAS Committee on Science in the Promotion of Human Welfare, October 4, 1969, minutes; and January 10, 1970, minutes.

19. "Science and Human Survival: Unlimited war is self-defeating and an alternative must be found by a new science of human survival," *Science* 134 (December 29, 1961), 2080–2083. As an outgrowth of the symposium on science and survival, the committee proposed establishing a special commission, similar to the Air Conservation Commission, "for the purpose of encouraging studies of new means for the protection of society that are free of the destructive character of modern warfare." The board invited the committee to spell out its plans in more detail, but on reconsideration, the committee chose to concentrate its efforts on other topics. See memorandum from the Committee on Science in the Promotion of Human Welfare to the board of directors, May 28, 1962, and Board of Directors, June 16–17, 1962, minute 30.

20. "The Integrity of Science: A Report by the AAAS Committee on Science in the Promotion of Human Welfare," *American Scientist* 53 (June 1965), 174–198.

21. Barry Commoner, *Science and Survival* (NY: Viking Press, 1966).

22. Committee on Science in the Promotion of Human Welfare, "Report of Activities in 1960," 2–3.

23. Board of Directors, March 17–18, 1962, agenda item 40 and minute 40.

24. *Air Conservation: The Report of the Air Conservation Commission of the American Association for the Advancement of Science* (Washington DC: American Association for the Advancement of Science, 1965).

25. Barry Commoner, "A Reporter at Large: The Environment," *The New Yorker*, June 15, 1987, 46–71.

26. Board of Directors, June 16–17, 1962, agenda item 39 and minute 39; October 13–14, 1962, agenda item 36 and accompanying proposal to the American Association for the Advancement of Science from the American Anthropological Association; and October 13–14, 1962, minute 36.

27. Board of Directors, December 27, 1962, agenda item 20.

28. Memorandum from the Committee on Science in the Promotion of Human Welfare to the board of directors, February 28, 1963.

29. "Science and the Race Problem: A Report of the AAAS Committee on Science in the Promotion of Human Welfare," *Science* 142 (November 1, 1963), 558–561.

30. Carleton Putnam, *Science* 142 (December 13, 1963), 1419; and Henry E. Garrett and Wesley C. George, *Science* 143 (February 28, 1964), 913–915.

31. Letter to the editor from the Committee on Science in the Promotion of Human Welfare, *Science* 143 (February 28, 1964), 915.

32. Margaret Mead, Theodosius Dobzhansky, Ethel Tobach, and Robert E. Light, eds., *Science and the Concept of Race* (NY: Columbia University Press, 1968).

33. Committee on Science in the Promotion of Human Welfare, October 4, 1969, minutes.

34. Ibid.; and Board of Directors, October 18–19, 1969, minute 26; see also 1970 Report of the Committee on Science in the Promotion of Human Welfare, November 16, 1970.

35. John Platt, "What We Must Do: A large-scale mobilization of scientists may be the only way to solve our crisis problems," *Science* 166 (November 28, 1969), 1115–1121.

36. Committee on Science in the Promotion of Human Welfare, November 21, 1970, minute 40.

37. Walter Modell, "Promotion of Human Welfare," *Science* 170 (December 18, 1970), 1254–1256.

38. Board of Directors, December 27, 1962, agenda item 33 and minute 31; and undated memorandum by Lawrence Cranberg, "An Ethical Practices Committee for Scientists?"

39. Dael Wolfle, "The AAAS as Seen by its Former Presidents," March 22, 1963, a summary of discussions at the March 2, 1963, meeting of the Board of Directors.

40. "Ethical Problems: An Invitation," *Science* 143 (January 31, 1964), 435.

41. "The Etiquette of Research and Publication," tab E to agenda item 26, Board of Directors, March 27–28, 1965. The other report — "Problems of Customs or Manners Arising in the Major Relationships of Scientists" — was tab F, same item.

42. Board of Directors, March 27–28, 1965, minute 26. The Committee on Cooperation Among Scientists asked to be dissolved and reported to the board that they believed AAAS would be better served by ad hoc committees to deal with special issues than by continuation of their committee. As examples of possible topics for ad hoc committees, they made three suggestions: (a) a committee to consider the problems of cross-disciplinary communication; (b) one on the education of foreign students attending U.S. universities; and (c) one on the arrangements that might be made to enable visiting scholars to make better use of the magnificent intellectual resources of history, art, public affairs, science, and other areas that existed in Washington, D.C. See Committee on Cooperation Among Scientists, February 17, 1965, minute 2. The board did not appoint any of the three proposed committees. However, I arranged several sessions on the third topic that brought together representatives of Washington, D.C., organizations that

would have been interested in a center for visiting scholars. Those meetings may have helped in planning and securing approval of the Woodrow Wilson Center of the Smithsonian Institution. See Dael Wolfle, "A Center for Advanced Study in the Nation's Capital," tab B for the meeting of the Board of Directors, October 19–20, 1968; see also Chapter 5 this volume.

43. Committee on Council Affairs, December 7, 1963, minute 16, and "Ethical Problems of Scientists," notes by T. C. Byerly and William Wildhack, sent to committee members, December 17, 1962.

44. Council, December 27 and 30, 1963, minute 16.

45. Committee on Council Affairs, March 29, 1965, agenda item 5 and minute 5.

46. "A Study of Scientists' Assessments of the Importance or Relevance of Certain Questions with Regard to the Ethics and Responsibilities of Scientists," preliminary report of the Council Study Committee on Ethics and Responsibilities of Scientists, Council, December 30, 1966, agenda item 8c.

47. Anatol Rapaport, chairman, Study Committee on Ethics and Responsibilities of Scientists, "A Study of Scientists' Views on Ethics and Responsibilities," a report of a survey, 1967.

48. Early in the course of AAAS involvement with the problems of military use of chemical herbicides and defoliants in Vietnam, members of the staff of the Science Policy Office of the Congressional Legislative Reference Service of the Library of Congress asked Don Price and me if they might serve as observers of the AAAS activities and write a history of the case. We agreed and gave them access to all relevant minutes, correspondence, and reports. They also had access to records from the Department of Defense, and, of course, the published materials. On that basis, Franklin Huddle of the Science Policy Office wrote a long and detailed account that includes some history of the development of herbicides and their military use, and then gives a factual account of the whole course of AAAS interest, uncertainty, debate, and final action. His paper is a critical and fair account of the difficulties AAAS experienced in deciding what to do and the disagreements among various participants in the whole history. It was published in one of the several volumes prepared at the request of the House of Representatives Committee on Science and Astronautics (as it was then named) as that committee was considering legislation to create the Congressional Office of Technology Assessment. See *Technical Information for Congress*, Report to the Subcommittee on Science, Research, and Development of the Committee on Science and Astronautics, U.S. House of Representatives, prepared by the Science Policy Research Division, Library of Congress, April 25, 1969, revised April 15, 1971, Part IV, Chapter 1, entitled "A Technology Assessment of the Vietnam Defoliant Matter, An Evaluation by a Scientific Organization," 531–591, and post-1970 supplement, 591–614.

49. Board of Directors, October 15–16, 1966, item 29 and minute 29; Committee on Council Affairs, December 27, 1966, minute 2 (a).

50. Council, December 30, 1966, minute 16.

51. *Technical Information for Congress* (note 46), 531.

52. Board of Directors, March 11–12, 1967, minute 28. Because the proposed study would involve policies and actions of a controversial nature that involved high levels of the U.S. government, the ad hoc committee was selected to include members who were of stature in science, well acquainted with AAAS policies and activities, and experienced in dealing with governmental policy issues. The full membership consisted of René Dubos,

professor, Rockefeller University, chairman; Joseph L. Fisher, president, Resources for the Future; Laurence M. Gould, professor, University of Arizona; Jean Mayer, professor, Harvard University School of Public Health; Walter Modell, director, clinical pharmacology, Cornell University Medical College; E. W. Pfeiffer, associate professor, University of Montana; Dixy Lee Ray, director, Pacific Science Center, Seattle; Paul B. Sears, emeritus professor, Yale University; and Athelstan F. Spilhaus, dean, Institute of Technology, University of Minnesota.

That list included three past or future presidents of AAAS (Gould, Sears, and Spilhaus), a future members of Congress (Fisher), a future university president (Mayer), and a future chairman of the Atomic Energy Commission and state governor (Ray).

53. Report of the AAAS Committee to Advise the Board of Directors Concerning Studies of Chemical and Biological Agents that Alter the Environment, July 1967.

54. Don K. Price to Robert S. McNamara, September 11, 1967.

55. John S. Foster, Jr., to Don K. Price, September 29, 1967.

56. W. B. House et al, *Assessment of Ecological Effects of Extensive or Repeated Use of Herbicides*, final report, August 15–December 1, 1967, Midwest Research Institute, sponsored by Advanced Research Projects Agency of the Department of Defense, ARPA Order No. 1086.

57. Board of Directors, March 9–10, 1968, minute 26.

58. H. Burr Steinbach to Dael Wolfle, May 28, 1968.

59. Spraying for defoliation increased from 8 square miles in 1962 to 2,320 square miles in 1967, and then declined to 1,980 in 1968. Spraying for crop destruction increased from 1 square mile in 1962 to 348 square miles in 1967, and 100 in 1968. See *Technical Information for Congress* (note 48), Table 3, 544.

60. Four agents were being used. Most widely used and known was Agent Orange, a 50:50 mixture of 2,4-D (2,4 dichlorophenoxyacetic acid) and 2,4,5-T (2,4,5-trichlorophenoxyacetic acid). Agent Blue (cacodylic acid, an arsenical compound) was used for destroying crops. Information on composition of each agent and their usage is given in *Technical Information for Congress* (note 48), Table 2, 539.

61. "On the Use of Herbicides in Vietnam, a Statement by the Board of Directors of the American Association for the Advancement of Science," *Science* 161 (July 19, 1968), 253–256. That statement was also published in *Scientist and Citizen* 10 (June–July 1968), 119–122, a journal which had published several earlier accounts of AAAS activities in this area and which published a special issue on chemical and biological agents in warfare (*Science and Citizen* 9, August–September 1967). In 1970 the name of the journal was changed to *Environment*.

62. Board of Directors, December 27, 1968, minute 30 as amended on December 30, 1968.

63. AAAS Herbicide Assessment Commission, progress report, June 1, 1970; and American Association for the Advancement of Science, Herbicide Assessment Commission, Woods Hole Conference, June 15–19, 1970. See also Board of Directors, June 20–21, 1970, minute 31. The cost of the field study was expected to be about $70,000. The board asked the Rockefeller and Sloan Foundations each to contribute $25,000 toward that cost, but both declined and AAAS bore the whole cost.

64. Philip M. Boffey, "Herbicides in Vietnam: AAAS Study Runs into a Military Roadblock," *Science* 170 (October 20, 1970) 42–52.

65. R. L. Johnson, Assistant Secretary of the U.S. Army to H. Bentley Glass, March 12, 1970.

66. Matthew Meselson and John Constable to the Honorable Ellsworth Bunker, November 12, 1970.

67. Philip M. Boffey, "Herbicides in Vietnam: AAAS Study Finds Widespread Devastation," *Science* 171 (January 8, 1971), 43–47. A summary of the symposium and its background materials was published in *Congressional Record*, 118 (No. 32, March 3, 1972), S3226–S3233. John Constable and Matthew Meselson gave a shorter report, with a number of photographic illustrations, in "The Ecological Impact of Large-Scale Defoliation in Vietnam," *Sierra Club Bulletin* 56 (April 1971), 4–9.

The Herbicide Assessment Commission was also responsible for work on the analysis, methods of assessment, and possible health effects of dioxin, a highly toxic pollutant in Agent Orange (see note 60). See Robert Baughman and Matthew Meselson, "An Improved Analysis for Tetrachlorodibenzo-p-dioxins," *Advances in Chemistry Series* 120 (1973), 92–104; Robert Baughman and Matthew Meselson, "An Analytical Method for Detecting TCDD (Dioxin): Levels of TCDD in Samples from Vietnam," *Environmental Health Perspectives*, experimental issue 5 (1973), 27–35; Patrick W. O'Keefe, Matthew S. Meselson, and Robert Baughman, "Neutral Cleanup Procedures for 2,3,7,8-Tetrachlorodibenzo-p-dioxin Residues in Bovine Fat and Milk," *Journal of the Association of Official Analytical Chemists* 6 (no. 3, 1978), 621–626; and Matthew Meselson, Patrick O'Keefe, and Robert Baughman, "The Evaluation of Possible Health Hazards from TCDD in the Environment," in *Symposium on the Uses of Herbicides in Forestry*, February 21–22, 1978 (Washington, DC: Department of Agriculture, 1978), 91–94.

The status in 1983 of evergreen forests in Vietnam that had been subjected to extensive defoliation during the war was described by Peter S. Ashton in "Regeneration in Inland Lowland Forests in South Viet-Nam One Decade After Aerial Spraying by Agent Orange as a Defoliant," *Bois de Forets des Tropiques* 211 (1er Trimestre, 1986), 19–34.

68. Harold J. Coolidge to Dael Wolfle, May 23, 1968.

69. Fred. H. Tschirley, "Defoliation in Vietnam: The ecological consequences of the defoliation program in Vietnam are assessed." *Science* 163 (February 21, 1969), 779–786.

70. Gordon H. Orians and E. W. Pfeiffer, "Ecological Effects of the War in Vietnam: Effects of defoliation, bombing and other military activities on the ecology of Vietnam are described," *Science* 168 (May 1, 1970), 544–554.

71. Boffey (note 67) and Terri Aaronson, "A Tour of Vietnam," *Environment* 13 (March 1971), 34–43.

72. E. W. Pfeiffer and Arthur H. Westing, "Land War, Three Reports," *Environment* 13 (November 1971), 2–15.

73. Boffey (note 67), 44.

74. Council, December 30, 1970, minute 5 (d).

75. *The Effects of Herbicides in South Vietnam* (Washington, DC: National Academy of Sciences, 1974).

76. Boffey (note 67), 43–44 and 47.

77. Board of Directors, June 10–11, 1967, minute 20.

78. Board of Directors, December 29, 1967, agenda item 22.

79. Board of Directors, December 29, 1967, minute 22; and March 9–10, 1968, minute 26.

80. Committee on Environmental Alteration, October 12, 1968, minutes; November 15–16, 1968, minutes; 1969 report, December 8, 1969; "Proposed Future Activities," a memorandum from the Committee on Environmental Alteration to the board of directors, March 1970; Board of Directors, March 7–8, 1970, minute 24; and October 17–18, 1970, minute 2 (c).

81. Board of Directors, April 20–21, 1969, minute 34; June 27–29, 1969, minute 29; March 7–8, 1970, agenda item 9; and June 20–21, 1970, agenda item 7.

82. Board of Directors, June 27–29, 1969, agenda item 10; June 20–21, 1970, agenda item 33 and minute 33.

83. Robert S. Morison, "Science and Social Attitudes: Growing doubts require that science be put more recognizably at the service of man," *Science* 165 (July 11, 1969), 150–156.

84. H. Bentley Glass, "Letter from the President," *AAAS Bulletin* 14 (September 1969), 1–2.

85. U.S. participants in the conference were Philip H. Abelson, Lewis M. Branscomb, Edward U. Condon, James D. Ebert, H. Bentley Glass, William T. Kabisch, Robert S. Morison, Mina Rees, Walter Orr Roberts, Jack P. Ruina, Carl Sagan, Athelstan F. Spilhaus, and Dael Wolfle. See Marcel Roche, "Science and the Future," *Science* 165 (August 8, 1969), 619–620.

86. Board of Directors, June 27–29, 1969, minute 24.

87. Board of Directors, October 18–19, 1969, tab A; and December 28–29, 1969, minute 25.

88. Report of the Committee of Young Scientists to the board of directors, March 7, 1970.

89. Board of Directors, March 7–8, 1970, minute 20; and Dael Wolfle to Committee of Young Scientists, March 30, 1970.

90. Committee of Young Scientists to the board of directors, June 20, 1970, and memorandum, "Selected Comments from COYS Responses to Dael Wolfle's letter of March 30, 1970" and "Summary of COYS Responses to Dael Wolfle's letter of March 30, 1970."

91. Board of Directors, June 20–21, 1970, minute 30; and October 17–18, 1970, supplementary information item m.

92. Board of Directors, April 20–21, 1969, minute 37; and June 27–29, 1969, agenda item 24 and minute 24.

93. Board of Directors, October 18–19, 1969, minute 23.

## Chapter 12

1. Several of these proposals were discussed by Carl M. York, in "Steps Toward a National Policy for Scientific Research," *Science* 172 (May 14, 1971), 643–648.

2. Elmer B. Staats, "Making the Science Budget for 1967," in Harold Orlans, ed., *Science Policy and the University* (Washington, DC: The Brookings Institution, 1968), 220.

3. William D. Carey, address at the Seventeenth Annual Conference on the Administration of Research, October 1963.

4. *National Patterns of R&D Resources; Funds and Manpower in the United States* (Washington, DC: National Science Foundation, published annually).

5. Arnold A. Strassenburg, "Supply and Demand for Physicists," *Physics Today* 23 (April 1970), 23–28.

6. Philip M. Boffey, "Recession in Science: Ex-Advisors Warn of Long-Term Effects," *Science* 168 (May 1, 1970), 555–557; and "A Time of Torment for Science," *Science News* 99 (January 2, 1971), 5–6.

7. Irvin Stewart, *Organizing Scientific Research for War: The Administrative History of the Office of Scientific Research and Development* (Boston: Little Brown, 1948), Chapter 13, "The Contract."

8. Rachel Carson, *Silent Spring* (Boston: Houghton Mifflin, 1962). In 1950 Rachel Carson was the recipient of a AAAS-Westinghouse Science Writing Award.

9. Ibid., 297.

10. *Air Conservation: The Report of the Air Conservation Commission of the American Association for the Advancement of Science* (Washington, DC: American Association for the Advancement of Science, 1965).

11. The 1947 ratings are from R. W. Hodge, P. M. Siegel, and P. H. Rossi, "Occupational Prestige in the United States," *American Journal of Sociology* 70 (November 1964), 286–302. The 1972 ratings are from *Science Indicators 1972* (Washington, DC: National Science Foundation, 1973), 97.

12. Philip H. Abelson, "Troubled Times for Academic Science," *Science* 168 (May 1, 1970).

13. Joint Committee on the Humanistic Implications of Science and Technology, sponsored by the American Association for the Advancement of Science, American Council of Learned Societies, and Social Science Research Council, May 16, 1970, minutes.

14. "AAAS (I)," "(II)," and "(III)," *Science* 172 (April 30, 1971), 453–458; (May 7, 1971), 542–547; and (May 14, 1971), 656–658.

15. H. Bentley Glass, "1970 Report of the Chairman of the Board of Directors," to the Council of the American Association for the Advancement of Science.

# INDEX

*Note: Items in which AAAS is part of the formal title are listed under the letter A. Those in which AAAS is implied—such as AAAS meetings, AAAS presidents, etc.—are listed as meetings of AAAS, presidents of AAAS, etc.*

AAAS awards and prizes, 99, 117-120, 189, 191
    policies concerning, 119-120
AAAS Anna Frankel Rosenthal Memorial Award for Cancer Research, 119
AAAS Ida B. Gould Memorial Award for Research on Cardiovascular Problems, 119
AAAS Prize for Behavioral Science Research, 118-119
AAAS-Westinghouse Science Writing Awards, 189
*AAAS Bulletin*, 6, 87
AAAS divisions. *See* Alaska (Arctic) Division of AAAS; Caribbean Division of AAAS; Pacific Division of AAAS; Southwestern and Rocky Mountain Division of AAAS
*AAAS Guide to Scientific Instruments*, 97
*AAAS Science Book List*, 167, 303 (note 48)
*AAAS Science Book List for Children*, 167
Abelson, Philip H., 60, 65, 85, 95-99, 213, 259
Abrams, Creighton, 244
academies of science, 3, 124-125, 152
Academy Conference (National Association of Academies of Science), 125, 152, 205
"Action Program" for science education, 151, 152, 155, 185
Adams, Roger, 7, 65, 92, 136
Adams, Scott, 62
Adams, Sherman, 21-22

administrative secretary of AAAS, 6
*Advancement of Science, The*, 81
affiliated societies, 3, 41-43, 58, 124, 132-134, 140
Agency for International Development, 213, 222
Agents Orange, White, and Blue, 243, 318 (notes 59 and 60)
*Air Conservation*, 233-234, 247
Alaska (Arctic) Division of AAAS, 4
Aldrin, Edwin, 65
Allen, George, 212
Allison Commission of 1884-86, 28
Alpert, Daniel, 191
amateurs, role in AAAS, 9
American Academy of Arts and Sciences, 95
American Anthropological Association, 234-235
American Association of Colleges of Teacher Education, 169
American Association of Physics Teachers, 149
American Association of School Administrators, 162, 179
American Association of University Professors, 19
American Chemical Society, 25-26, 43, 102, 107, 109, 149, 170, 207
American Council of Learned Societies, 169, 192, 261
American Council on Education, 36, 38, 102, 113
American Geographical Society, 223
American Geological Institute, 109
American Geophysical Union, 109

American Institute of Biological Sciences, 43, 48, 59, 66, 202, 207, 283 (note 6)
American Institute of Physics, 207, 237
*American Journal of Science*, 44
American Library Association, 165
*American Men of Science*, 77, 289 (note 24)
American Museum of Natural History, 220
American Orthopsychiatric Association, 23
American Psychological Association, 103, 294 (note 6)
*American Scientist*, 75, 78, 80, 123
American Society for the Diffusion of Useful Knowledge, 164
American Society of Newspaper Editors, 200
American Society of Photogrammetry, 109
American Society of Range Management, 223
American Society of Zoologists, 43
American University, 73, 101, 103
Anderson, Clinton, 113
Arches of Science Award, 208-210
Arden House Statement, 48-53, 55-57, 58, 69, 91, 193, 208, 260
*Aridity and Man*, 222
arid lands conferences, 214-216, 221-224, 261
*Arid Lands Research Newsletter*, 221
Arizona Academy of Science, 223
Arthur D. Little Foundation, 219
Asimov, Isaac, 167, 209
Assessment of Biological Effects of Repeated Use of Herbicides, 242
Association of American Colleges, 169
Association of American Geologists and Naturalists, 2
Association of the Bar of the City of New York, 22
Astin, Allen V., 24
Atomic Energy Commission, 84, 219, 222
  radiation standards, 23-24

atomic radiation, biological effects of, 110, 229
Atwood, Mildred, 73-74, 88
Auger, Pierre, 224-225

Bache, Alexander Dallas, 13
Baez, Albert V., 163
Bailey, Liberty Hyde, 44
Baird, Spencer F., 13
Baitsell, George, 74, 76, 88
Baker, William O., 30, 61, 203
Ballard, B. G., 60
Barnard, F. A. P. (Frederick), 39
Bascom, Willard, 192-193, 195
basic research, symposium on, 29-31
*Basic Research: A National Resource*, 34
Bateson, Gregory, 24
Bateson, Mary Catherine 24, 212
Battelle Memorial Institute, 185
Beadle, George W., 29, 142, 203
Begle, Edward G., 171
Behnke, John, 57, 94, 192, 216
Bell, Alexander Graham, 71
Bell, David, 45
Bentley, Arthur F., 118-119
Berkner, Lloyd V., 27
Berl, Walter, 64, 65-69, 207
Berlin, Irving, 88
Bevan, William, 260
biological effects of atomic radiation, 110, 229
Biological Sciences Curriculum Study, 160, 171
Blackwelder, Richard E., 91-93
Blakeslee, Albert S., 42
Blakeslee, Alton, 199
board of directors of AAAS
  chairman of, 6
  as Executive Committee of the Council, 5, 127
  organization and responsibilities of, 127-129, 135-137
Boeing Company, 209
Boffey, Philip, 85, 213
Bok, Bart, 225
Bolling, Richard, 204

Index

*Books of the Traveling High School Library*, 166
Borlaug, Norman, 65
Botanical Society of America, 43
Bow, Frank T., 113
Bowles, Frank, 113
Bowman, Isaiah, 14-15, 211, 223
Brademas, John, 203
Brain, Lord, 62
branches of AAAS, 4-5. *See also* Lancaster Branch of AAAS
Brenner, Sydney, 61
Brewster, Kingman, 66
British Association for the Advancement of Science, 58, 62, 81, 211, 224-225, 311 (note 49)
  joint meeting with AAAS, 249-250, 251, 252
Broadcast Music, Inc., 192, 201
Brode, Wallace R., 85, 92, 94, 106, 206, 237
Bronk, Detlev W., 21-22, 26, 42, 52, 90-92, 98, 106, 110, 183, 224, 237
Bronowski, Jacob, 57-58, 62
Brookings Institution, 7, 102, 112, 113, 121, 203-204
Brown, Frank, 200
Brownell, Samuel, 163-164
Brussels Worlds Fair, 195
building, AAAS, 39, 43, 101-114
  Board of Zoning Adjustment, 105, 109, 110
building fund of AAAS, 102, 103
Bunker, Ellsworth, 244
Bunting, Mary I., 29
Bureau of the Budget, 35-37, 255
Burrill, Meredith F., 92
Bush, Vannevar, 13-15, 17, 136, 297 (note 18)

Cain, Stanley, 204
Calder, Ritchie, 209
Caldwell, Otis W., 6, 14
Calkins, Robert, 113
Cambel, Ali, 204

Campbell, Frank, 72, 74, 101-102
Campbell, Louise, 196
Capra, Frank, 209
Carey, William, 16-17, 255
Caribbean Division of AAAS, 4
Carleton College, 199
Carlson, Anton J., 7, 48-49
Carlson, Edward, 196
Carmichael, Leonard, 60, 196
Carnegie Corporation of New York, 32, 153, 156, 160, 168, 171, 172, 194
Carnegie Foundation for the Advancement of Teaching, 149
Carnegie Institution of Washington, 26, 95, 102, 112, 113
Carson, Rachel *(Silent Spring)*, 227, 257, 258
Cattell, Jacques, 5, 72-74, 77
Cattell, James McKeen, 5-6, 71-72, 82, 97, 99, 287 (note 4)
Cattell, Josephine Owen, 72-74
Cattell, Ware, 72, 81
centennial meeting of AAAS, 43-46
certification of teachers, 169-173, 186
chairman of the board of directors of AAAS, 6
Chamber of Commerce, 151
charter of AAAS, 39
*Chemical and Engineering News*, 77-78
chemical defoliants and herbicides in Vietnam, 207, 240-247, 317 (note 48)
Chisholm, Brock, 46
Christenson, Theodore, 103
Christmas Course of Lectures Adapted to a Juvenile Auditory, 183
classification of research, 19-23
Clay, Lucius, 29
Cleveland, Newcomb, 117-118
Clifford, Clark, 44-45
Cloud, Preston, 60
Cobb, W. Montague, 54-55
Cohn, Victor, 195, 199
Cold War, 17, 18, 151, 193, 257
Coleman, John, 192

Colgate, Craig, 196
Columbia Broadcasting System, 192-193, 201, 207
Columbia University Faculty Club, 220
Commission on Environmental Alteration, 241, 247, 248, 317-318 (note 52)
Commission on Industrial Competitiveness, 28
Commission on Population and Reproduction Control, 236, 261
Commission on Professionals in Science and Technology, 26
Commission on Science Education, 33, 95, 160-163, 173, 177-178, 179, 182, 184, 185, 252, 261, 302 (notes 36 and 37). *See also* Cooperative Committee on the Teaching of Science and Mathematics; science education
Committee of Young Scientists, 251-252
Committee on Scientific Freedom and Responsibility, 24
Committee on Cooperation Among Scientists, 113, 230, 236-239, 316 (note 42)
Committee on Council Affairs, 121, 129-130, 133, 139, 239, 240-241, 260
Committee on Desert and Arid Zone Research, 221, 261
Committee on Environmental Alteration, 207, 236, 242, 247-248, 252, 261
Committee on Governance, 135, 139, 252, 260
Committee on Investment and Finance, 143, 147
Committee on AAAS Meetings, 57-62, 65-66, 239, 252
Committee on Membership Development, 58
Committee on Nominations and Elections, 84, 137, 140, 141
Committee on Public Understanding of Science, AAAS, 123, 195, 201, 207, 230, 237, 252
Committee on Research Grants, 125
Committee on Science in the Promotion of Human Welfare, 207, 228-236, 239, 242, 247, 248, 252, 253, 261
Committee on the Social Aspects of Science, Interim, 227-230
Committee on Un-American Activities (Thomas Committee), 45, 90, 283 (note 14)
Commoner, Barry, 230-234, 248
Commonwealth Foundation, 149
communication with scientists
"Scientists' and Engineers' TV Journal", 122-123
*See also* meetings; *Science*
communism, 18, 20, 257
competency measures, 178-179
Conant, James B., 7, 15, 17, 65, 136, 209, 237
Condon, Edward U., 44-45, 52-53, 90-93, 142, 191, 237, 283 (notes 11 and 14), 291-292 (note 59)
Conference on Scientific Manpower, 61, 152
congressional testimony, 26-27
"Conquest" television programs, 192-193, 201, 207
Constable, John D., 244
constituencies of AAAS, 2, 41
constitution and bylaws of AAAS, 6, 127-129, 130, 134, 137, 139
Cook, Robert E., Jr., 244
Coolidge, Harold, 245
cooperation with government agencies, 24-25
Cooperative Committee on the Teaching of Science and Mathematics, 149-153, 159, 161-163, 169, 170, 135, 299 (note 3), 301-302 (note 30). *See also* Commission on Science Education; science education

Corcoran, William W., 64
Cornell, Douglas, 51
Cornell University, 177
Corner, George W., 87-88
Council for the Advancement of Negroes in Science, 53-54
Council for the Advancement of Science Writing, 199, 200, 206, 208
Council of Chief State School Officers, 97, 158, 167, 171
council of AAAS, organization and responsibilities of, 5, 50-51, 127-140, 260
Cousteau, Jacques, 220
Cranberg, Lawrence, 139
Crow, James F., 29
Curtis, Thomas, 204

Daddario, Emilio Q., 203
Dana, James Dwight, 3, 39
Dasbach, Joseph, 163
Davidson, Martin, 79
Davies, Clarence, 28, 92
Davis, Jefferson, 64
Davis, Watson, 205-206, 209
Day, Edmund E., 15-17
Deason, Hilary, 164-169, 196, 198
Deere and Company, 223
Dees, Bowen, 163
Dees, Sarah, 213
Department of Commerce, 222
Department of Defense, 243, 245
Department of Health, Education and Welfare, 21-23
Department of Labor, 151
Department of Science, 27-28, 29
Department of State, 215, 219, 243
Department of the Interior, 222
Dewey, John, 118
DeWitt, Nicholas, 151
Dickson, David, 213
Dickson, Leonard Eugene, 117
Dingwall, Ewen, 196
divisions of AAAS, 3-4, 44, 125, 130-135
  *See also* individual divisions

Dixon, James P., 234
Douglas, Paul, 73
Douglass, Andrew T., 4
Doxiadis, A., 65
Dubos, René, 183, 204, 209, 241, 317-318 (note 52)
Duisberg, Peter, 214
Dunn, L. C., 110
Dunning, John, 106
Du Pont Company, 185, 223
Dupree, A. Hunter, 27
DuShane, Graham, 74, 80, 83, 94, 95, 98

Earth Day, 228, 257
Edison, Thomas A., 71
editor of *Science*, 53, 89-96
Editorial Board of *Science*, 74, 76, 89, 90, 93
EDO Foundation, 219
Ehrman, Libert, 35
Eisenhower, Dwight D., 45-46, 107, 196, 256, 259
Eisley, Loren, 65
electorate of AAAS, 135-140, 260, 298 (note 37)
elementary school students, science programs for, 177-182
Elementary Science Study, 179-182
eligibility for federal grants, 20-23
  reviewer blacklist, 257
Eli Lilly Company, 119
Elliott, Carl, 156, 157
El Paso Natural Gas Company, 223
Elsey, George M., 44-45
Emery, Alden, 26
Emory University, 159
Engineering Manpower Commission, 25
*Enough of Pessimism*, 98
environmental problems and education, 185, 207, 233-234, 235-236, 240-247, 257-259
ethical problems of scientists, 237-240, 258
"Etiquette of Research and Publication, The," 237-238

Executive Committee of the Council (board of directors), 5, 127
executive officer of AAAS, 53, 56, 95, 135, 143
expenditures for research, federal, 1, 11, 13, 30, 235, 259
Eyring, Henry, 29, 232

Faraday, Michael, 66, 183
Faulkner, Waldron, 105, 107-109, 111-112
Fawcett, C. B., 211
federal grants, eligibility rules for, 20-23
Fellows of AAAS, 3, 139
fellowships for teachers, 162
finances of AAAS. *See* receipts and expenditures of AAAS
Finger, Grayce, 80
Flanagan, Dennis, 75, 208
Ford Foundation, 34, 66, 68, 76, 113, 194, 202, 222
Foster, John S., Jr., 242-243
Fraley, Pierre, 200
Friendly, Alfred, 200
Fulbright, James W., 113
*Future of Arid Lands, The,* 215
"Future of Science, The" (address by Bentley Glass), 250

Gallant, Joseph, 167
Gamow, George, 209
Gardner, John W., 153
General Electric Company, 82, 151, 154
General Motors Corporation, 185
general secretary of AAAS, 5, 6, 127, 150
George Sarton Memorial Lecture, 61
Gerard, Ralph W., 15, 61
Gibson Island. *See* Gordon Research Conferences
Gitlin, Irving, 46
Glass, H. Bentley, 74, 82, 85, 91, 93, 97, 167, 171, 209, 212, 250, 261
Glennan, T. Keith, 27
Goddard, Robert H., 126
Goddard, Mrs. Robert H., 126
Godfrey, Kneeland, 191, 206-207

Goethe, Charles Matthias, 124-125
Gofman, John, 23
Golden, William T., 143, 147
Goodall, Jane, 62
Gordon, Neil, 115-116
Gordon Research Conferences, 7, 115-117, 295 (notes 1 and 7), 296 (note 8)
Gould, Laurence M., 195, 199
governance of AAAS, 5-6, 127-147
government agencies, AAAS cooperation with, 24-28
government responsibility for science, 1, 30
Graham, George, 203
grants, federal, eligibility rules for, 20-23
grants to students, 125-126
Gravel, Mike, 23
Gray, Asa, 39
Gray, George W., 209
Greenberg, Daniel S., 81, 83, 85, 86, 213
Greenwalt, Crawford, 30
Gregory, Sir Richard A., 211
Gregory, Richard L., 212
Gropius, Walter (Architects Collaborative), 105-107
growth of AAAS, 9-11, 71, 81, 146, 213
Grunbaum, James, 79
*Guide to Science Reading, A,* 167
Gustavson, Reuben G., 15
Guthrie, James, 64

Hackerman, Norman, 60
Hafstad, Lawrence R., 110
Hale, George Ellery, 66, 103
Hammond, Allen L., 98
Handler, Philip, 60, 65, 117
Hanson, Arthur, 39, 98, 106-108
Hanson, Elisha, 107
Hardin, Garrett, 248
Harris, Milton, 204
Harris, Ray L., 191
Haskins, Caryl, 68-69, 95, 110, 113
Heatwole, Thelma, 191, 206
Heffner, Richard, 122, 195, 201

Hendricks, Sterling B., 60
Henry, Joseph, 13, 39
Herbicide Assessment Commission, 244-248, 252, 261, 319 (note 67)
herbicide use in Vietnam, 242-248
Herblock (Herbert Block), 45
Herron, Dudley, 173
Hesburgh, Theodore, 61
Hickel, Walter J., 24
Hill, A. V., 62
Hill, Lister, 156-157
Hiscocks, E. S., 60
Hobby, Oveta Culp, 21
Hodge, Carl O., 215, 222
Hodges, Luther P., 196
Hogness, John, 117
Holiday Science Lectures, 183-184, 186, 187, 206
Holton, Gerald, 252-253
Hoover, Herbert, 18
Hornig, Donald, 242
Horsfall, Frank, 203
Horsky, Robert, 113
Houston, William V., 29
Howard, Leland O., 44, 71, 101
Hoyle, Fred, 209
Hubbard, Gardiner Greene, 71
Humphrey, Hubert H., 27, 68, 113
Hunter College, 159
Hutzel, John, 47

income tax reform, 34-38
Industrial Science Award, 119
*Inexpensive Science Library, An,* 167
institutes for teachers of science and mathematics, 122, 151, 160, 162, 173-176, 187
*Integrity of Science, The,* 232-233
International Advisory Committee on the Public Understanding of Science, 225
*International Clearinghouse on Science and Mathematics Curriculum Developments, Report of the,* 185
International Conference on Arid Lands in a Changing World, 221-224

International Federation of Associations for the Advancement of Science, 222-225, 313 (note 49)
International Geographic Union, 223
International Geophysical Year, 197, 199
International Oceanographic Congress, 214, 216-221
International Oceanographic Foundation, 219
international scientific meetings, AAAS policies for, 216-217
International Symposium on Communication and Social Interaction in Primates, 62
International Union of Geodesy and Geophysics, 218
Inter-Society Committee for a National Science Foundation, 15-17
Ivy, A. C., 48-49

Jackson, Frederick, 171
Jastrow, Robert, 60
Johnson, Lyndon B., 27, 113, 255
Joint Commission on the Training of Teachers of Science and Mathematics, 169-170
Jones, Howard Mumford, 192
Jones, R. V., 176
*Journal of Geophysical Research,* 95
*Journal of the American Medical Association,* 97
junior high school students, science education for, 182
Junior Scientists' Assembly, 152, 205

Kabisch, William, 23, 175, 206, 234-235, 260
Kaempffert, Waldemar, 209
Kalinga Prize, 208-209
Kanwar, J. S., 223
Keener, Gladys M., 52-53, 74, 91-92
Kefauver, Estes, 27
Kelly, Harry C., 29, 60, 163-164
Kennedy, John F., 82, 203
Kettering, Charles F., 7, 142
Kilgore, Bill, 14-17

Kilgore, Harley M., 14-15
Killian, James R., 30, 258
King, Robert W., 34
Klopsteg, Paul E., 34-38, 84, 106, 137, 142, 143, 147, 230, 237
Knowles, John H., 24
Kovda, Viktor, 217, 223
Krul, W. F. J. M., 216

Lancaster Branch of AAAS, 5, 8, 190
Langer, Elinor, 86
Lanham, Fritz, 18
Lark-Horovitz, Karl, 150, 224
Lasker, Gabriel, 54
Latin American Conference on Arid Lands, 222
Laurence, William, 45
Lawrence Berkeley Radiation Laboratory, 23
Lazarsfeld, Paul, 60
Leake, Chauncey D., 64, 137, 225, 237
Leakey, Louis S. B., 65
Lee, Milton O., 92
Lewenstein, Bruce V., 307 (note 2)
Libby, Willard F., 110
*Life* reports of AAAS meetings, 191
Linowitz, Sol, 65
Livermore, Arthur, 163, 179
Livingston, Burton E., 6, 7
Livingston, Robert B., 163, 179
local branches of AAAS, 4-5, 8, 190
Lomonosov, 219-220
Lonsdale, Dame Kathleen, 62, 65
Lauwerys, Joseph, 175
Lowbeer, Hans, 174
Lowe, David, 224
loyalty and security issues, 17-24
Luck, J. Murray, 29

MacArthur, Donald, 242
MacDougal, Daniel T., 4
Macleod, Robert, 60
Magee, Warren, 39, 73, 91, 97, 98, 105-107
Magnuson, Bill, 14-17
Magnuson, Warren G., 14-16, 196
Major, Robert, 60, 175

Malone, Thomas F., 203
Margolis, Howard, 83, 85
Marston, Robert Q., 23
Massachusetts Institute of Technology, 249
Mathematical Association of America, 149
Mather, Kirtley F., 15, 51, 142-143, 224, 283 (note 14)
Mathias, Charles, 204
May, Catherine, 204
Mayo Clinic, 199
Mayor, John, 153, 158, 163, 170, 171, 185
McCarley, Orin, 153
McCarran-Walter Act, 18
McCarthy, Joseph, 18, 22, 257
McCarthyism, 17-19, 23
McClure, Stewart, 157
McCormack, James, 27
McElheney, Victor, 85, 213
McGill, William J., 86
McNamara, Robert, 242
Mead, Margaret, 24, 38, 54-55, 64, 65, 137, 140-141, 167, 197, 193, 209, 212, 235, 251
Meeting, Charlotte, 77, 79, 82
meetings of AAAS
  annual, registration at, 10, 47, 51, 54, 66, 67, 261; social events at, 48, 64-65, 286 (note 67); television and radio coverage of, 46, 66-68, 207; time of, 63-64
  audiotapes, 68
  centennial, 43-46
  dissent and disruption at, 259
  First International Congress of Oceanography, 216-220; television and radio coverage at, 312 (note 30)
  *Life* reports on, 191
  "Moving Frontiers of Science," 59-61
  national, 2, 41-69, 190-191, 261
  Parliament of Science, 28-29
  policies, 59-63
  press coverage of, 190-191

Symposium on Basic Research, 29-31
membership classes of AAAS, 3, 5, 139
membership drive, AAAS, 43
Mendeleev, Dmitri, 66
Merton, Robert, 65, 239
Meselson, Matthew, 244-247
Metreaux, Rhoda, 167, 198
Meyerhoff, Howard A., 14, 16, 25-26, 31, 39, 47, 52-53, 55, 57, 58, 72, 74, 76, 77, 88, 89, 90-93, 94, 97, 101, 105, 106, 109, 150, 281 (note 45), 285 (note 41)
Michels, John, 71
Michelsen, Börge, 224
Michigan State University, 157, 178, 182
Midwest Research Institute, 242
Miles, Walter, 14, 224
Military Air Transport Service, 219
Miller, Donald H., Jr., 208
Mills, Wilbur, 15, 36
Minneapolis *Star and Tribune*, 199
Mitchell, James, 27, 196, 203, 237
Modell, Walter, 235
Moe, Henry Allen, 15
Mohole Project, 192
Monsanto Company, 192-193
Moon issue of *Science*, 96-97
Moore, John A., 163, 185
Morison, Robert, 250
Mosher, Charles A., 204
Moulton, Forest R., 6, 42, 47, 53, 72-73, 77, 99, 101-105, 112, 113, 120, 213
Moulton, Harold, 7
"Moving Frontiers of Science" symposium series, 59-61
Murphy, Ellen, 77
Murphy, Walter, 26
Muskie, Edmund S., 23

National Academy of Sciences-National Research Council, 44, 51, 65, 83, 95, 105, 107, 151, 174, 192, 204, 214, 216, 219, 225, 242, 245, 247

arid lands committee, 221
genetic recombination concerns of, 117
policy on government support of unclassified research, 21-22
symposium on basic research, 30
National Aeronautics and Space Administration, 27, 96
National Association for Research in Science Teaching, 149
National Association for the Advancement of Colored People, 54
National Association of Manufacturers, 151
National Association of Science Writers, 194, 197, 198, 200, 208
National Association of State Directors of Teacher Education and Certification, 170-173, 186
National Association of State Universities and Land Grant Colleges, 36
National Broadcasting Company, 207
National Capital Planning Commission, 111-112
National Citizens' Committee for Educational Television, 201
National Council for the Accreditation of Teacher Education, 173
National Council of Teachers of Mathematics, 169
National Defense Education Act of 1958, 156, 157-153, 167, 186, 203
National Education Association, 149, 151
National Geographic Society, 71, 102, 107, 190
National Industrial Conference Board, 56
National Institutes of Health, 23, 117, 125, 235, 257, 314 (note 1)
national meetings of AAAS. *See* meetings, national
National Paint, Varnish and Lacquer Association, 111-112
National Patent Council, 18
National Publishing Company, 82

National Science Board, 14-17, 25
National Science Fair, 205
National Science Foundation, 2, 24-25, 31, 44, 61, 62, 99, 121-123, 143, 147, 154, 182, 185, 186, 196, 200, 222, 225, 256, 314 (note 1)
  arid lands meetings, support of, 215
  establishment of, 13-17, 279 (note 4)
  foreign scientists' visits, support for, 173-174, 176-179
  oceanography conference, support of, 219
  overhead payments, 32, 34
  public understanding of science, support of, 207
  rules for grant eligibility, 20-22
  science education improvements, support of, 150-152, 160-162, 164, 166, 168, 169
  student grant, support, 125
  teacher qualifications, study of, 172
National Science Seminars, 205-206
National Science Teachers Association, 150, 165, 169
National Society for Medical Research, 48
*Nature,* 75, 99, 211
Nature Conservancy, 121
Negus, Sidney, 190-191
Nelson, Bryce, 23, 85
Nevins, Alan, 192
Newcomb Cleveland Prize, 117-118, 189, 191
Newell, Homer B., 60
New Mexico Academy of Science, 223
New Mexico Institute of Mining and Technology, 215
*New Roads to Yesterday,* 190
News and Comment section of *Science,* 57, 82-86, 87
New York Philharmonic Orchestra concerts, 208
*New York Times,* 45, 170, 231
Nichols, Rodney W., 242

Northwest Bell Telephone Company, 209
Noyes, W. Albert, 15, 29
Nussbaum, Hans, 75, 107, 108, 143

objectives of AAAS, 8-9, 48-50, 59, 65, 107, 120, 249-253, 261-262
*Oceanography,* 220
Office of International Science, AAAS, 225-226
Office of Naval Research, 219, 256, 314 (note 1)
Office of Scientific Research and Development, 1, 32, 256
Office of Technology Assessment, 240
Oppenheimer, Robert, 30
Organization for Economic Cooperation and Development, 174, 175, 212
Organization of American States, 222
Orians, Gordon, 245
Ormes, Robert V., 77
Osborne, Fairfield, 46
Ost, David, 173
overhead costs, AAAS policy on recovery of, 32-34

Pacific Division of AAAS, 3-4, 44, 125
Pacific Science Center, 162, 184, 197, 208-209
Park, Thomas, 78, 94, 133, 137, 195, 237
Parkinson, Francis, 175
Parks, W. George, 116, 117
Parliament of Science, 28-29, 195, 215, 229
Patnaik, B., 208
Payne, Fernandus, 97
Pederson, Bethsabe, 82
Pennsylvania State University, 155-157
permanent secretary of AAAS, 5, 6, 127
Pfeiffer, E. W., 240, 241, 245, 246, 247, 252
Phi Beta Kappa Address (annual meeting), 61, 259
Philip Hauge Abelson Prize, 99

Physical Sciences Study Committee, 160
Piel, Gerard, 60, 75-79, 208, 209
Pierce, Franklin, 64
Pigman, Ward, 227-228, 252, 253
Piori, Emanuel R., 27, 29
Platt, John, 236
Polanyi, Michael, 60
*Politics of Pure Science, The,* 85
popular science
   books, 189-190
   lectures, 189-190
   *See also* Committee on Public Understanding of Science
Powell, Adam Clayton, 204
Powell, John Wesley, 65, 223
president of AAAS, 5, 140-142, 185, 271-274, 299 (note 43)
president-elect of AAAS, 6, 185
President's Committee on Scientists and Engineers, 151, 158
press coverage of AAAS. *See* meetings; *Science*
Press, Frank, 200
prestige of science, 254
Price, Derek J. deSolla, 60
Price, Don K., 27, 37-38, 60, 139, 242, 249
prizes, AAAS, 117-120
   *See also* AAAS awards and prizes
"Problems of Customs or Manners Arising in the Major Relationships of Scientists," 237
Prostak, Arnold, 138
public attitudes toward science, 197-198, 250, 309 (note 37)
Public Health Service, 21, 23, 229, 257
Publications Committee, 76, 89, 96-98, 120
publication of committee reports, 228-233
public understanding of science, 189-210, 307 (note 2)
   television and radio coverage for, 191-201
   *See also* Committee on Public Understanding of Science; International Advisory Committee on the Public Understanding of Science; Parliament of Science

Quie, Albert, 203

Rabinowitch, Eugene, 209
race, issues involved in scientific study of, 234-235
Rackley, Ralph, 163-164
Rand McNally Company, 185
Rapaport, Anatol, 239
receipts and expenditures of AAAS, 10, 11, 146
recognition of exceptional teachers, 155, 157, 186
Rees, Mina, 29, 140-141, 147
regional consultants on education, 154, 155
regional divisions of AAAS. *See* divisions of AAAS
research, federal expenditures for, 1, 11, 13, 30, 235, 259
research, government classification of, 19-23
research grants by AAAS, 123-126
Revelle, Roger, 199, 203, 214, 219-220
Rice, Stuart, 35-36, 282 (note 67)
Rieser, Leonard, 135, 163
Ripley, S. Dillon, 113
Roberts, Walter Orr, 203, 207, 251, 259
Roche, Marcel, 250
Rockefeller Foundation, 32, 76, 194, 197, 198, 208, 215, 219, 222
Rockefeller University, 30, 68
Rogers, Will, 93
Rogers, William P., 244
Rohweder, Ralph, 48-49
Roller, Duane, 55, 74, 77, 87-88, 92, 93-94, 292 (note 68)
Romer, Alfred, 139
Roosevelt, Franklin D., 1, 13
Rosenthal Foundation, 119
Ross, Charles, 45
Rubey, William W., 137, 217
Ruina, Jack, 248

Saltonstall, Leverett, 113
San Francisco State University, 159
SAPA. *See* "Science—A Process Approach"
Sapolsky, Harvey, 249
satellites and sputniks, public attitudes toward, 197, 203
scientists, high school students' images of, 198
Scheele, Leonard A., 22
Scherago, Earl J., 79, 97
Scherer, Paul A., 27, 32, 78, 92, 143, 146-147
Schmitt, Francis O., 60, 200
School Mathematics Study Group, 154, 160, 171
Schramm, Wilbur, 197
Schuss Yucca spoof, 75
Schwarzschild, Martin, 204
science, government responsibility for, 1, 30
*Science*, 5, 6, 9-10, 19-20, 22, 23, 27, 34, 44, 46, 53, 71-99, 120, 261, 287 (note 2)
    advertising in, 78-80, 87, 256, 260, 290 (note 32)
    complaints about, 87-88
    editor of, 89-96
    editorial policies, 84-89
    editorials in, 84, 97-99
    News and Comment section, 57, 82-86, 87
    readers' judgment of, 86-87
    press coverage of, 190-191
    special issues of, 96-97
"Science and Human Survival," 232, 233, 315 (note 19)
"Science and Human Welfare," 231
science and mathematics education, 149-187, 193, 199
    conferences on, 154, 158, 162, 177, 187
    *See also* science education
science and mathematics teachers
    consultants for, 154-157, 186
    institutes for teachers of science and mathematics, 122, 151, 160, 162, 173-176, 187
    Joint Commission on the Training of Teachers of Science and Mathematics, 169-170
    special teachers in elementary schools, 158-159
    *See also* science teachers; teachers
*Science and Government Report*, 85
Science and Public Policy Study Group (SPPSG), 249
*Science and Public Policy* (Steelman Report), 150
science and society bibliographies, 185, 306 (note 95)
*Science and Survival*, 232
"Science and the Future," 249-250
"Science—A Process Approach (SAPA)," 163, 178-182, 186, 187
    publisher of, 182
*Science Books* (later, *Science Books & Films*), 169, 303 (note 48)
Science Center for Instructional Materials and Processing, 157
Science Curriculum Improvement Study, 179-182
science education, 149-187
    "Action Program," 151
    curriculum, AAAS. *See* "Science—A Process Approach"
    environmental problems and, 185, 207, 233-234, 235-236, 240-247, 257-259
    university consultants for, 155-157
    *See also* Commission on Science Education; Cooperative Committee on the Teaching of Science and Mathematics; Science Teaching Improvement Program
*Science Education News*, 153, 163, 185
*Science Illustrated*, 76
*Science in Communist China*, 62

Index

*Science in Japan,* 63
Science Materials Center of Fairfax County, VA, 157
science news column for newspapers, 207
science policy
   AAAS support of studies of, 249
   problems with, 258-259
Science Press Printing Company, 83
Science Service, 151, 198, 205-208
science teachers
   education of, 159-173, 186
   institutes for, 122, 151, 160, 162, 173-176, 187
Science Teaching Improvement Program, 152-159, 160, 161, 162, 169, 185
*Science—The Endless Frontier,* 13-14
*Science 80-86,* 81, 290 (note 42)
*Scientific American,* 75-79, 80, 123, 208
*Scientific Estate, The,* 249
scientific freedom vs. Cold War fears, 18-24
Scientific Manpower Commission, 25-26
*Scientific Monthly, The,* 5, 6, 9, 10, 53, 74-76, 80-82, 101, 103
"Scientists' and Engineers' TV Journal," 122-123
Scoggins, Margaret, 164
Scudder, Stevens, and Clark, 146
Seaborg, Glenn, 84-85, 117, 209
Sears, Mary, 217, 220
Sears, Paul B., 78, 110, 163, 199, 237
Seattle University, 209
Seattle World's Fair, 195-197, 209
sections of AAAS, 1, 41-43, 130-135
   secretaries of, 135
security and loyalty issues, 17-24
segregation (Atlanta meeting), 53-55
Seitz, Frederick, 83, 242
*Selected List of Career Guidance Publications,* 168
Semenov, Vadim S., 65
seminars for foreign scientific attachés, 121-122, 204
seminars for members of Congress, 27, 121, 202-204, 206

Seymour, Dan, 202
Shapley, Harlow, 14-15, 42-44, 46, 57, 58, 61, 65, 69, 103, 112, 113, 136, 199, 200, 209, 224, 237
Sherburne, Edward G., 122-123, 183, 191, 194-197, 200, 201, 204, 206, 207
Siddons, Frederick P. H., 106, 143
Sigma Xi, 201
Sigma Xi Lecture (annual meeting), 61, 259
Simpson, George Gaylord, 61
Singer, Fred, 60
Singer, Maxine, 117
Sinnott, Edmund W., 46, 142
Skinner, B. F., 65, 203
Sklar, Michael, 192-193
Skolnikoff, Eugene, 249
Sloan, Alfred P., 29-30
Sloan Foundation, 30, 32, 201, 202, 219, 249
Smiley, Terah L., 222
Smithsonian Institution, 73, 101, 102, 112, 113, 126
Snow, Sir Charles, 61
Snyder, Laurence H., 110, 156, 237
social events at AAAS annual meetings, 48, 64-65, 286 (note 67)
Social Science Research Council, 261
Sociopsychological Prize. *See* AAAS awards and prizes; AAAS Prize for Behavioral Science Research
Soll, Dieter, 117
Southwestern and Rocky Mountain Division of AAAS, 3-4, 205, 214-216, 221-223, 261
*Soviet Science,* 62
Special Committee on Council Activities and Organization, 129-130
Special Committee on Oceanic Research, 218-219, 221, 223
Spilhaus, Athelstan, 196, 199
*SPPSG Newsletter* (later *Public Policy*), 249
Staats, Elmer, 16-17, 255
Stafford, Robert, 204
Stakman, Elvis, 224

Standard Oil of New Jersey, 223
Stanford Linear Accelerator,
  announcement of, 30
Stanton, Frank, 192, 202
Steelman, John R., 44, 150
Steere, William, 60
Stephenson, Charles S., 73
Stratton, Julius A., 22
Straus, William, 74
students
  AAAS grants for, 125-26
  images of scientists, 198
  performance of, using new science
    curricula, 179-181
Study Committees of Council, 129,
  225, 239-240, 248
Study Committee on Cooperation with
  Developing Countries, 225
Study Committee on Ethics, 239-240
Study Committee on the Use of
  Natural Areas as Research
  Facilities, 120-121
Suits, C. G., 15
Sullivan, Walter, 60
Survey Research Center, 197-198
Swann, W. F. G., 5
Swarbrick, James, 215
symposia, AAAS, 6, 62-63, 120, 232-
  235, 245-246, 248, 249
*Symposium on Basic Research,* 28, 29-31,
  34, 215
symposium volumes, AAAS, 6-7

Tamplin, Arthur, 23
tax liability, AAAS as tax-exempt organization, 38-40
Taylor, Raymond L., 47, 48, 51, 54,
  58, 63, 65, 91, 94, 106-107, 125
teachers
  certification of, 169-173, 186
  fellowships for, 162
  recognition of exceptional in-
    dividuals, 155, 157, 186
television and radio programs
  for public understanding of science,
    191-201, 312 (note 30)
  for scientists, 122-123
  *See also* meetings, television and
    radio coverage of
Textronics Incorporated, 209
Theobold Smith Award in Medical
  Sciences, 119
Thimann, Kenneth V., 247
Thomas Alva Edison Foundation, 193
Thompson, Frank, 36
Thornhill, Daniel, 65
Tichenor, Phillip J., 197
Timken Roller Bearing Company, 123
Traveling Libraries of Science, 24-25,
  32, 158, 160, 163-169, 186,
  187, 196, 198, 256
treasurer of AAAS, 5, 32, 135, 142-147
Truman, Harry S, 1, 16-17, 44-45, 46,
  150, 259
Trumbull, William, 260
Trytten, M. H., 26
Tschirley, Fred H., 245
Turner, Joseph, 98
Twist, Sumner, 116

Ubell, Earl, 51-53, 199
Ullman, Ai, 203
*Understanding,* 200-201, 206
UNESCO (United Nations Education-
  al, Scientific, and Cultural Or-
  ganization), 150, 174, 212, 213,
  214-216, 221-225
UNESCO Advisory Committee on
  Marine Sciences, 217-219
Union of Biological Societies, 149
United Nations, 217, 219, 243
university consultants for science and
  mathematics teachers, 155-158
University of Arizona, 223
University of Minnesota, 199
University of Nebraska, 155-157
University of Oregon, 155-156
University of Tennessee, 159
University of Texas, 155-157
University of Washington, 151, 168,
  184
University of Wisconsin, 153, 177
Urey, Harold, 14, 115

U.S. Commission on Government Security, 23
U.S. Office of Education, 151, 156, 159, 162, 167, 169, 182
U.S. Science Exhibit, 196, 209

Vail, Richard, 45
Valentine, Willard L., 73-74, 77, 80, 88, 96, 101, 103, 212-213
Vetter, Betty, 26
Viall, William P., 171-172
vice presidents for AAAS sections, 135
Vietnam, chemical defoliants and herbicides in, 207, 240-247, 317 (note 48)
Visiting Foreign Lecturers Program, 160, 173-176, 187, 212
Von Frisch, Karl, 209

Walbesser, Henry, 179
Walcott, Gregory, 307 (note 7)
Wallen, Eugene, 154
*Wall Street Journal,* 199
Walsh, John, 74, 83
Warren, Earl, 22, 24
Warren, Shields, 110
Washington, George, 112
*Washington Post,* 45, 114, 199-200
Waterman, Alan T., 17, 22, 85, 137, 138, 199
Weaver, Warren, 19, 28, 30, 49, 52-53, 55, 76, 77, 78, 90-93, 110, 167, 195, 199, 200, 201, 208, 209, 228, 260
Weis, Jessica, 204
Weisbrod, Burton, 38
Weiss, Paul, 42, 60, 183-184
Westing, Arthur W., 244, 246, 247
Westinghouse Company, 151, 189

Westinghouse Science Talent Search, 126
"What We Must Do," 236
Whitaker, Douglas M., 15
White, Edward, 62
White, Gilbert, 214, 216, 221, 223
White, Lynn, 237-238
*Who's Who in America,* 7, 95
Wiesner, Jerome, 60, 82
Wigner, Eugene, 232
Wildhack, William, 239
Wilkes, Daniel, 191
William P. Lipscomb Company, 109
Wilson, Logan, 113
Wilson, Robert L., 30
Wolfle, Dael, 15-17, 18, 55-56, 74, 77, 78, 82, 92, 98, 103, 107, 152, 173, 196, 199, 206, 209, 260
Woodrow Wilson Center, 113
Woodrow Wilson Memorial Commission, 113
World Meteorological Association, 222
"World of the Mind, The," 192, 193, 201
Wrather, William E., 92, 142-143, 145
Wright, Louis, 113
Wright, Sewall, 42
Writers' Guild of America, 201, 206

Xerox Corporation, 178-182, 185

Yerkes, Robert M., 42
Yochelson, Ellis, 139
Yoshido, M., 217
Youth Science Activities Committee, 206

Zacharias, Jerrold, 60
Znaniye, 212

**564935**

3 1378 00564 9358

JAN 23 1991

THE LIBRARY
UNIVERSITY OF CALIFORNIA
San Francisco
(415) 476-2335

**THIS BOOK IS DUE ON THE LAST DATE STAMPED BELOW**

Books not returned on time are subject to fines according to the Library Lending Code. A renewal may be made on certain materials. For details consult Lending Code.